Metals Challenged by Neutron and Synchrotron Radiation

Special Issue Editor
Klaus-Dieter Liss

MDPI • Basel • Beijing • Wuhan • Barcelona • Belgrade

MDPI

Special Issue Editor
Klaus-Dieter Liss
Guangdong Technion—Israel Institute of Technology (GTIIT)
China

Editorial Office
MDPI AG
St. Alban-Anlage 66
Basel, Switzerland

This edition is a reprint of the Special Issue published online in the open access journal *Metals* (ISSN 2075-4701) from 2015–2017 (available at: http://www.mdpi.com/journal/metals/special_issues/metals-challenged-neutron-synchrotron-radiation).

For citation purposes, cite each article independently as indicated on the article page online and as indicated below:

Author 1; Author 2. Article title. *Journal Name*. **Year**. Article number/page range.

Image courtesy of Elsevier

ISBN 978-3-03842-679-0 (Pbk)
ISBN 978-3-03842-680-6 (PDF)

Cover Picture: Traces within a Materials Oscilloscope, representing timelines in an azimuthalangle/ time plot. Here, the evolution of the α-101 reflection ring (horizontal axis: angle) of a titanium aluminide intermetallics is streaked in time (vertical axis) upon heating to 1573 K (bottom half figure) and subsequent plastic deformation at high-temperature (top half figure). See Editorial for References.

Table of Contents

About the Special Issue Editor vii

Preface to "Metals Challenged by Neutron and Synchrotron Radiation" ix

Klaus-Dieter Liss
Metals Challenged by Neutron and Synchrotron Radiation
Reprinted from: *Metals* **2017**, 7(7), 266; doi: 10.3390/met7070266 1

Gang Chen, Klaus-Dieter Liss and Peng Cao
An *in situ* Study of NiTi Powder Sintering Using Neutron Diffraction
Reprinted from: *Metals* **2015**, 5(2), 530–546; doi: 10.3390/met5020530 9

Yu Xiao, Feng Xu, Xiaofang Hu, Yongcun Li, Wenchao Liu and Bo Dong
In situ Investigation of Titanium Powder Microwave Sintering by Synchrotron Radiation
Computed Tomography
Reprinted from: *Metals* **2016**, 6(1), 9; doi: 10.3390/met6010009 24

Jun Ma, Aijun Li and Huiping Tang
Study on Sintering Mechanism of Stainless Steel Fiber Felts by X-ray Computed Tomography
Reprinted from: *Metals* **2016**, 6(1), 18; doi: 10.3390/met6010018 33

Hikari Nishijima, Yo Tomota, Yuhua Su, Wu Gong and Jun-ichi Suzuki
Monitoring of Bainite Transformation Using in Situ Neutron Scattering
Reprinted from: *Metals* **2016**, 6(1), 16; doi: 10.3390/met6010016 42

**Petra Erdely, Thomas Schmoelzer, Emanuel Schwaighofer, Helmut Clemens, Peter Staron,
Andreas Stark, Klaus-Dieter Liss and Svea Mayer**
In Situ Characterization Techniques Based on Synchrotron Radiation and Neutrons Applied for the
Development of an Engineering Intermetallic Titanium Aluminide Alloy
Reprinted from: *Metals* **2016**, 6(1), 10; doi: 10.3390/met6010010 52

**Andreas Stark, Marcus Rackel, Aristide Tchouaha Tankoua, Michael Oehring, Norbert Schell,
Lars Lottermoser, Andreas Schreyer and Florian Pyczak**
In Situ High-Energy X-ray Diffraction during Hot-Forming of a Multiphase TiAl Alloy
Reprinted from: *Metals* **2015**, 5(4), 2252–2265; doi:10.3390/met5042252 79

**Klaus-Dieter Liss, Ken-Ichi Funakoshi, Rian Johannes Dippenaar, Yuji Higo, Ayumi Shiro,
Mark Reid, Hiroshi Suzuki, Takahisa Shobu and Koichi Akita**
Hydrostatic Compression Behavior and High-Pressure Stabilized β-Phase in γ-Based Titanium
Aluminide Intermetallics
Reprinted from: *Metals* **2016**, 6(7), 165; doi: 10.3390/met6070165 91

Takuo Okuchi, Akinori Hoshikawa and Toru Ishigaki
Forge-Hardened TiZr Null-Matrix Alloy for Neutron Scattering under Extreme Conditions
Reprinted from: *Metals* **2015**, 5(4), 2340–2350; doi: 10.3390/met5042340 113

Darren J. Goossens
Monte Carlo Modelling of Single-Crystal Diffuse Scattering from Intermetallics
Reprinted from: *Metals* **2016**, 6(2), 33; doi: 10.3390/met6020033 122

Hiroshi Suzuki, Rui Yamada, Shinki Tsubaki, Muneyuki Imafuku, Shigeo Sato, Tetsu Watanuki, Akihiko Machida and Junji Saida
Investigation of Elastic Deformation Mechanism in As-Cast and Annealed Eutectic and Hypoeutectic Zr–Cu–Al Metallic Glasses by Multiscale Strain Analysis
Reprinted from: *Metals* **2016**, *6*(1), 12; doi: 10.3390/met6010012 135

Takeshi Egami, Yang Tong and Wojciech Dmowski
Deformation in Metallic Glasses Studied by Synchrotron X-Ray Diffraction
Reprinted from: *Metals* **2016**, *6*(1), 22; doi: 10.3390/met6010022 146

Gu-Qing Guo, Shi-Yang Wu, Sheng Luo and Liang Yang
How Can Synchrotron Radiation Techniques Be Applied for Detecting Microstructures in Amorphous Alloys?
Reprinted from: *Metals* **2015**, *5*(4), 2048–2057; doi: 10.3390/met5042048 158

Gu-Qing Guo, Shi-Yang Wu, Sheng Luo and Liang Yang
Detecting Structural Features in Metallic Glass via Synchrotron Radiation Experiments Combined with Simulations
Reprinted from: *Metals* **2015**, *5*(4), 2093–2108; doi: 10.3390/met5042093 166

Eui Pyo Kwon, Shigeo Sato, Shun Fujieda, Kozo Shinoda, Ryosuke Kainuma, Kentaro Kajiwara, Masugu Sato and Shigeru Suzuki
Characterization of Deformation Behavior of Individual Grains in Polycrystalline Cu-Al-Mn Superelastic Alloy Using White X-ray Microbeam Diffraction
Reprinted from: *Metals* **2015**, *5*(4), 1845–1856; doi: 10.3390/met5041845 180

Kouhei Ichiyanagi and Kazutaka G. Nakamura
Structural Dynamics of Materials under Shock Compression Investigated with Synchrotron Radiation
Reprinted from: *Metals* **2016**, *6*(1), 17; doi: 10.3390/met6010017 191

Soo Yeol Lee, E-Wen Huang, Wanchuck Woo, Cheol Yoon, Hobyung Chae and Soon-Gil Yoon
Dynamic Strain Evolution around a Crack Tip under Steady- and Overloaded-Fatigue Conditions
Reprinted from: *Metals* **2015**, *5*(4), 2109–2118; doi: 10.3390/met5042109 207

Soo Yeol Lee, Huamiao Wang and Michael A. Gharghouri
Twinning-Detwinning Behavior during Cyclic Deformation of Magnesium Alloy
Reprinted from: *Metals* **2015**, *5*(2), 881–890; doi: 10.3390/met5020881 216

Michael J. Demkowicz and Jaroslaw Majewski
Probing Interfaces in Metals Using Neutron Reflectometry
Reprinted from: *Metals* **2016**, *6*(1), 20; doi: 10.3390/met6010020 224

Sara J. Callori, Christine Rehm, Grace L. Causer, Mikhail Kostylev and Frank Klose
Hydrogen Absorption in Metal Thin Films and Heterostructures Investigated in Situ with Neutron and X-ray Scattering
Reprinted from: *Metals* **2016**, *6*(6), 125; doi: 10.3390/met6060125 241

Kengo Kidena, Minami Endo, Hiroki Takamatsu, Masahito Niibe, Masahito Tagawa, Kumiko Yokota, Yuichi Furuyama, Keiji Komatsu, Hidetoshi Saitoh and Kazuhiro Kanda
Resistance of Hydrogenated Titanium-Doped Diamond-Like Carbon Film to Hyperthermal Atomic Oxygen
Reprinted from: *Metals* **2015**, *5*(4), 1957–1970; doi: 10.3390/met5041957 261

Klaus-Dieter Liss, Ken-Ichi Funakoshi, Rian Johannes Dippenaar, Yuji Higo, Ayumi Shiro, Mark Reid, Hiroshi Suzuki, Takahisa Shobu and Koichi Akita
Correction: Liss, K.-D., et al. Hydrostatic Compression Behavior and High-Pressure Stabilized β-Phase in γ-Based Titanium Aluminide Intermetallics. Metals 2016, 6, 165
Reprinted from: *Metals* **2017**, 7(9), 353; doi: 10.3390/met7090353 273

About the Special Issue Editor

Klaus-Dieter Liss is a Professor at the Guangdong Technion – Israel Institute of Technology in Shantou, China (since 2017) holding an academic appointment as Full Professor at the Technion, Israel. He also is a Honorary Professor at the school of Mechanical, Materials, Mechatronic and Biomedical Engineering at the University of Wollongong, Australia (since 2014). Starting in 1983, he studied General Physics at the Technische Universität München, Germany (Dipl. Phys., 1990) and defended a thesis at the RWTH Aachen, Germany (Dr. rer. nat., 1995) supported by a doctoral scholarship at the Institut Laue-Langevin (ILL), France. His career in neutron-scattering started in 1986, spending continuously several months per year at the ILL, complemented by high-energy synchrotron radiation since 1991. In 1994, Liss started as a postdoctor at the European Synchrotron Radiation Facility (ESRF) in France, followed by a full position as beamline scientist and beamline responsible from 1995 to 2001. He took one semester sabbatical leave, lecturing diffraction theories at the Friedrich-Alexander-Universität Erlangen-Nürnberg, Germany in 1998–1999. After a short time in Germany at GKSS Research Center, DESY and Technische Universität Hamburg Harburg (2001-2003) he moved as a Senior Researcher to the Bragg Institute of the Australian Nuclear Science and Technology Organisation, ANSTO in 2004, where he enjoyed the inaugural Senior Research Fellowship (2007-2010) of the organization. In 2013-2014 he spent a sabbatical year at the Japan Atomic Energy Agency in Tokai.

In the recent years, Liss' research focuses on neutron and synchrotron diffraction methods for the investigation of thermo-mechanical processes mainly on metals, including in situ and time-resolved measurements and pioneering experiments to study phase transformations, microstructure evolution, order and disorder, and defect kinetics. Current research themes encompass materials processing under extreme conditions, fast and ultrafast time response in materials, and novel and enhanced quantum beam sources for the application in materials science and physics.

Klaus-Dieter Liss held presidency of the Materials Australia New-South-Wales Branch (2014-2017) and is National Councilor in that society (since 2011). He is Fellow of the Australian Institute of Physics and Member of the TMS Phase Transformations Committee. As Editor-In-Chief he created the journal Quantum Beam Science and is member of the Editorial Board of Metals, as well as Advanced Engineering Materials. He has been Guest Editor for MRS Bulletin, June 2016 and for MRS Proceedings. In 2012 and 2014, Liss chair-organized the MRS Fall Meeting symposia on applications of neutron and synchrotron radiation, reading tutorials, as well as at TMS. He organized similar symposia at PRICM-8 (2013) and THERMEC (2011, 2013, 2016).

Preface to "Metals Challenged by Neutron and Synchrotron Radiation"

Neutron and Synchrotron radiation methods have matured to become powerful techniques for the study of a vast range of materials, including metals. The characterization methods comprise the categories of diffraction, spectroscopy and imaging, which themselves can alter greatly in detail, to include hundreds of variants, problems and sample environments. In a similar way, their applications to metals and hard condensed matter materials cover disciplines spanning engineering, physics, chemistry, materials science and their derivatives such as geology, energy storage, etc.

The present book, "Metals Challenged by Neutron and Synchrotron Radiation" is a first compilation in Metals of 20 original and review works on research utilizing or designing those state-of-the-art techniques at modern facilities. The Editorial reviews the context of and identifies thematic links between these papers, grouping them into five interwoven themes, namely Sintering Techniques and Microstructure Evolution, Titanium Aluminides and Titanium Alloys Under Extreme Conditions, Metallic Glass and Disordered Crystals, In Situ and Time-Resolved Response to Mechanical Load and Shock, and Thin Films and Layers.

This book represents a good cross-section of the status quo of neutron and synchrotron radiation with respect to questions in the metallurgical field, which by far is not exhaustive. Nor are the methods and other materials, which motivated me to the creation of a new sister-journal, entitled Quantum Beam Science. With this, I would like to thank all authors, reviewers and contributors behind the scene for the creation of this work, presenting to you a piece of interesting reading and reference literature.

Klaus-Dieter Liss
Special Issue Editor

metals

MDPI

Editorial

Metals Challenged by Neutron and Synchrotron Radiation

Klaus-Dieter Liss [1,2]

[1]   Australian Nuclear Science and Technology Organisation, New Illawarra Road,
      Lucas Heights, NSW 2234, Australia; kdl@ansto.gov.au or liss@kdliss.de; Tel.: +61-2-9717-9479
[2]   School of Mechanical, Materials, Mechatronic & Biomedical Engineering, University of Wollongong,
      Wollongong, NSW 2522, Australia

Received: 28 June 2017; Accepted: 29 June 2017; Published: 11 July 2017

1. Introduction and Scope

In the past one and a half decades, neutron and synchrotron radiation techniques have come to the forefront as an excellent set of tools for the wider investigation of material structures and properties [1,2], becoming available to a large user community. This holds especially true for metals, which are a fascinating class of materials with both structural and functional applications. With respect to these application classes, metals are used to engineer bridges and automotive engines as well as to exploit magnetic and electric properties in computer storage, optics, and electronics. Both neutron sources and synchrotrons are large user facilities of quantum-beam installations [3] with the implementation of a common accelerator or nuclear reactor-based source, often serving over 50 beamlines simultaneously and even more end stations. Up to a few thousand experiments are undertaken yearly, utilizing specialized beam conditions, sample environments, and detection systems. Their variations range across spectroscopy, diffraction, small-angle scattering, and inelastic scattering for sample sizes ranging from nanometers to meters. Examples of such installations can be found in the Topical Collection *Facilities of Metals'* sister journal *Quantum Beam Science* [3].

The scope of the present Special Issue in *Metals* comprises articles on research case studies on individually selected systems. Fields of interest range from engineering, through materials design, to fundamental materials science, including non-exclusively, strain scanning, texture analysis, phase transformation, precipitation, microstructure reconstruction, crystal defects, atomic structures (both crystalline and amorphous), order and disorder, kinetics, time-resolved microstructure evolution, local structure correlations, phonons, deformation and transformation mechanisms, response to extreme conditions, local and integrated studies, both within the bulk and at interfaces. Regarding the breadth of the discipline, this contribution is not exhaustive by far, but stimulates important and evolving studies throughout the metals community.

2. Contributions

Twenty articles have been published in the present Special Issue of *Metals*, encompassing the fields of sintering techniques, titanium aluminides and titanium alloys, metallic glass and disorder in crystals, and thin layers and interfaces. This grouping is not hard-bound, and thematic links can be established beyond them, which shall be emphasized in the following presentation.

2.1. Sintering Techniques and Microstructure Evolution

In situ investigations for following morphology and phase constitution in sintering and annealing processes have been enabled by the high flux of modern neutron and X-ray beams. While wide-angle diffraction allows for the recording of the time-resolved crystallographic structural evolution, small-angle scattering determines precipitations on the nanometer scale, complemented by tomographic imaging into micrometer-sized features.

Such in situ phase evolutions are presented by Chen et al. upon vacuum sintering of Ni-Ti powder compacts [4]. The authors present neutron diffraction patterns of both Ti-Ni and TiH_2-Ni mixtures taken every 34 s upon a heating ramp to 1373 K, unraveling the appearance and evolution of intermediate and final product phases. Hydrogen and its release not only influence the residual pore size of the product, but also highly affects the intermetallic diffusion and reaction during the sintering process.

Sub-micrometer-resolution synchrotron X-ray computed tomography has been used by both Xiao et al. [5] and Ma et al. [6] for in situ investigations of microwave sintering metals. Xiao et al. [5] work on titanium powder, which was sintered at 1173 K and data recorded in 900 s time steps. The reconstructed data allows for the extraction of phenomena such as particle densification together with neck formation and growth. The objects of Ma et al. [6] were packed stainless-steel fiber felts sintered at 1273 K–1473 K. Similarly, the neck forming between two crossing sintering fibers were evaluated. The data is interpreted analytically with atom migration rates due to atom and vacancy diffusion processes and the surface energy of the neck. Because of the complicated heating process in microwave sintering driven by electro-magnetic wave interaction, anomalous sintering rates result from the observations.

In situ investigations on microstructure and phase evaluations from the bulk of material are most important to develop modern high-strength, transformation-induced plasticity steels. Here, Nishijima et al. [7] present small-angle neutron scattering simultaneously taken with dilatometry in order to determine the precipitation and evolution of ferrite in austenite, after quenching from 1173 K to 573 K and subsequent holding. Furthermore, the quantitative phase evolution is monitored in wide-momentum-transfer neutron diffraction, revealing not only phase fractions but also strain in the lattice parameters.

Such combined techniques open a pathway for new alloy development in a larger sense.

2.2. Titanium Aluminides and Titanium Alloys under Extreme Conditions

Already two of the examples presented in Section 2.1 emphasize the importance of titanium and its alloys [4,5]. Much focus on alloys designed for extreme operating conditions is on the titanium aluminide intermetallics, complemented by a nice feature of a Ti-Zr solid solution.

Neutron and synchrotron in situ characterization methods have contributed tremendously to the development of modern γ-based titanium aluminides, as reviewed by Erdely et al. [8]. The Ti-Al phase diagram containing *fcc*-based γ- and *hcp*-based α_2-phase at operating temperature is not only complex by nature; moreover, alloying is undertaken to provide a ductile, *bcc*-based β-phase at processing temperatures above 1600 K. All phases can be ordered and fully or partially disordered, adding to the complexity, while the prevailing structure strongly influences the mechanical and thermal properties. The very particular neutron scattering behavior of titanium, possessing a negative scattering amplitude—in contrast to most other elements—is uniquely exploited to determine the atomic order in such phases. Superstructure peaks, uncloaking the order, are extremely strong while the main reflections, seen with X-rays, literally disappear, forming a so-called null matrix. Therefore, synchrotron X-ray diffraction can determine the overall crystal structure, and neutron scattering can determine the atomic order therein very sensitively. Similar to the above on steels, Erdely et al. [8] utilize small- and wide-range-scattering to determine in situ fingerprints of microstructure and phase evolution in reciprocal space. State-of-the-art, real-time, in situ studies during plastic deformation at 1573 K, where the morphology of intensity distribution along the two-dimensional Debye-Scherrer rings is streaked in time, reveal crystal deformation, dynamic recovery, and dynamic recrystallization processes. The paper not only reviews many of the pioneering methods cited therein, but also gives comprehensive reference to the thermo-mechanical properties and the development of such intermetallics for industrial application.

A greater emphasis on plasticity is presented by Stark et al. [9], in a particular case study using the abovementioned method to investigate γ-based titanium aluminum alloyed with niobium. In Ti-42Al-8.5Nb (at. %), the ductile β- and potentially ordered β_o-phase is abundant above ~ 1400 K, evolving even further at higher temperatures. In their experiment, two-dimensional in situ synchrotron high-energy X-ray diffraction is undertaken upon plastic deformation to 40% in 100 s, and is continuously recorded with 4 s time frames. The azimuthal intensity distribution leads to a detailed texture analysis and an illustration of their evolution, revealing intensity fluctuations that indicate dynamic recrystallization. It is revealed that phase composition, recrystallization, and the amount of deformation influence the texture significantly. The study demonstrates the potential of in situ diffraction techniques in a so-called Materials Oscilloscope [10] for the optimization of hot-forming multi-phase alloys.

More often, titanium aluminides are processed under extreme conditions, including high pressure, which gave myself and others, Liss et al. [11], the motivation to investigate a γ-based titanium-aluminide under high pressure and temperature. We used a 15 MN press to impose on a volume containing a 1.5 mm^3 sample inside a heating element while synchrotron X-ray diffractograms were recorded. Pressurizing at room temperature reveals the equation of states, introduction of crystallographic disorder, and continuous phase transformation of γ in favor of α/α_2. Regarding the high ordering energy of γ, this transformation is favored despite increasing the volume per atom, which is counterintuitive under pressure. Heating at 10 GPa to melting reveals shifting of the phase transformation temperatures. Moreover, the β-phase is found at high pressure and temperature in 3- and 2-phase fields. Such a ductile β-phase opens processing windows under high pressure, while it would not be abundant at operating conditions under ambient pressure, reducing problems of creep under constant load.

Related to this topic is the making of a titanium-zirconium null-matrix material, reported by Okuchi et al. [12], which is being developed as a neutron-transparent material for gaskets and high-pressure cells in neutron scattering experiments under extreme conditions, as outlined above. In a null-matrix material, coherent neutron scattering is suppressed by alloying elements of positive and negative scattering amplitude, Zr and Ti, respectively, in a solid solution. The composition is chosen such that the average coherent neutron scattering length, which determines the strength of the Bragg peaks, is equal to zero. The forging method presented here results in finer lamellar microstructures, improving the mechanical properties of the material while still exposing excellent neutron-optical properties.

Interestingly, the scattering process of the titanium-zirconium null-matrix material produced by Okuchi et al. [12] relates to the investigations summarized by Erdely et al. [8], because titanium-aluminides equally form a null-matrix when fully disordered. In their study, the authors use the latter to determine the aforementioned crystallographic order parameter upon disorder, rounding up this section.

2.3. Metallic Glass and Disordered Crystals

Crystallographic disorder, as described in Section 2.2., can be detected by diffuse neutron or X-ray diffraction, which is presented in an article by Goossens [13]. As ordered intermetallics undergo disorder, for example, atoms randomly swap sites and their species on a given lattice site are determined by their random probability as given by their concentration, i.e., they become a solid solution, and their coherent scattering differences contribute to incoherence and scatter into an all solid angle. If there is remaining order, i.e., short-range order existing just over a few unit cells, this diffuse scattering peaks at particular locations in a reciprocal lattice. There exist numerous kinds of disorder, such as anti-site swaps, vacancies, interstitials, spin-orientation, magnetism, two-body correlations, and thermal vibrations—phonons. Such local order and defects occur through a variety of complex metals, including intermetallics, alloys, and quasicrystals, and can strongly influence mechanical as well as functional properties of the material. Detailed modeling of the diffuse scattering can be a challenge, and is the object of Goossen's contribution [13], realized here by an inverse Monte-Carlo method. Crystal structures and defects are simulated minutely in a computer program and the corresponding diffraction pattern is calculated and iterated by randomization of the defects step by step, until a realistic, averaged diffraction pattern is obtained. The author shows in three-dimensional reciprocal space the response to different defect contributions and crystal symmetries in various configurations.

A further degree of disorder is achieved when atoms do not sit on crystal lattice sites at all, which is the case in amorphous materials, and in the present case of metallic glass. Within the context of X-ray, neutron, or even electron beam diffraction, the concept of Bragg diffraction, based on crystal planes, breaks down. As Bragg diffraction is a discrete Fourier series of an infinite periodic structure, the discrete points Q_i forming the reciprocal lattice, and coefficients expressed by structure factors S_i, amorphous disordered systems have to be treated in a continuous Fourier integral with a continuous scattering function $S(Q)$. The back-transformation from an amorphous diffraction pattern leads to the pair distribution function or radial distribution function, as briefly introduced by Egami et al. [14]. Similarly, electrons scatter from atoms and form diffraction patterns, which can be obtained by measuring the extended X-ray absorption fine structure, known as EXAFS. In this process, electrons are emitted from an element-specific atom by tuning the X-ray energy across the absorption edge, scattering at surrounding atoms, and forming an interference pattern. Similarly, Fourier back-transformation leads to local atom distribution functions.

Emphasis is given not only on the structure, but also on the strain response to elastic and plastic deformation. Suzuki et al. [15] report on elastic deformation mechanisms in Zr-Cu-Al metallic glasses under as-cast and heat-treated conditions. Like in any object under uniaxial tensile loading, atom distances expand in the loading direction and normally contract transversely, reflecting in changes and anisotropy of the diffraction patterns. Peak shifts can be observed and evaluated both in reciprocal space mapping and after back transformation to direct space. A salient feature of this paper is that elastic response, i.e., the Young modulus, depends on the length of the scale observed. It differs for its nearest neighbors, for long-range atomic average, and for macroscopic distances. The latter can contain pseudo-elastic strain, which is localized plastic strain, where atoms in building blocks change their morphology in a reversible manner, leading to a larger overall strain. Diffraction processes detect only elastic strain, such that this pseudo-elastic effect is not seen directly. As the method can distinguish nearest-neighbor distances, second-nearest-neighbors etc., it is sensitive to separating strain response on different length scales. Suzuki et al. [15] apply these aspects to determine the stress behavior in various eutectic, hypoeutectic, and heat-treated glasses, and reveal strongly- and weakly-bonded regions and clusters. By the way of elastically perturbing the system, the authors have shown how to obtain structural information on various length scales, not otherwise possible to reveal from diffraction and macroscopic methods alone.

Likewise, Egami et al. [14] treat the atomic response of metallic glass to perturbations by a uniaxial mechanical load, breaking the symmetry of the system and extracting further information on the amorphous microstructure over length-scales. Non-affine atomic rearrangements below the yield stress have been found to be localized and reversible, leading to the abovementioned pseudo-elasticity. As an analysis method, the authors present the difference pair-distribution function, showing residuals between the loading and transverse direction, and allowing the extraction of the non-affine response of deformation in so-called localized shear-transformation zones. Increasing temperature leads to avalanche collapses of the latter, resulting in creep.

The structure of metallic glass is further discussed by Guo et al., particularly how synchrotron radiation [16] and combined simulations [17] are employed to determine icosahedral and icosahedral-like clusters. Both techniques, diffraction and absorption spectroscopy, EXAFS, are used to index Voronoi clusters in a tessellation. Because the determined popular clusters in Zr-30Pd (at. %) glass share configuration similarities with the Zr_2Pd quasi-crystalline phase, the latter precipitates upon annealing before transformation into a fully crystalline phase.

All contributions lead to the conclusion that a metallic glass is largely made up of various kinds of close-packed building fragments which link together in an amorphous network. There are localized hard centers, with softer regions in-between, which can reversibly deform in a non-affine way and lead to pseudo-elasticity. Diffraction and scattering methods primarily probe for atomic length-scales, but when coupled with response to an applied uniaxial stress, it can deliver representative structural information on larger, intermediate length-scales, which is not otherwise accessible.

2.4. In Situ and Time-Resolved Response to Mechanic Load and Shock

The concept of real-time and in situ studies has already been touched in the previous chapters, where examples of thermally driven reactions and transformations, as well as response to mechanical stress, were investigated. In particular, the brilliance of synchrotron radiation allows one to focus the investigation under highly advanced resolution, be it in time or spatial.

The deformation behavior of individual grains in polycrystalline Cu-Al-Mn super-elastic alloy is presented by Kwon et al. [18], who use a 15×15 μm^2 beam of white high-energy X-rays in the transmission of a 200 μm thick tensile specimen with a 400 μm grain size. When scanned over the grains and their boundaries, the recorded white-beam Laue diffraction patterns minutely reveal grain orientation, as well as gradients therein. In particular, streaking of the Laue spots occurs according to the gradients in the illuminated volume. On top of this, the Laue spots have been analyzed by an energy-dispersive detector, altogether resolving locally reciprocal space in three dimensions, namely orientation and depth. Data is taken in situ while the specimen is being loaded, giving local strain feedback to the applied stress. It is found that the resulting strain distribution not only varies from grain to grain, but, furthermore, orientation and lattice-spacing gradients evolve within one grain. Reversible phenomena during super-elastic deformation have been observed and are interpreted by a reversible martensitic transformation. In this way, super-elasticity shows parallels to the non-affine pseudo-elastic deformation described in Section 2.3 on metallic glasses.

Ultra-high time resolution is reviewed by Ichiyanagi et al. [19], investigating the dynamic structural response to shock waves in a laser-pump X-ray-probe setup. High-power lasers are used to induce pressure waves into the 10 GPa range within nanoseconds, synchronized with a given time shift to 100 ps probing X-ray pulses. Repeating the experiment with different time delays results in scanning the temporal evolution of the structural changes. From an X-ray optical point of view, 'pink' beams covering a wide bandwidth are used to match enough reciprocal-space acceptance for given crystal orientations. Similar to the abovementioned Laue method, streaking of the reflection spots determines lattice gradients in a CdS single crystal at each snapshot in time. In contrast to quasi-static compression, the structural phase transition at 3.25 GPa was not observed on nanosecond dynamic loading-unloading, revealing an over-compressed state with an incubation time greater than 10 ns. Furthermore, temporal response is demonstrated for polycrystalline and amorphous materials. Different behavior between single-crystalline and polycrystalline specimens allow for the elucidation of nucleation and growth mechanisms of the pressure-induced phase transformations.

Back to more conventional time-scales, Lee & Huang et al. [20] and Lee & Wang et al. [21] employ neutron scattering on cyclic loading for fatigue testing a crack tip in stainless steel under steady and overloaded fatigue conditions. They also study the twinning-detwinning behavior in magnesium alloy under reversed loading. The crack tip analysis was performed by scanning a 1 mm^2 neutron beam along the crack-propagation direction, revealing internal lattice strains which concentrate at the tip. A comparison of the overloaded with the steady fatigued specimen shows large compressive stresses behind the crack tip, suggesting a mechanism for crack-growth retardation after overloading. The magnesium study considers load partitioning between the different crystallographic grain orientations and texture evolution in Mg-8.5Al (mass %) alloy in four scenarios, compression followed by reverse tension and vice versa, respectively starting with extrusion texture and a reoriented texture. In both cases, extension twinning is activated in the first deformation step, basically flipping the basal pole from transverse to axial in the first case, and axial to transverse in the second case. There are asymmetries and non-linearities in unloading and reverse deformations, which are discussed in detail with the stress states of the different plane families.

These contributions round up important insights into the potential of in situ deformation studies, comprising lattice-orientation resolved strain and texture. The knowledge, measurements, and modeling of these inherently relate to the mechanical and physical properties of a material.

2.5. Thin Films and Layers

Besides the bulk studies presented here, metals and materials in the form of thin films play a highly important role in modern technology, in the form of functional materials, such as magnetic nano-layers in data storage, catalysts, optical components, and electronic devices, and even beyond hard condensed matter physics in biological systems, such as peptide layers or cell membranes.

Neutron and X-ray reflectometry and derived grazing-incident diffraction are important methods where evanescent waves are used to probe depths into the films from atomic distances to micrometers. When these depths are scanned by varying the incident angle to the surface, or more quantitatively by varying the scattering vector, accurate depth profiles are obtained. Demkowicz and Majewski [22] review the basic scattering theory and concepts behind this method, starting from the very basic concepts of diffraction, the definition of scattering vector, evanescent wave functions, and interaction potential expressed by atomic scattering lengths. Both neutron and X-ray scattering theories are very similar, differing mostly on the different scattering amplitudes, resulting in distinct interaction potentials, which are also expressed in refractive index. First of all, the thickness of the layers leads to thickness fringes, which is an important measure, particularly for in situ studies. For example, in magnetic metallic multilayers, fringes and intensities minutely depend on the spin orientations and help to understand, e.g., exchange couplings through a non-magnetic layer under applied conditions. In addition, the sharpness and the inter-diffusion of atomic layers can be revealed from the overall reflection curves.

A very nice application of the capabilities of neutron and X-ray reflectometry is on hydrogen absorption in metallic thin films and heterostructures, presented by Callori et al. [23]. Safe hydrogen storage not only plays a paramount role in chemical energy storage, but hydrogen can also alter physical properties, such as magnetism and electric conductivity in a functional device. In contrast to characterization by X-rays, neutron scattering is the ultimate probe for atomic structures containing hydrogen. This is because of the large neutron scattering lengths of hydrogen. Moreover, the isotope ^{1}H possesses a negative coherent scattering length of -3.7406 fm while its heavier counterpart deuterium, ^{2}H or D, diffracts with 6.671 fm. Mixing the two in pre-determined concentrations allows for contrast variation, including the null-matrix effect as discussed in Section 2.2 and other contrast matching. It shall be noted that hydrogen also possesses a nuclear spin, which in principle can be polarized for complementary analysis, but is not part of the present study. In contrast, X-ray scattering is poorly sensitive to hydrogen atoms, as they only possess one scattering electron which most often is not even localized. Callori et al. [23] demonstrate that the refractive index varies in the Pd-H system, related to the scattering-length-density changes due to mixing negative H with positive Pd scatterers, reducing the critical angle for total reflection in a reflectivity curve, from which the H concentration is obtained. In metallic crystal lattices, hydrogen diffuses on interstitial sites, which can be barely filled, to be fully occupied in a stoichiometric metal hydride, straining the original metal lattice. The authors present data for several Fe-Nb multilayers before and after hydrogen loading, showing shifts of the thickness fringes due to swelling of the layer thicknesses. On top of this, polarized neutron reflectivity is sensitive to the magnetism in the system carried by the Fe layers, showing that its ferromagnetism is not affected by hydrogen loading the Fe/Nb multilayers. All three effects, namely refractive index, thickness fringes, and magnetic polarization, are obtained simultaneously. The results are further discussed with X-ray diffraction results and various other methods to quantify phase transformations, anisotropic expansion, and magnetism. Such investigations not only play an important role in the microstructure engineering of hydrogen-storage devices, but also for the functional properties of sensors.

Hydrogenated diamond-like carbon film with titanium doping is a lubricant in spacecraft applications, and is investigated by Kidena et al. [24] with respect to resistance to hyperthermal oxygen exposure, as found in the lower orbits around Earth. Thin films of 400 nm have been investigated by a multitude of quantum-beam methods, including Rutherford back-scattering and elastic recoil detection analysis of He, X-ray photo-electron spectroscopy, and synchrotron X-ray absorption fine structure (NEXAFS) measurements. The results of these methods lead to the determination of film thickness, layer profile, and hydrogen content. It has been found that the material is excellently stable under low-orbit conditions, as bombardment with oxygen forms a stable, 5 nm thick titanium oxide layer, having no influence on lubrication but protecting the degradation of hydrogen.

The examples given are not only important pieces of research, but also demonstrate the unique capabilities of neutron and synchrotron radiation for the investigation of thin films, coupled to complementary methods.

3. Conclusions and Outlook

A variety of interwoven topics have been compiled in the present Special Issue of *Metals*. The materials range from bulk metals to thin films, under extreme conditions in space, temperature, pressure, and shock, with applications as structural materials, sensors, as well as in data storage, energy storage, and neutron optics. The methods of investigation and niche applications of neutron and synchrotron radiation are similarly widespread, covering their diversity which includes spectroscopic methods, diffraction, polarization analysis, reflectometry, contrast matching, and coherent and incoherent scattering.

There is a large overlap of questions and scientific concepts shared by the various research communities, such as response to stress in glass and crystals, non-affine super-plasticity, order and disorder, phase transformations and chemical reactions, localization, thermal equilibrium, and materials under extreme conditions. The reader will benefit from cross-disciplinary work in this widespread field designated by the word *Metals*. Moreover, it is demonstrated that one probe alone, i.e., synchrotron radiation or neutron scattering, is not sufficient for an in-depth understanding of the relevant problems. Furthermore, cross-disciplinary application of different quantum beams can benefit from synergies beyond them. Therefore, as a next step addressing various classes of materials beyond the metals alone, a new sister journal to *Metals* has been created, entitled *Quantum Beam Science* [3], to open broader opportunities for future challenges by neutron and synchrotron radiation in *Metals*.

Conflicts of Interest: The author declares no conflict of interest.

References

1. Vogel, S.C. A Review of Neutron Scattering Applications to Nuclear Materials. *Int. Sch. Res. Not.* **2013**, *2013*, e302408. [CrossRef]
2. Liss, K.D.; Chen, K. Frontiers of synchrotron research in materials science. *MRS Bull.* **2016**, *41*, 435–441. [CrossRef]
3. Liss, K.D. Quantum Beam Science—Applications to Probe or Influence Matter and Materials. *Quantum Beam Sci.* **2017**, *1*, 1. [CrossRef]
4. Chen, G.; Liss, K.D.; Cao, P. An in situ Study of NiTi Powder Sintering Using Neutron Diffraction. *Metals* **2015**, *5*, 530–546. [CrossRef]
5. Xiao, Y.; Xu, F.; Hu, X.; Li, Y.; Liu, W.; Dong, B. In situ Investigation of Titanium Powder Microwave Sintering by Synchrotron Radiation Computed Tomography. *Metals* **2016**, *6*, 9. [CrossRef]
6. Ma, J.; Li, A.; Tang, H. Study on Sintering Mechanism of Stainless Steel Fiber Felts by X-ray Computed Tomography. *Metals* **2016**, *6*, 18. [CrossRef]
7. Nishijima, H.; Tomota, Y.; Su, Y.; Gong, W.; Suzuki, J. Monitoring of Bainite Transformation Using in Situ Neutron Scattering. *Metals* **2016**, *6*, 16. [CrossRef]

8. Erdely, P.; Schmoelzer, T.; Schwaighofer, E.; Clemens, H.; Staron, P.; Stark, A.; Liss, K.D.; Mayer, S. In Situ Characterization Techniques Based on Synchrotron Radiation and Neutrons Applied for the Development of an Engineering Intermetallic Titanium Aluminide Alloy. *Metals* **2016**, *6*, 10. [CrossRef]

9. Stark, A.; Rackel, M.; Tchouaha Tankoua, A.; Oehring, M.; Schell, N.; Lottermoser, L.; Schreyer, A.; Pyczak, F. In Situ High-Energy X-ray Diffraction during Hot-Forming of a Multiphase TiAl Alloy. *Metals* **2015**, *5*, 2252–2265. [CrossRef]

10. Liss, K.D. Structural Evolution of Metals at High Temperature: Complementary Investigations with Neutron and Synchrotron Quantum Beams. In *Magnesium Technology 2017*; Solanki, K.N., Orlov, D., Singh, A., Neelameggham, N.R., Eds.; The Minerals, Metals & Materials Series; Springer International Publishing: Cham, Switzerland, 2017; pp. 633–638.

11. Liss, K.-D.; Funakoshi, K.-I.; Dippenaar, R.J.; Higo, Y.; Shiro, A.; Reid, M.; Suzuki, H.; Shobu, T.; Akita, K. Hydrostatic Compression Behavior and High-Pressure Stabilized β-Phase in γ-Based Titanium Aluminide Intermetallics. *Metals* **2016**, *6*, 165. [CrossRef]

12. Okuchi, T.; Hoshikawa, A.; Ishigaki, T. Forge-Hardened TiZr Null-Matrix Alloy for Neutron Scattering under Extreme Conditions. *Metals* **2015**, *5*, 2340–2350. [CrossRef]

13. Goossens, D.J. Monte Carlo Modelling of Single-Crystal Diffuse Scattering from Intermetallics. *Metals* **2016**, *6*, 33. [CrossRef]

14. Egami, T.; Tong, Y.; Dmowski, W. Deformation in Metallic Glasses Studied by Synchrotron X-Ray Diffraction. *Metals* **2016**, *6*, 22. [CrossRef]

15. Suzuki, H.; Yamada, R.; Tsubaki, S.; Imafuku, M.; Sato, S.; Watanuki, T.; Machida, A.; Saida, J. Investigation of Elastic Deformation Mechanism in As-Cast and Annealed Eutectic and Hypoeutectic Zr–Cu–Al Metallic Glasses by Multiscale Strain Analysis. *Metals* **2016**, *6*, 12. [CrossRef]

16. Guo, G.-Q.; Wu, S.-Y.; Luo, S.; Yang, L. How Can Synchrotron Radiation Techniques Be Applied for Detecting Microstructures in Amorphous Alloys? *Metals* **2015**, *5*, 2048–2057. [CrossRef]

17. Guo, G.-Q.; Wu, S.-Y.; Luo, S.; Yang, L. Detecting Structural Features in Metallic Glass via Synchrotron Radiation Experiments Combined with Simulations. *Metals* **2015**, *5*, 2093–2108. [CrossRef]

18. Kwon, E.P.; Sato, S.; Fujieda, S.; Shinoda, K.; Kainuma, R.; Kajiwara, K.; Sato, M.; Suzuki, S. Characterization of Deformation Behavior of Individual Grains in Polycrystalline Cu-Al-Mn Superelastic Alloy Using White X-ray Microbeam Diffraction. *Metals* **2015**, *5*, 1845–1856. [CrossRef]

19. Ichiyanagi, K.; Nakamura, K.G. Structural Dynamics of Materials under Shock Compression Investigated with Synchrotron Radiation. *Metals* **2016**, *6*, 17. [CrossRef]

20. Lee, S.Y.; Huang, E.-W.; Woo, W.; Yoon, C.; Chae, H.; Yoon, S.-G. Dynamic Strain Evolution around a Crack Tip under Steady- and Overloaded-Fatigue Conditions. *Metals* **2015**, *5*, 2109–2118. [CrossRef]

21. Lee, S.Y.; Wang, H.; Gharghouri, M.A. Twinning-Detwinning Behavior during Cyclic Deformation of Magnesium Alloy. *Metals* **2015**, *5*, 881–890. [CrossRef]

22. Demkowicz, M.J.; Majewski, J. Probing Interfaces in Metals Using Neutron Reflectometry. *Metals* **2016**, *6*, 20. [CrossRef]

23. Callori, S.J.; Rehm, C.; Causer, G.L.; Kostylev, M.; Klose, F. Hydrogen Absorption in Metal Thin Films and Heterostructures Investigated in Situ with Neutron and X-ray Scattering. *Metals* **2016**, *6*, 125. [CrossRef]

24. Kidena, K.; Endo, M.; Takamatsu, H.; Niibe, M.; Tagawa, M.; Yokota, K.; Furuyama, Y.; Komatsu, K.; Saitoh, H.; Kanda, K. Resistance of Hydrogenated Titanium-Doped Diamond-Like Carbon Film to Hyperthermal Atomic Oxygen. *Metals* **2015**, *5*, 1957–1970. [CrossRef]

metals

MDPI

Article

An *in situ* Study of NiTi Powder Sintering Using Neutron Diffraction

Gang Chen [1,2,*], Klaus-Dieter Liss [3,4] and Peng Cao [1,*]

[1] Department of Chemical and Materials Engineering, the University of Auckland, Private Bag 92019, Auckland 1142, New Zealand

[2] State Key Laboratory of Porous Metal Materials, Northwest Institute for Nonferrous Metal Research, Xi'an 710016, Shaanxi, China

[3] Australian Nuclear Science and Technology Organisation, New Illawarra Road, Lucas Heights, NSW 2234, Australia; kdl@ansto.gov.au

[4] Quantum Beam Science Directorate, Japan Atomic Energy Agency, 2-4 Shirakata-Shirane Tokai-mura, Naka-gun, Ibaraki-ken 319-1195, Japan

* Authors to whom correspondence should be addressed; mychgcsu@163.com (G.C.); p.cao@auckland.ac.nz (P.C.); Tel.: +86-29-8623-1095 (G.C.); Fax: +86-29-8626-4926 (G.C.).

Academic Editor: Hugo F. Lopez

Received: 27 February 2015; Accepted: 27 March 2015; Published: 3 April 2015

Abstract: This study investigates phase transformation and mechanical properties of porous NiTi alloys using two different powder compacts (*i.e.*, Ni/Ti and Ni/TiH$_2$) by a conventional press-and-sinter means. The compacted powder mixtures were sintered in vacuum at a final temperature of 1373 K. The phase evolution was performed by *in situ* neutron diffraction upon sintering and cooling. The predominant phase identified in all the produced porous NiTi alloys after being sintered at 1373 K is B2 NiTi phase with the presence of other minor phases. It is found that dehydrogenation of TiH$_2$ significantly affects the sintering behavior and resultant microstructure. In comparison to the Ni/Ti compact, dehydrogenation occurring in the Ni/TiH$_2$ compact leads to less densification, yet higher chemical homogenization, after high temperature sintering but not in the case of low temperature sintering. Moreover, there is a direct evidence of the eutectoid decomposition of NiTi at *ca.* 847 and 823 K for Ni/Ti and Ni/TiH$_2$, respectively, during furnace cooling. The static and cyclic stress-strain behaviors of the porous NiTi alloys made from the Ni/Ti and Ni/TiH$_2$ compacts were also investigated. As compared with the Ni/Ti sintered samples, the samples sintered from the Ni/TiH$_2$ compact exhibited a much higher porosity, a higher close-to-total porosity, a larger pore size and lower tensile and compressive fracture strength.

Keywords: NiTi; powder sintering; dehydrogenation; neutron diffraction

1. Introduction

NiTi alloys have excellent properties including unique shape memory effect (SME), superelasticity, good biocompatibility and great energy absorption, which have been attracting attention from multiple areas such as medical devices, energy absorbers, actuators and mechanical couplings [1,2]. Powder metallurgy (PM) is a simple, energy-saving and widely used route to produce NiTi alloys [3]. Additionally, powder sintering is an effective technique to produce various porous structures, which are beneficial to bone tissue ingrowth and also provide an effective way of reducing stiffness of the implant [4].

Elemental powder sintering to fabricate porous NiTi alloys has been tremendously successful recently [4–10]. Interestingly, TiH$_2$ powder was frequently used in NiTi powder sintering in previous studies [4,10–17] due to its cleansing effect of dehydrogenation, which lowers oxygen content and potentially promotes chemical homogenization and densification [18,19]. There is no doubt that the use of TiH_2 favors final phase homogenization after high temperature sintering in the previous reports [4,10–17]. However, our most recent results [10,17,20,21] and the report from Robertson and Schaffer [14] disclosed a discouraging densification and a much larger porosity when using TiH_2 powder. As such, the use of such powder cannot guarantee densification promotion in all NiTi studies, although it does show densification in some other alloys, e.g., pure Ti, Ti-6Al-4V, Ti-5Al-2.5Fe and TiAl [19,22–27]. This might be caused by other factors simultaneously affecting the sintering process and thus the densification. These factors include TiH_2 particle size in Refs. [11,12,28,29] and the binders used in the reports [4,16]. Our recent results [17,20,21] also pointed out that it is the dehydrogenation of TiH_2 powder that increased the porosity of sample and then hindered its densification, when compared with that using similar particle size of Ti powder.

The process of TiH_2 dehydrogenation has been studied for many years [17,19,20,25,27,30–36]. However, most of the studies are conducted in either argon or air atmosphere [15,19,32,33,35]. With respect to the atmosphere, the dehydrogenation usually takes place in the temperature range from 523 to 973 K (250 to 700 °C), which possibly causes the concern of TiH_2 oxidation. On the other hand, some studies, e.g., Refs. [31,34], were performed in vacuum, effectively avoiding the oxidation issue. In spite of this, the diffraction instrument used is laboratory low-intensity X-ray diffraction systems [34], which normally require several minutes to one hour to achieve a complete scan for phase analysis and the achieved data is normally semi-accurate. Such "long"-time scanning properly leads to delayed or missing information. These technical limitations can be tackled with high-energy neutron diffraction under vacuum, which is able to penetrate bulk metals, and this type of diffraction has been successfully employed for *in situ* studies for sintering mechanism and reactions [20,36]. The beam intensities allow information from bulk material to be followed on short time scales (less than 60 s), while undergoing an *in situ* heating/cooling cycle to observe phase transformations. Furthermore, due to the strong incoherent neutron scattering from hydrogen, neutron diffraction can also track the development of hydrogen concentration during dehydrogenation [20].

Since dehydrogenation of TiH_2 involving in the reaction procedure of powder sintering, this reactive process is thought to be more intricate and different from the case of Ni/Ti blend. To the best of our knowledge, no report has elaborated the reactive sintering mechanism using Ni/TiH_2 blend involving dehydrogenation of TiH_2 and the mechanism investigation of TiH_2 decomposition under vacuum. Bearing in mind, it is of great importance to investigate the combination of dehydrogenation of TiH_2 and newly born Ti and Ni sintering hereafter and the comparative study of mechanical properties of as-fabricated NiTi alloys using Ni/Ti and Ni/TiH_2 powder blends. In this study, it is the first time to observe and study the combined phase transformation processes of dehydrogenation of TiH_2 and the subsequent reactions between new-born Ti and Ni particles using *in situ* neutron diffraction under vacuum as a comparison of the Ni/Ti blend. Further, the systematic mechanical comparison was investigated in terms of pore size, porosity, pore shape and pore size distribution. Therefore, this study is an additional and supplemental report to our recent results in Refs. [17,20].

2. Experimental Section

The mean particle size of Ti, TiH_2 and Ni raw powders used in this study was 32.2, 24.6 and 16.4 μm, respectively. Powder mixtures of Ni/Ti and Ni/TiH_2 were gently mixed in a ball mill for 10 h. Both powder mixtures had a nominal composition of 51 at.% Ni and 49 at.% Ti.

After mixing, powder mixtures were pressed into cylindrical discs of 12 mm diameter with three heights (*i.e.*, 4, 10 and 20 mm for microstructural characterization, neutron diffraction measurement and compression test, respectively) and tensile testing bars (15 mm in gauge length and 2 mm in thickness) in a single-action steel die under 250 MPa pressure. Stearic acid lubricant was slightly applied to the compaction die wall. Subsequently, the 4- and 20-mm-thick green compacts and tensile bars were sintered in a vacuum furnace at 3×10^{-3} Pa, while the 10-mm-thick green compacts were sintered in a high temperature vacuum furnace (5×10^{-4} Pa) equipped on the WOMBAT for *in situ* neutron diffraction measurements. The WOMBAT is a high-intensity diffractometer at the Australian Nuclear Science and Technology Organization (ANSTO), which uses monochromatic neutrons and is equipped with a two-dimensional area detector [37]. The basic technical information of WOMBAT is detailed in Refs. [20,36]. The sintering profile with a heating rate of 5 K/min will be shown in Section 3.2. The heating process was designed into two stages where the first stage is for dehydrogenation of TiH_2 powders, while the second one is to perform final sintering at a temperature of 1373 K (1100 °C) for 2 h, followed by furnace cooling.

A free Rietveld program *MAUD* was chosen to analyze the full powder-diffraction pattern using the Rietveld method, which is to obtain quantitative values of the phase fractions throughout the *in situ* experiments [20]. To determine the phase fractions, each 1-D diffraction pattern was subsequently fed into the Rietveld analysis as a function of time. The analysis was began with a well-fitted analysis file in *MAUD*, which was then used for recursive fitting of the following data files. The batch running was repeated several times with different starting values and constraints to start the iterating process until there was a consistently good fitting throughout the entire run.

Open porosity and sintered density were measured by the Archimedes method as specified in the ASTM B962-08 standard. Pore size distribution analysis was conducted using a pore-size distribution analyzer (GaoQ PDSA-20) using the bubble-point method as per the ASTM F316-03 standard [38]. Microstructures of the as-sintered compacts were observed using an environmental scanning electron microscope (ESEM, FEI Quanta 200F, FEI, Houston, TX, USA) equipped with an energy dispersive X-ray spectrometer (EDX, Oxford Instruments, Oxfordshire, UK). Phase constituents were determined using X-ray diffraction (XRD, Bruker D2 Phaser, Bruker, Karlsruhe, Germany). Differential scanning calorimetry (DSC, Netzsch 404 F3, Netzsch, Selb, Germany) was used to determine the various reactions of compacts during sintering with a heating rate of 5 K/min under flowing argon gas.

The tensile properties of the as-sintered NiTi tensile bars were measured on an Instron 3367 universal machine with a cross-head speed of 0.5 mm/min at ambient temperature. The tensile bars were tensioned approaching to its fracture strength. The compressive properties of the 20-mm-thick samples after 1373 K sintering were measured on an MTS 810 universal machine with a load rate of 0.6 kN/s at room temperature. An alignment cage ensured the parallelism of all samples during testing. The ends of compression cylindrical samples (machined into 10.5-mm diameter and 15-mm height) were polished and smoothed using sand papers, and finally the ends were greased before compression tests. Cyclic experiments were performed to study possible deformation and superelasticity. The cylindrical samples were first compressed until a significant deflection of the linear elastic deformation portion on the stress-strain curve was obtained or the stress level approached to its fracture strength. After that they were unloaded to zero stress and the subsequent cycle followed.

3. Results

3.1. Microstructure

Differential scanning calorimetry (DSC) measurements were conducted to investigate the phase evolution for each compact. Figure 1 shows the DSC curves of the Ni/Ti, Ni/TiH_2 and pure TiH_2 compacts after 250 MPa compaction with a heating rate of 5 K/min. According to Figure 1, a broad exothermic peak can be seen at *ca.* 1036 K for the Ni/Ti compact, which is followed by an endothermic peak developing with an onset temperature at 1143 K. With increasing temperature, this is immediately followed by an apparent exothermic peak at around 1240 K. The final peak is an endothermic peak whose temperature is 1417 K. As discussed in Ref. [20], the four peaks correspond to formation of intermetallic phases (e.g., NiTi, Ni_3Ti and $NiTi_2$, *etc.*), eutectic reaction to generate liquid Ti-rich phase, combustion reaction between molten Ti-rich and Ni-rich phases, and another eutectic reaction between NiTi and Ni_3Ti phases, respectively. In contrast, the dehydrogenation of TiH_2 is a thermally endothermic process [39]. Therefore, the first two endothermic peaks for Ni/TiH_2 and TiH_2 compacts correspond to the dehydrogenation, which ranges from ~630 to 920 K. However, the following peaks for the Ni/TiH$_2$ are less manifest as compared with the Ni/Ti compact. This is due to the fact that the dehydrogenation peaks may overlap with the following reaction peaks [20].

The X-ray diffraction (XRD) results are presented in Figure 2 for both compacts sintered at 1373 K. It can be seen that the main sintered phase is austenitic B2 NiTi in both cases, with the existence of martensitic B19′, secondary NiTi$_2$, Ni$_3$Ti and Ni$_4$Ti$_3$. The existence of these phases in the as-sintered samples is further confirmed in the ESEM micrographs and EDX analysis (Figure 3). It should be noted that the amount of Ni$_4$Ti$_3$ phase is too little to be detected by EDX. The needle-like structural phase is determined to be Ni$_3$Ti in both samples (Figure 3b,d), which is due to the eutectoid reaction of NiTi → NiTi$_2$ + Ni$_3$Ti during cooling [20]. However, it is interesting to observe that the amount of secondary phases of the Ni/TiH$_2$ sintered sample is less compared than that of the Ni/Ti sintered based on the XRD (Figure 2) and energy dispersive X-ray (EDX) results (Figure 3). This means that the final chemical homogeneity of the Ni/TiH$_2$ sintered is higher than that of the Ni/Ti sintered sample.

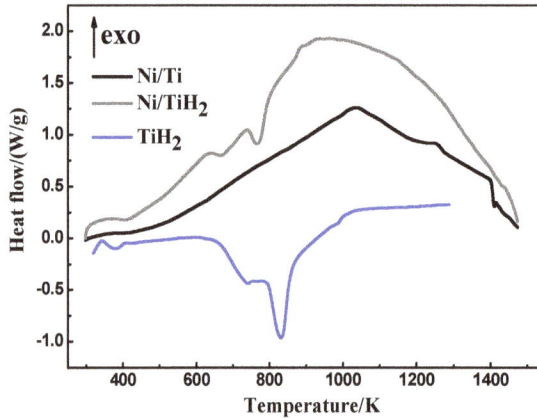

Figure 1. Differential Scanning Calorimetry (DSC) curves of Ni/Ti, Ni/TiH$_2$ and TiH$_2$ compacts with a heating rate of 5 K/min.

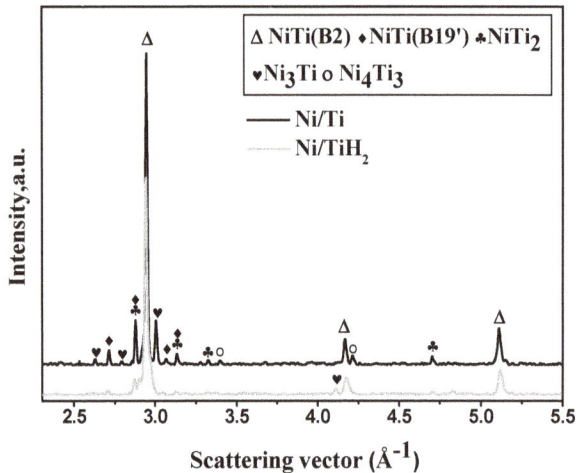

Figure 2. X-ray Diffraction (XRD) patterns of the samples after being sintered at 1373 K.

Table 1 summarizes the basic data of both sintered compacts from 4-mm-thick green samples. It can be figured that the dimension exhibits shrinkage for both sintered samples in terms of either radial or axial direction. Moreover, the shrinkage of the Ni/Ti is larger than that of the Ni/TiH$_2$ after sintering, with the concomitant higher density for former case. In addition to the shrinkage and density, the open porosity and close-to-total porosity ratio are significantly different from each other. For instance, the close-to-total porosity ratio of the Ni/Ti sintered sample is 89.6% ± 3.4%, while it is only 12.2% ± 0.8% in the case of Ni/TiH$_2$.

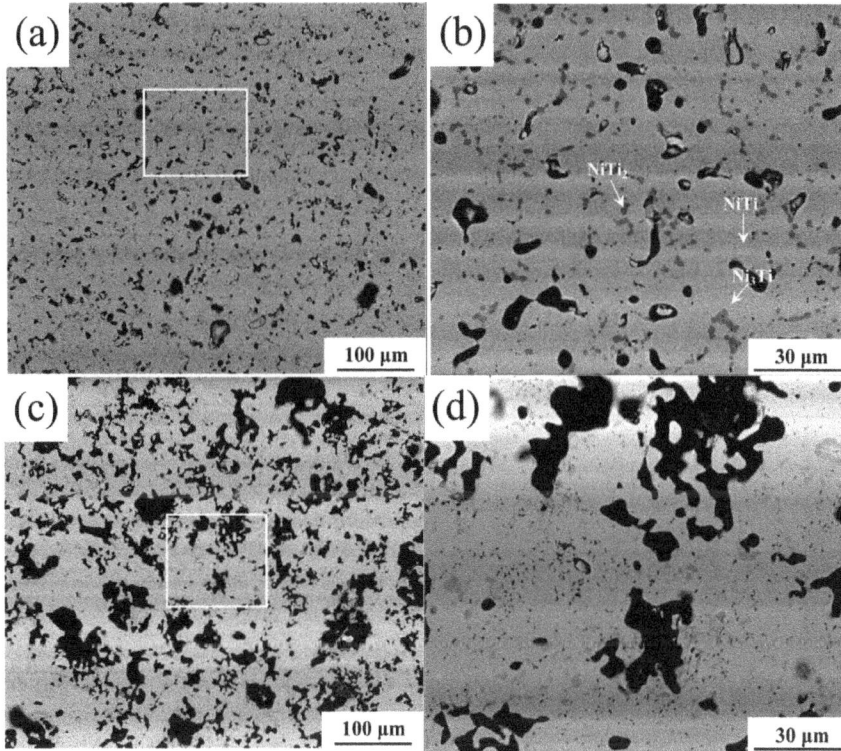

Figure 3. Back-scattered electron images of samples sintered from the Ni/Ti compact at (**a**) 1373 K, (**b**) enlarged square area in (**a**); sintered from the Ni/TiH$_2$ compact at (**c**) 1373 K, (**d**) enlarged square area in (**c**).

Table 1. Characteristics of the 1373 K sintered porous NiTi samples.

Sample	Shrinkage/%		Density/g·cm^{-3}	Open porosity/%	Close-to-total porosity ratio/%
	axial	radial			
Ni/Ti	10.47 ± 1.23	6.49 ± 0.62	5.81 ± 0.11	1.0 ± 0.1	89.6 ± 3.4
Ni/TiH$_2$	5.93 ± 0.49	4.21 ± 0.37	4.47 ± 0.07	26.9 ± 2.9	12.2 ± 0.8

3.2. In situ Neutron Diffraction

Figure 4 presents the neutron diffraction patterns of the Ni/Ti and Ni/TiH$_2$ compacts collected as a function of time in the 2D plot. The intensity is displayed by the grey scale values as a function of scattering vector Q (Q = $4\pi/\lambda$·sinθ) on the abscissa and time on the ordinate. It is focused on the dehydrogenation process and its effect on the phase transformation of the Ni/TiH_2 compact as compared with the Ni/Ti compact.

From Figure 4a, it can be seen when the temperature approaches *ca.* 840 K, the intensities of intermetallic phases (*i.e.*, B2 NiTi, Ni_3Ti, $NiTi_2$ and Ni_4Ti_3) start to establish as a result of the intensity decrease of elemental Ni and Ti in the Ni/Ti compacts. Afterwards, the intensities of elemental Ni and Ti gradually decrease and it is almost nil at about 1076 K, while these intermetallic phases largely increase. Until the temperature increases to 1163 K, the peaks of some secondary phases (Ni_3Ti and Ni_4Ti_3) almost disappear while the NiTi and $NiTi_2$ phases still remain with temperature increase even when holding at 1373 K. Additionally, it is interesting to note that the intensities of previously disappeared Ni_3Ti and Ni_4Ti_3 phases re-emerge when the furnace was cooled to *ca.* 847 K. This phenomenon has also been discussed in our recent reports [20,36]. It is due to the eutectoid reaction ($NiTi \rightarrow NiTi_2 + Ni_3Ti$) taking place at *ca.* 903 K during furnace cooling [20,36,40–42]. Additionally, it is obvious that the peaks are significantly shifted in position, which is attributed to thermal expansion of crystal lattice when the temperature is relatively high [43]. Moreover, the Mo peaks come from the Mo wires holding the samples in the instrument.

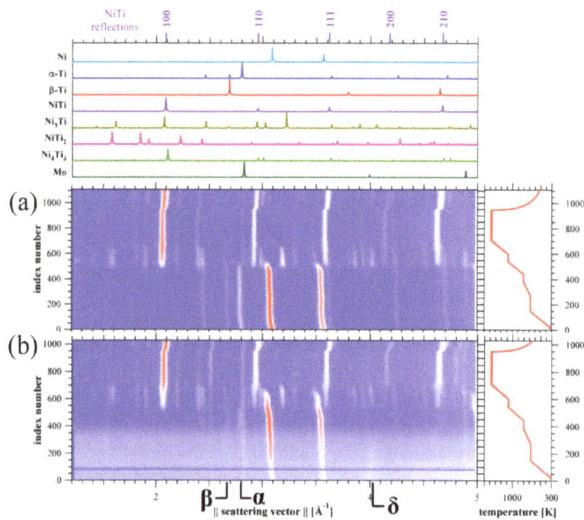

Figure 4. Neutron diffraction patterns as a function of time while temperature is ramped 1373 K from (a) Ni/Ti and (b) Ni/TiH_2.

In contrast, several differences can be seen between the Ni/TiH_2 compact (Figure 4b) and Ni/Ti compact (Figure 4a) in the heating and cooling process. First, when involving TiH_2 sintering, the initial background is much more significant compared to the Ni/Ti case (Figure 2a). Second, the temperature to establish intensities of intermetallic phases (*i.e.*, B2 NiTi, $NiTi_2$ and Ni_3Ti) is nearly 100 K higher than the Ni/Ti (Figure 4b *cf.* Figure 4a). Third, there is no Ni_4Ti_3 phase formed during sintering in the Ni/TiH_2 case and the intensities of secondary phases are weaker as compared to the Ni/Ti sample. The initial pattern background is caused by the strong incoherent neutron scattering from hydrogen atoms in TiH_2. Then, it gradually decreases with the temperature till ~923 K when it is thought the dehydrogenation of δ-Ti(H) is almost complete. It is noteworthy that both α-Ti(H) and β-Ti(H) phases appeared during decomposition of TiH_2 below 780 K, which is consistent with the recent study by Jiménez *et al.* [33]. Several intermetallic phases (*i.e.*, B2 NiTi, Ni_3Ti and $NiTi_2$) start to form when the temperature reaches ~975 K concomitant with the intensity decrease of elemental Ni and Ti. After this, the intensities of these phases continue to increase until the temperature rises to ~1350 K when the peaks of Ni, Ti and Ni$_3$Ti phases disappear. There only exist B2 NiTi and minor NiTi$_2$ phases when holding at 1373 K. It is similar with the case of Ni/Ti compact that the intensity of Ni$_3$Ti phase starts to re-establish when it was cooled to ~823 K.

With a particular focus on the dehydrogenation process of TiH$_2$, it can be seen from Figure 4b that the starting constituent includes δ-Ti(H) phase, and with increasing temperature another two hydrogen-containing solid solutions, *i.e.*, α-Ti(H) and β-Ti(H) phases, establish their intensities. The α-Ti(H) and β-Ti(H) phase has hcp and bcc structure, respectively, and hydrogen atoms sit randomly on the tetrahedral sites of both phases [44]. When the temperature approaches *ca.* 695 K, the intensity of δ-Ti(H) phase completely vanishes. Afterwards, the β-Ti(H) and α-Ti(H) phases totally transfer to α-Ti phase at ~780 K.

3.3. Pore-Size Distribution

The use of bubble-point method is to measure the pore-size distribution of both green and sintered samples, which can determine the pore-throat size in the pore tunnel as specified in the American Society of Testing Materials (ASTM) F316-03 standard. As presented in Figure 5a, most pores of the green Ni/Ti compact are in the range of 2.5~7.5 µm accounting for about 80% and only few pores are larger than 15.0 µm or smaller than 2.0 µm. In contrast, the original pore size in the green Ni/TiH$_2$ compact (Figure 5b), which mostly positions less than 5.0 µm, is smaller compared to the green Ni/Ti compact. However, after 1373 K sintering the pore size can be split into two main ranges for each sample, which are 2.5~20.0 and 4.0~20.0 µm for the Ni/Ti and Ni/TiH$_2$, respectively. This means pore-size distribution is broader and pores become larger after sintering. Such phenomenon is significantly obvious in the Ni/TiH$_2$ case that pores are previously positioned below 5.0 µm as shown in Figure 5b, while most of them enlarge to the range between 5 and 10 µm, accounting *ca.* 50% porosity, after sintering.

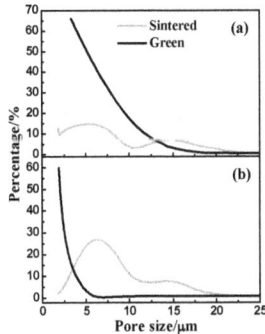

Figure 5. Pore-size distribution of green and 1373 K sintered samples from (**a**) Ni/Ti and (**b**) Ni/TiH$_2$.

3.4. Mechanical Properties

3.4.1. Static Tensile Test

Figure 6 displays typical stress-strain curves of the NiTi bars being sintered at 1373 K from both compacts. However, both sintered samples exhibited typical brittle fracture behaviors. As presented in Table 2, the fracture tensile strength of the Ni/Ti sintered sample (549.4 ± 9.6 MPa) is much higher than that of the Ni/TiH$_2$ (160.2 ± 7.3 MPa). Accordingly, the fracture strain of the former sample (4.6% ± 0.2%), which is expectable for porous NiTi, is much higher compared to the later sample (0.9% ± 0.1%). Nevertheless, both Ni/Ti sintered bars demonstrated quasi linear elastic deformation behavior. In contrast, the Young's modulus of both samples is quite similar but significantly lower than that of the wrought NiTi alloys (~70 GPa) [45].

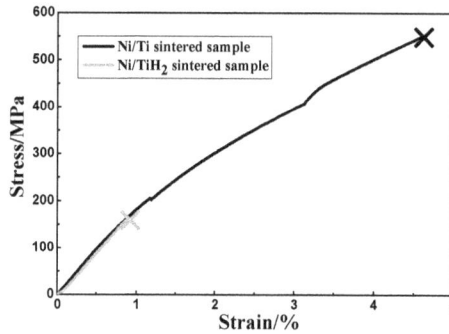

Figure 6. Tensile stress-strain curves for the 1373 K sintered NiTi parts made from Ni/Ti and Ni/TiH$_2$.

Table 2. Static tensile properties of the as-sintered NiTi alloys.

Sample	Fracture tensile strength/MPa	Fracture strain/%	Young's modulus/GPa
Ni/Ti	549.4 ± 9.6	4.6 ± 0.2	18.9 ± 1.1
Ni/TiH$_2$	160.2 ± 7.3	0.9 ± 0.1	18.0 ± 0.9

3.4.2. Cyclic Compressive Test

To investigate the porosity effect on the compressive properties, a total of five cycles was applied to each sintered sample. The cyclic compressive samples were compressed to 500, 800 and 1200 MPa, respectively, for the 1373 K-sintered samples from the Ni/Ti compact and then completely unloaded. In the Ni/TiH$_2$ case, the compressive load changes to 300, 500 and 800 MPa, respectively, since the tensile strength of the Ni/TiH$_2$ sintered sample is much lower than the Ni/Ti sintered sample (Figure 6). Figure 7 shows the strain curves as a function of time for the compressive cycles. The Ni/TiH$_2$ sintered sample failed during the third cycle with a fracture strain of 7.14% under 800 MPa stress (Figure 7b). By contrast, the Ni/Ti sintered sample could withstand all the five cycles under both 500 and 800 stresses only, except the 1200 MPa load, where the sample collapsed at the third cycle (Figure 7a).

There can be seen several interesting aspects of superelasticity originating from these curves. First, the residual strain increases with the compressive stress. On the other hand, it is noteworthy that the residual strain of the Ni/Ti sintered sample is less compared with the case of Ni/TiH$_2$ under the identical compressive load. For instance, the residual strain is 0.75% for the Ni/Ti sintered sample while it is 1.96% in the latter case. Additionally, the maximum strain slightly rises with the cycle number obviously for the higher compressive stress.

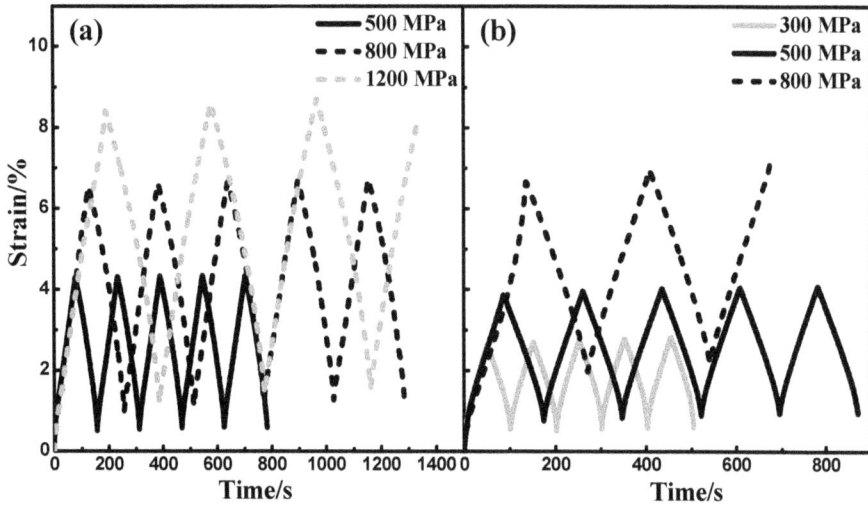

Figure 7. Compressive load-unload-recovery cycles under different compressive stresses for the samples after sintering at 1373 K (a) the Ni/Ti compact and (b) the Ni/TiH$_2$ compact. A total of five cycles was applied to each sample.

4. Discussion

4.1. Microstructural Evolution

4.1.1. Dehydrogenation process

Several *in situ/ex situ* studies have been focused on the thermal decomposition of TiH$_2$ [4–35,46,47]. However, it should be noted that the *ex-situ* XRD and TEM investigations may suffer from instant information loss in terms of the phase transformation during the heating process [19,27,30,46,47]. Additionally, although *in situ* high temperature XRD and X-ray synchrotron/neutron diffraction techniques were applied, their results may still be of concern. First, some experiments were conducted in argon atmosphere [32,33,35], which possibly causes the oxidation problem and may mislead the result. Moreover, other reports using vacuum atmosphere may result in instant information loss or delay due to the fact that XRD scanning required a long time (usually several minutes to one hour to achieve a complete scan) [31,34]. To our best knowledge, it is the first time using the neutron diffraction technique to *in situ* investigate the dehydrogenation process of TiH$_2$. This means could not only solve the long-time scanning problem (needed below 60 s), but also involve vacuum furnace to effectively avoid the oxidation issue.

According to the Ti-H phase diagram (Figure 8), titanium hydride appears as δ, β and α-phase has *ca.* 50~66.7 at.%, 0~50 at.% and 0~8.5 at.% of hydrogen content [48,49], respectively. In our case, the initial titanium hydride phase includes δ phase as shown in Figure 4b. Based on this phase diagram and the neutron diffraction pattern in Figure 4b, it can be concluded that its dehydrogenation could take place as follows: $\delta \rightarrow \delta + \alpha \rightarrow \delta + \beta + \alpha \rightarrow \beta + \alpha \rightarrow \alpha$. This finding is with great agreement with the report in Ref. [33]. Attributed to the strong incoherent neutron scattering from hydrogen atoms, there is an obvious background during the initial heating. In spite of this, the hydrogen release progresses with the temperature and time concomitant with the background slash. The dehydrogenation temperature range in this study occurs between 573 and 1073 K as presented in our DSC curves (Figure 1) and neutron diffraction pattern (Figure 4b), which is consistent with previous studies [15,30,32,33,50]. As a result, the background evolution is consistent with the process of dehydrogenation during heating and finally almost disappears at about 780 K, Figure 4b.

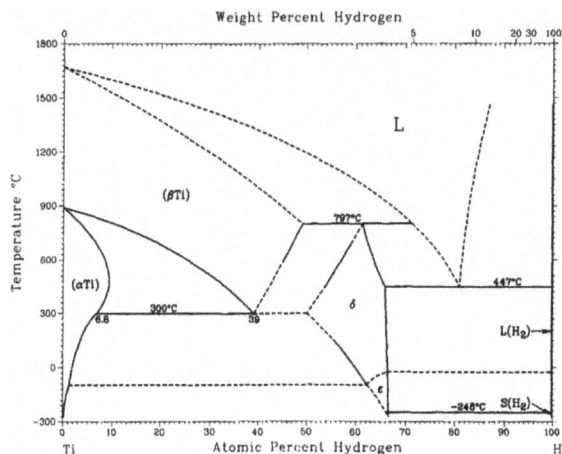

Figure 8. Ti-H binary phase diagram redrawn from Ref. [48].

4.1.2. Pore and Phase Evolution

As discussed in our previous reports [4,10,17,20] together with other studies [5,14,51–56], the pores present in the final sintered samples can be originated from the following four sources: (1) original pores in the green compact, (2) Kirkendall pores formed due to the different diffusion rates between Ni and Ti or newly born Ti elements, (3) pores occurred by the following phase transformation or alloying and (4) large pores caused by liquid phase sintering (LPS). It has been proved in Refs. [17,20] that dehydrogenation in the Ni/TiH_2 compact causes porosity increase during sintering and then the diffusion distance between Ni and new-born Ti particles enlarges, which is thought to delay sequential alloying and increase pore size and porosity in the Ni/TiH_2 sample. In contrast, LPS has two opposite effects on densification. On the one hand, it would favor densification since it promotes diffusion due to the presence of liquid [57,58]. On the other hand, however, it could give rise to swelling because of pores leaving behind [5]. We recall the microstructure images (Figure 3), density and porosity data (Table 1), and pore-size distribution (Figure 5), it seems the combination of the two factors, which are dehydrogenation and LPS, leads to the fact that the density of Ni/Ti sintered sample is much higher compared to the case of Ni/TiH_2 although the relative density of both green compacts is similar (*i.e.*, it is 73.0% and 71.2% for the Ni/Ti and Ni/TiH_2 compact, respectively).

The Rietveld quantitative analysis from the neutron diffraction data, shown in Figure 9, further supports the discussion above. Figure 9 displays the weight fraction of various intermetallic phases for both compacts during sintering and furnace cooling. It can be confirmed that the whole sintering process of the Ni/TiH_2 compact below 1373 K is postponed compared to the Ni/Ti compact (Figure 9b *cf*. Figure 9a). Nevertheless, at the final holding stage at 1373 K, the amount of B2 NiTi phase is slightly lower for the Ni/TiH_2 compact (94.3 wt.%) than that in the Ni/Ti compact (96.2 wt.%). However, such situation occurs oppositely after furnace cooling, because the final B2 phase amount of the Ni/TiH_2 compact after cooling (87.3 wt.%) is higher as compared to the Ni/Ti compact (81.3 wt.%). This observation has been reported in our recent result [20] that there is a eutectoid reaction NiTi $\rightarrow$ $Ni_3Ti + NiTi_2$ happened at around 903 K, which means the B2 NiTi phase decomposed into Ni_3Ti and $NiTi_2$ phases during cooling and thus gives rise to the phase amount change accordingly. All the amount of secondary phases such as $NiTi_2$, Ni_3Ti and Ni_4Ti_3 in the Ni/Ti sintered sample is higher than that in the Ni/TiH_2 sintered sample, which is consistent with the XRD results (Figure 2). This can further confirm that the dehydrogenation from TiH_2 activates titanium surface and thus enhances final chemical homogenization.

4.2. Fracture, Superelasticity and Modulus

With regard to the strength of a material, it is dependent on the weakest portion in the material. Normally, porosity, pore size and pore shape have a significant effect on the strength of porous NiTi alloys. For instance, a more severe stress concentration may arise from a sharp edge of the pores. Furthermore, a larger pore size and/or higher porosity result in more reduction in the effective load-carrying cross section [10]. These factors all result in the strength drop the porous NiTi alloys [59]. Recalling the fracture tensile strength and Young's modulus (Figures 6 and 7, Table 2), these values of the Ni/Ti sintered sample are higher compared with the Ni/TiH_2 sintered sample.

On the one hand, from a fracture mechanics point of view, the material fails when the stress intensity factor K ($= Y\sigma\sqrt{\pi a}$) reaches its fracture toughness [60]. In this respect, the "a" represents the pore size and pore-size distribution, while the "Y" is a collective parameter of pore shape and orientation in a porous material. In this study, the mean pore size of the Ni/Ti sintered sample is significantly smaller than did in the Ni/TiH_2 case, Figures 3 and 5. However, the ESEM micrographs (Figure 3) show that the pore shape is similar in both samples. This implies that the average "Y" value is analogous in both cases, while the "a" value gives rise to a higher stress intensity factor K for the Ni/TiH_2 sintered sample. As such, the Ni/Ti sintered sample demonstrated a higher fracture stress, as compared to the case of Ni/TiH_2. Alternatively, this means the use of TiH_2 powder leads to lower fracture strength caused by larger pore size and lower densification (Table 1) although it shows higher chemical homogenization (Figure 9).

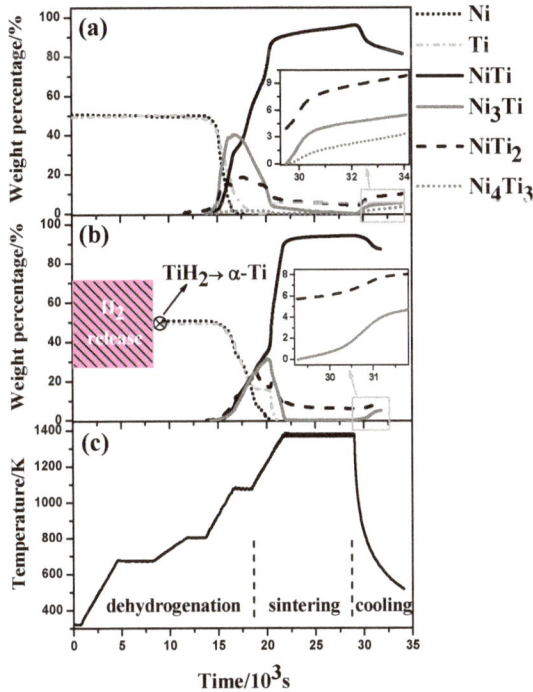

Figure 9. Weight fractions of the detected phases as a function of time (temperature) during *in situ* scan as achieved by Rietveld refinement analysis upon heating and cooling (a) the Ni/Ti compact, (b) the Ni/TiH_2 compact, and (c) heating and cooling profile as a function of time.

On the other hand, compressive tests show the typical superelasticity properties of sintered NiTi alloys, which are attributed to the stress-induced martensitic transformation [2]. With increasing the cycle number, the accumulated residual strain increases and then levels off to a constant value (Figure 7). This phenomenon has been discussed regarding to the general shape memory "training process" [20,61]. The Young's modulus of the Ni/Ti sintered sample is greater than that of the Ni/TiH$_2$ sintered sample (Table 2). First, as shown in Table 1 the close-to-total porosity ratio is 89.6% $\pm$ 3.4% and 12.2% $\pm$ 0.8% for the Ni/Ti and Ni/TiH$_2$ sintered compacts, respectively. Normally, higher ratio of close-to-total porosity would give rise to higher elastic modulus [60,62]. Second, the higher density of the Ni/Ti sintered sample would result in higher elastic modulus than did the Ni/TiH$_2$ sintered sample after 1373 K sintering as shown in Table 1. Additionally, it should be noted that the final phases present in the sintered compacts also affect the elastic modulus. Recalling Figure 9 that the Ni/TiH$_2$ sintered compact contains 8.0 wt.% NiTi$_2$ phase while the Ni/Ti sintered sample has 9.8 wt.% NiTi$_2$. More amount of NiTi$_2$ phase also causes higher elastic modulus for the Ni/Ti sintered sample [20].

5. Summary

In this report, porous NiTi alloys from Ni/Ti and Ni/TiH$_2$ powder compacts were produced by introducing a conventional press-and-sinter method. The microstructure and mechanical properties of sintered samples were investigated and compared with involving the use of TiH$_2$ powder. The following conclusions can be drawn from this study.

(1) B2 NiTi phase is the dominant phase identified in both samples after being sintered at 1373 K holding for two hours together with the presence of some minor secondary phases.

(2) Dehydrogenation from TiH$_2$ leads to a lower density, a much higher porosity, a larger pore size but higher final chemical homogenization after sintering as compared with the Ni/Ti compact.

(3) The use of TiH$_2$ powder causes lower fracture strength and lower elastic modulus compared with the Ni/Ti sintered sample.

Acknowledgments: We acknowledge the financial support from Ministry of Business Innovation and Employment (MBIE), New Zealand. Gang Chen thanks the China Scholarship Council (CSC) for providing him a doctoral scholarship. We also acknowledge the support of the Bragg Institute, Australian Nuclear Science and Technology Organization (ANSTO), in providing the neutron research facilities used in this work. The authors would like to thank Australian Institute of Nuclear Science and Engineering (AINSE) Ltd for providing financial assistance (award No. P2716) to enable work on WOMBAT to be conducted. The authors also appreciate the funding from Shaanxi Science and Technology Co-ordination and Innovation Project (2014KTZB01-02-04).

Conflicts of Interest: The authors declare no conflict of interest.

References

1. Duering, T.W.; Pelton, A.R. *Materials Properties Handbook: Titanium Alloys*; ASM International, the Materials Information Society: Materials Park, OH, USA, 1994.
2. Yamauchi, K.; Ohkata, I.; Tsuchiya, K.; Miyazaki, S. *Shape Memory and Superelastic Alloys: Technologies and Applications*; Woodhead Publishing: Cambridge, UK, 2011; p. 390.
3. Elahinia, M.H.; Hashemi, M.; Tabesh, M.; Bhaduri, S.B. Manufacturing and processing of NiTi implants: A review. *Prog. Mater. Sci.* **2012**, *57*, 911–946. [CrossRef]
4. Chen, G.; Cao, P.; Wen, G.; Edmonds, N.; Li, Y. Using an agar-based binder to produce porous NiTi alloys by metal injection moulding. *Intermetallics* **2013**, *37*, 92–99. [CrossRef]
5. Whitney, M.; Corbin, S.F.; Gorbet, R.B. Investigation of the mechanisms of reactive sintering and combustion synthesis of NiTi using differential scanning calorimetry and microstructural analysis. *Acta Mater.* **2008**, *56*, 559–570. [CrossRef]
6. Sadrnezhaad, S.K.; Hosseini, S.A. Fabrication of porous NiTi-shape memory alloy objects by partially hydrided titanium powder for biomedical applications. *Mater. Des.* **2009**, *30*, 4483–4487. [CrossRef]
7. Tosun, G.; Ozler, L.; Kaya, M.; Orhan, N. A study on microstructure and porosity of NiTi alloy implants produced by SHS. *J. Alloys Compd.* **2009**, *487*, 605–611. [CrossRef]

8.   Whitney, M.; Corbin, S.F.; Gorbet, R.B. Investigation of the influence of Ni powder size on microstructural evolution and the thermal explosion combustion synthesis of NiTi. *Intermetallics* **2009**, *17*, 894–906. [CrossRef]
9.   Liu, X.; Wu, S.; Yeung, K.W.K.; Xu, Z.S.; Chung, C.Y.; Chu, P. Superelastic porous NiTi with adjustable porosities synthesized by powder metallurgical method. *J. Mater. Eng. Perform.* **2012**, *21*, 2553–2558. [CrossRef]
10. Chen, G.; Cao, P.; Edmonds, N. Porous NiTi alloys produced by press-and-sinter from Ni/Ti and Ni/TiH$_2$ mixtures. *Mater. Sci. Eng. A* **2013**, *582*, 117–125. [CrossRef]
11. Li, B.-Y.; Rong, L.-J.; Li, Y.-Y. Stress–strain behavior of porous Ni-Ti shape memory intermetallics synthesized from powder sintering. *Intermetallics* **2000**, *8*, 643–646. [CrossRef]
12. Li, B.-Y.; Rong, L.-J.; Li, Y.-Y. The influence of addition of TiH$_2$ in elemental powder sintering porous Ni-Ti alloys. *Mater. Sci. Eng. A* **2000**, *281*, 169–175. [CrossRef]
13. Bertheville, B.; Neudenberger, M.; Bidaux, J.E. Powder sintering and shape-memory behaviour of NiTi compacts synthesized from Ni and TiH$_2$. *Mater. Sci. Eng. A* **2004**, *384*, 143–150. [CrossRef]
14. Robertson, I.M.; Schaffer, G.B. Swelling during sintering of titanium alloys based on titanium hydride powder. *Powder Metall.* **2010**, *53*, 27–33. [CrossRef]
15. Wu, S.; Liu, X.; Yeung, K.W.K.; Hu, T.; Xu, Z.; Chung, J.C.Y.; Chu, P.K. Hydrogen release from titanium hydride in foaming of orthopedic NiTi scaffolds. *Acta Biomater.* **2011**, *7*, 1387–1397. [CrossRef] [PubMed]
16. Chen, G.; Wen, G.A.; Cao, P.; Edmonds, N.; Li, Y.M. Processing and characterisation of porous NiTi alloy produced by metal injection moulding. *Powder Injection Moulding Int.* **2012**, *6*, 83–88.
17. Chen, G.; Cao, P. NiTi powder sintering from TiH$_2$ powder: An *in situ* investigation. *Metall. Mater. Trans. A* **2013**, 1–4.
18. Wang, H.; Fang, Z.Z.; Sun, P. A critical review of mechanical properties of powder metallurgy titanium. *Int. J. Powder Metall.* **2010**, *46*, 45–57.
19. Wang, H.T.; Lefler, M.; Fang, Z.Z.; Lei, T.; Fang, S.M.; Zhang, J.M.; Zhao, Q. Titanium and titanium alloy via sintering of TiH$_2$. *Key Eng. Mater.* **2010**, *436*, 157–163. [CrossRef]
20. Chen, G.; Liss, K.-D.; Cao, P. *In situ* observation and neutron diffraction of NiTi powder sintering. *Acta Mater.* **2014**, *67*, 32–44. [CrossRef]
21. Chen, G. Powder Metallurgical Titanium Alloys (TiNi and Ti-6Al-4V): Injection Moulding, Press-and-Sinter, and Hot Pressing. Ph.D. Thesis, The University of Auckland, Auckland, New Zealand, 2014.
22. Azevedo, C.R.F.; Rodrigues, D.; Beneduce Neto, F. Ti-Al-V powder metallurgy (PM) via the hydrogenation–dehydrogenation (HDH) process. *J. Alloys Compd.* **2003**, *353*, 217–227. [CrossRef]
23. Robertson, I.M.; Schaffer, G.B. Comparison of sintering of titanium and titanium hydride powders. *Powder Metall.* **2010**, *53*, 12–19. [CrossRef]
24. Ivasishin, O.M.; Eylon, D.; Bondarchuk, V.I.; Savvakin, D.G. Diffusion during powder metallurgy synthesis of titanium alloys. *Defect Diffus. Forum* **2008**, *277*, 177–185. [CrossRef]
25. Zhang, J.M.; Yi, J.H.; Gan, G.Y.; Yan, J.K.; Du, J.H.; Liu, Y.C. Research on dehydrogenation and sintering process of titanium hydride for manufacture titanium and titanium alloy. *Adv. Mater. Res.* **2013**, *616–618*, 1823–1829.
26. Ivasishin, O.M.; Savvakin, D.G.; Froes, F.; Mokson, V.C.; Bondareva, K.A. Synthesis of alloy Ti-6Al-4V with low residual porosity by a powder metallurgy method. *Powder Metall. Metal Ceram.* **2002**, *41*, 382–390. [CrossRef]
27. Bhosle, V.; Baburaj, E.G.; Miranova, M.; Salama, K. Dehydrogenation of nanocrystalline TiH$_2$ and consequent consolidation to form dense Ti. *Metall. Mater. Trans. A* **2003**, *34*, 2793–2799. [CrossRef]
28. Li, B.Y.; Rong, L.J.; Li, Y.Y. Porous NiTi alloy prepared from elemental powder sintering. *J. Mater. Res.* **1998**, *13*, 2847–2851. [CrossRef]
29. Li, B.-Y.; Rong, L.-J.; Li, Y.-Y.; Gjunter, V.E. An investigation of the synthesis of Ti-50 at. Pct Ni alloys through combustion synthesis and conventional powder sintering. *Metall. Mater. Trans. A* **2000**, *31*, 1867–1871. [CrossRef]
30. Bhosle, V.; Baburaj, E.G.; Miranova, M.; Salama, K. Dehydrogenation of TiH$_2$. *Mater. Sci. Eng. A* **2003**, *356*, 190–199. [CrossRef]
31. Sandim, H.R.Z.; Morante, B.V.; Suzuki, P.A. Kinetics of thermal decomposition of titanium hydride powder using *in situ* high-temperature X-ray diffraction (HTXRD). *Mater. Res.* **2005**, *8*, 293–297. [CrossRef]

32. Liu, H.; He, P.; Feng, J.C.; Cao, J. Kinetic study on nonisothermal dehydrogenation of TiH_2 powders. *Int. J. Hydrog. Energy* **2009**, *34*, 3018–3025. [CrossRef]

33. Jiménez, C.; Garcia-Moreno, F.; Pfretzschner, B.; Klaus, M.; Wollgarten, M.; Zizak, I.; Schumacher, G.; Tovar, M.; Banhart, J. Decomposition of TiH_2 studied *in situ* by synchrotron X-ray and neutron diffraction. *Acta Mater.* **2011**, *59*, 6318–6330. [CrossRef]

34. Farhana, H.N.; Wang, Y.; Noor, M.M.; Chan, S.I. Static X-ray scans on the titanium hydride (TiH_2) powder during dehydrogenation. *Adv. Mater. Res.* **2013**, *795*, 124–127. [CrossRef]

35. Jiménez, C.; Garcia-Moreno, F.; Pfretzschner, B.; Kamm, P.H.; Neu, T.R.; Klaus, M.; Genzel, C.; Hilger, A.; Manke, I.; Banhart, J. Metal foaming studied *in situ* by energy dispersive X-ray diffraction of synchrotron radiation, X-ray radioscopy, and optical expandometry. *Adv. Eng. Mater.* **2013**, *15*, 141–148. [CrossRef]

36. Chen, G.; Liss, K.-D.; Cao, P. *In situ* observation of phase transformation of powder sintering from Ni/TiH2 using neutron diffraction. In *TMS 2014 Supplemental Proceedings*; John Wiley & Sons, Inc.: Hoboken, NJ, USA, 2014; pp. 967–973.

37. Studer, A.J.; Hagen, M.E.; Noakes, T.J. Wombat: The high-intensity powder diffractometer at the opal reactor. *Phys. B Condens. Matter* **2006**, *385–386*, 1013–1015. [CrossRef]

38. Yu, J.; Hu, X.; Huang, Y. A modification of the bubble-point method to determine the pore-mouth size distribution of porous materials. *Sep. Purif. Technol.* **2010**, *70*, 314–319. [CrossRef]

39. Viswanathan, B.; Murthy, S.S.; Sastri, M.V.C. *Metal Hydrides: Fundamentals and Applications*, 1st ed.; Springer: Berlin, Germany, 1999; p. 189.

40. Duwez, P.; Taylor, J.L. The structure of intermediate phases in alloys of titanium with iron, cobalt, and nickel. *Trans. AIME* **1950**, *188*, 1173–1176.

41. Poole, D.M.; Hume-Rothery, W. The equilibrium diagram of the system nickel-titanium. *J. Inst. Metals* **1954**, *83*, 473–480.

42. Gupta, S.P.; Mukherjee, K.; Johnson, A.A. Diffusion controlled solid state transformation in the near-equiatomic Ti-Ni alloys. *Mater. Sci. Eng.* **1973**, *11*, 283–297. [CrossRef]

43. Liss, K.-D.; Bartels, A.; Schreyer, A.; Clemens, H. High-energy X-rays: A tool for advanced bulk investigations in materials science and physics. *Textures Microstruct.* **2003**, *35*, 219–252. [CrossRef]

44. Predel, B. H-Ti (Hydrogen-Titanium). In *Ga-Gd-Hf-Zr*; Madelung, O., Ed.; Springer: Berlin/Heidelberg, Germany, 1996; Volume 5f, pp. 1–2.

45. Greiner, C.; Oppenheimer, S.M.; Dunand, D.C. High strength, low stiffness, porous NiTi with superelastic properties. *Acta Biomater.* **2005**, *1*, 705–716. [CrossRef] [PubMed]

46. Mandrino, D.; Paulin, I.; Škapin, S.D. Scanning electron microscopy, X-ray diffraction and thermal analysis study of the TiH_2 foaming agent. *Mater. Charact.* **2012**, *72*, 87–93. [CrossRef]

47. Paulin, I.; Donik, Č.; Mandrino, D.; Vončina, M.; Jenko, M. Surface characterization of titanium hydride powder. *Vacuum* **2012**, *86*, 608–613. [CrossRef]

48. Okamoto, H. H-Ti (Hydrogen-Titanium). *J. Phase Equilib. Diffus.* **2011**, *32*, 174–175. [CrossRef]

49. Fukai, Y. *The Metal-Hydrogen System, Basic Bulk Properties*, 2nd ed.; Springer: Berlin/Heidelberg, Germany, 2005; p. 497.

50. Igharo, M.; Wood, J.V. Compaction and sintering phenomena in titanium-nickel shape memory alloys. *Powder Metall.* **1985**, *28*, 131–139. [CrossRef]

51. Biswas, A. Porous NiTi by thermal explosion mode of SHS: Processing, mechanism and generation of single phase microstructure. *Acta Mater.* **2005**, *53*, 1415–1425. [CrossRef]

52. Otsuka, K.; Ren, X. Physical metallurgy of Ti-Ni-based shape memory alloys. *Prog. Mater. Sci.* **2005**, *50*, 511–678. [CrossRef]

53. Laeng, J.; Xiu, Z.; Xu, X.; Sun, X.; Ru, H.; Liu, Y. Phase formation of Ni–Ti via solid state reaction. *Phys. Scr.* **2007**, *2007*, 250. [CrossRef]

54. Bansiddhi, A.; Dunand, D.C. Shape-memory NiTi foams produced by replication of NaCl space-holders. *Acta Biomater.* **2008**, *4*, 1996–2007. [CrossRef] [PubMed]

55. Li, H.; Yuan, B.; Gao, Y.; Chung, C.Y.; Zhu, M. High-porosity NiTi superelastic alloys fabricated by low-pressure sintering using titanium hydride as pore-forming agent. *J. Mater. Sci.* **2009**, *44*, 875–881. [CrossRef]

56. Wen, C.E.; Xiong, J.Y.; Li, Y.C.; Hodgson, P.D. Porous shape memory alloy scaffolds for biomedical applications: A review. *Phys. Scr.* **2010**, *2010*, 014070. [CrossRef]

Metals **2015**, *5*, 530–546

57. German, R.M. *Powder Metallurgy Science*; Metal Powder Industries Federation: Princeton, NJ, USA, 1998.
58. German, R.; Suri, P.; Park, S. Review: Liquid phase sintering. *J. Mater. Sci.* **2009**, *44*, 1–39. [CrossRef]
59. Ashby, M.F.; Evans, A.; Fleck, N.A.; Gibson, L.J.; Hutchinson, J.W.; Wadley, H. *Metal Foams: A Design Guide*; Butterworth-Heinemann: Boston, MA, USA, 2000.
60. Anderson, T.L. *Fracture Mechanics Fundamentals and Applications*, 3rd ed.; CRC Press: Boca Raton, FL, USA, 2005.
61. Nemat-Nasser, S.; Guo, W.-G. Superelastic and cyclic response of NiTi SMA at various strain rates and temperatures. *Mech. Mater.* **2006**, *38*, 463–474. [CrossRef]
62. Gibson, L.J.; Ashby, M.F. *Cellular Solids: Structure and Properties*, 2nd ed.; Cambridge University Press: Cambridge, UK, 1999.

metals

MDPI

Article

In situ Investigation of Titanium Powder Microwave Sintering by Synchrotron Radiation Computed Tomography

Yu Xiao [1], Feng Xu [1,*], Xiaofang Hu [1], Yongcun Li [2], Wenchao Liu [1] and Bo Dong [1]

[1] CAS Key Laboratory of Mechanical Behavior and Design of Materials, Department of Modern Mechanics, University of Science and Technology of China, Hefei 230026, China; xiaoyuxy@mail.ustc.edu.cn (Y.X.); huxf@ustc.edu.cn (X.H.); liuwc@mail.ustc.edu.cn (W.L.); dongbo@mail.ustc.edu.cn (B.D.)

[2] Department of Mechanics, Taiyuan University of Technology, Taiyuan 030024, China; liyongcun@tyut.edu.cn

*   Correspondence: xufeng3@ustc.edu.cn; Tel.: +86-551-6360-0564; Fax: +86-551-6360-6459

Academic Editor: Klaus-Dieter Liss
Received: 3 October 2015; Accepted: 23 December 2015; Published: 4 January 2016

Abstract: In this study, synchrotron radiation computed tomography was applied to investigate the mechanisms of titanium powder microwave sintering *in situ*. On the basis of reconstructed images, we observed that the sintering described in this study differs from conventional sintering in terms of particle smoothing, rounding, and short-term growth. Contacted particles were also isolated. The kinetic curves of sintering neck growth and particle surface area were obtained and compared with those of other microwave-sintered metals to examine the interaction mechanisms between mass and microwave fields. Results show that sintering neck growth accelerated from the intermediate period; however, this finding is inconsistent with that of aluminum powder microwave sintering described in previous work. The free surface areas of the particles were also quantitatively analyzed. In addition to the eddy current loss in metal particles, other heating mechanisms, including dielectric loss, interfacial polarization effect, and local plasma-activated sintering, contributed to sintering neck growth. Thermal and non-thermal effects possibly accelerated the sintering neck growth of titanium. This study provides a useful reference of further research on interaction mechanisms between mass and microwave fields during microwave sintering.

Keywords: microwave sintering; microstructure; synchrotron radiation computed tomography

1. Introduction

As an advanced material preparation method, microwave sintering has been extensively investigated because of its numerous advantages, such as high heating rate, overall and even heating, and material microstructure improvement. This technique is initially applied to nonmetallic material sintering. Microwaves were initially believed to be unable to heat metal materials because of the reflection at surfaces. Since its first application performed by Roy *et al.* in 1999, microwave sintering of metal, alloy, and metallic glassy powder has been investigated [1–4]. This technique can promote particle size uniformity and densification rate; it can also improve the macro-properties of materials to a greater extent than conventional sintering. For example, Roy *et al.* [1] confirmed that the modulus of Fe–Ni rupture is 60% higher than that of conventional specimens after 10–30 min of microwave treatment. These excellent advantages are attributed to the combination of thermal and non-thermal effects. However, the definite mechanisms of microwave sintering remain unclear. Janney *et al.* [2] indicated that the activation energy is lower in microwave sintering than in conventional sintering. Conversely, Saitou K *et al.* [3] demonstrated that the microwave field does not affect activation energy during sintering. Other theories, including eddy currents [5], crystallization enhancement [6], and micro-focusing and polarization effects [7], have been proposed. However, direct experimental evidence is difficult to obtain; as such, these theories cannot be easily verified because of the extreme experimental conditions required during microwave sintering.

The macro-properties of materials are determined on the basis of their microstructures which are driven by mechanisms of the sintering kinetics mechanisms. Relevant microwave sintering mechanisms can be revealed by continuous *in situ* tracking of the internal microstructure evolution. After determining all the microstructure characteristics, microwave heating mechanisms and sintering kinetics can be further analyzed. However, studies adopting conventional testing techniques have yet to observe microwave sintering *in situ* because of the experimental limitations of conventional testing techniques, such as optical microscopy, transmission electron microscopy, and scanning electron microscopy. Therefore, *in situ* and non-destructive testing techniques [8], such as synchrotron radiation computed tomography (SR-CT) [9], should be developed.

SR-CT is a novel testing technology that can achieve non-destructive and real-time 3D observations in extreme environments, such as high or low temperature, high pressure, and intense radiation. For example, Bale *et al.* [10] conducted a real-time study of microstructure behavior under mechanical loading at >1600 °C. When applied in sintering studies, SR-CT can continuously obtain accurate experimental data of surface and internal 3D microstructure evolution during sintering without interrupting the process and destroying the sample. Quantitative analysis can then be conducted on the basis of SR-CT experimental data. With these excellent advantages, SR-CT is quite suitable for the *in situ* investigation of sintering.

In this study, SR-CT was applied to investigate titanium powder microwave sintering *in situ*. 2D and 3D images at different sintering times were reconstructed. In the images of internal microstructure evolution, sintering phenomena included sintering neck growth, powder densification, and pore closing, which commonly occur in conventional sintering. Unique phenomena, such as particle smoothing, rounding, and short-term growth, contacted particles being isolated, which rarely occur in conventional sintering, were also observed. These phenomena may be attributed to eddy current loss, micro-focusing, and interfacial polarization effect. Our experimental data revealed the sample microstructure characteristics, such as surface area and sintering neck size. The kinetic mechanisms of sintering neck growth were quantitatively analyzed and compared with those of other metal microwave sintering techniques. Our results revealed that sintering kinetic behaviors were different from those of aluminum in early and intermediate periods. The cause of this difference was investigated, and the corresponding sintering mechanisms were analyzed. Results showed that the sintering behaviors were quite different during microwave sintering because of the heterogeneity between titanium and aluminum. Eddy current loss, other heating mechanisms, and non-thermal effects in microwave sintering were also observed.

2. Experimental Section

The microwave sintering experiment was conducted in a BL13W1 beam line at Shanghai Synchrotron Radiation Facility (Shanghai, China); the beam line is a third-generation light source with excellent features, including high intensity, high brilliance, high polarization rate, and quasi-coherence. The charge-coupled device resolution was 0.74 µm per pixel. Chemically pure titanium powder with an average particle size of approximately 23 µm and 99.9% purity (Aladdin Biochemical Technology Co., Ltd., Shanghai, China) was used. Acid pickling was conducted to remove oxides on the particle surfaces. The powder was dried in a vacuum oven at 120 °C and then loosely poured into a capillary tube with an inner diameter of approximately 0.3 mm and a height of approximately 10 mm. Some copper particles were fixed on the outer capillary surface as marking points to easily locate the same region at different sintering times. The sample was then introduced to a specially designed microwave sintering furnace with a multimode cavity of 2.45 GHz and an output power of 0–3 kW. An SiC susceptor was used to preserve heat and to accelerate the increase in sample temperature because of the small sample size and large space of the multimode cavity chamber. The top of the sample at the far end of the susceptor was chosen as the test region for the SR-CT experiment to reduce the influence of the susceptor (Figure 1). The temperature was measured using a thermo tracer (type TH5104; NEC Corp., Tokyo, Japan) under the following parameters: temperature measurement range from $-10\,°C$ to $1500\,°C$, accuracy of $\pm1.0\%$ (full scale), and emissivity of 0.6. The temperature increased to $900\,°C$ within approximately 20 min and remained constant for 60 min. Argon was used as a protective atmosphere. The experimental facility is shown in Figure 1. We compared the proposed technique with a previously described aluminum microwave sintering [11] that requires the same parameters as the microwave sintering furnace.

The furnace chamber must be an open space to allow the rotation system (Figure 1) to dive into the furnace because sample rotation is necessary to obtain computed tomography data. Therefore, the protective atmosphere must be constantly pumped into the furnace. The gas flow rate was set at a very slow limit to avoid excessive heat loss from the hole of the rotation system.

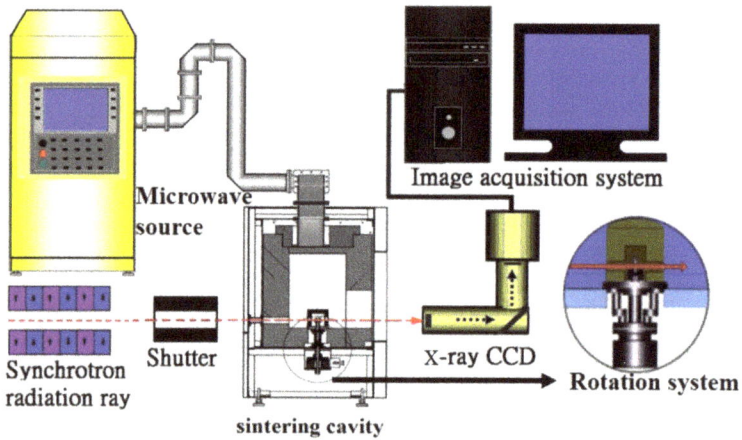

Figure 1. Schematic of *in situ* synchrotron radiation computed tomography (SR-CT) system of microwave sintering.

3. Results

Figure 2 shows the 3D images of microstructure evolution at different times during microwave sintering. Sintering phenomena, such as particle densification and sintering neck formation and growth, were observed. Further analysis was conducted on the basis of 2D images to present internal microstructure evolution and to obtain morphological parameters.

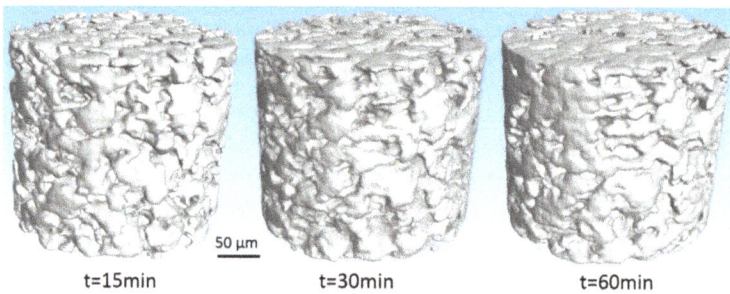

t=15min   t=30min   t=60min

Figure 2. 3D images of the microstructure at different sintering times.

The images shown in Figure 3 were reconstructed by using the filtered back projection algorithm. The cross-section images of the same internal areas at different sintering times were obtained on the basis of marking particles present on the sample and related algorithms. Grayscale ranges from 0–255. As the grayscale value approaches 255, X-ray is increasingly absorbed; the relative density is also high. This trend indicates that white regions represent particles and black regions represent pores. These cross-section images also reveal typical sintering phenomena similar to conventional sintering: (1) the particles in the circle contacted with each other while sintering neck formed and grew and (2) the pores marked with red asterisks closed and their size decreased.

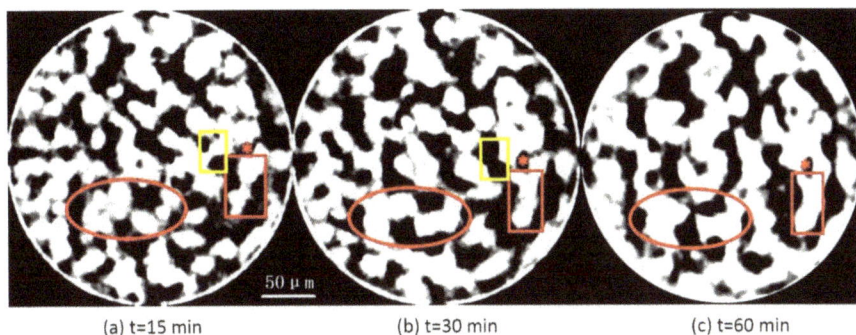

(a) t=15 min (b) t=30 min (c) t=60 min

Figure 3. Microstructure of the same cross section at different sintering times of 15 min (**a**), 30 min (**b**) and 60 min (**c**).

Special microwave sintering phenomena which were similar and different with other kinds of metals and rarely seen in conventional sintering were captured and shown in Figure 3.

(1) The particles in the red rectangle grow more spherically and smoothly within a short time as the corners and burrs on the surface disappear. A similar process can also be observed in the circle. This phenomenon may be attributed to the loss of eddy current on the particle surface; as a result, mass diffusion is accelerated. Furthermore, non-thermal effects, such as interfacial polarization between grain surface and pores, and micro-focusing effects at the corner and burrs likely contribute to particle smoothing and rounding processes. Therefore, the proposed process was much faster than conventional sintering.

(2) The two large particles in the red circle shown in Figure 3b come in contact with each other when the sintering neck is formed. However, the two particles are isolated from each other instead of sintering together (Figure 3c) as a consequence of the tensile and pressure from other particles. The particles in the yellow rectangle also exhibit similar behaviors. This phenomenon may have been caused by the micro-focusing effect. During microwave distribution, local microwave fields are disproportionately strong in certain regions because of the focusing influences of the microstructure, such as sintering neck, particle boundaries, and rough surfaces. Thus, non-uniform energy deposition occurs and temperature remarkably increases. The connecting region of two particles may be melted by the local high temperature attributed to the micro-focusing effect and then be broken by the tensile and pressure from other particles.

(3) Several small particles in the red circle shown in Figure 3a quickly grow together into two large particles within 15 min. This phenomenon may have been caused by the acceleration of mass diffusion as a result of the micro-focusing effect at the sintering neck regions.

These unique microwave sintering phenomena can be rarely observed in conventional sintering but can be detected in other microwave sintering experiments of metal and ceramic-metal mixtures; these phenomena are attributed to non-thermal effects. Phenomenon (1) also occurs in the microwave sintering of aluminum [11]. This phenomenon indicates that some common mechanisms occur during microwave sintering of different kinds of metals, such as titanium and aluminum. However, phenomena (2) and (3) were not observed in the microwave sintering of aluminum; therefore, different mechanisms are associated with the microwave sintering of different metals. These two phenomena were captured in the microwave sintering of the Al–SiC mixed system, which showed different microwave sintering mechanisms from those of pure aluminum [12]. The morphological parameters were quantitatively analyzed to further investigate the different microwave sintering mechanisms of titanium compared with other metal of aluminum.

The particles on the same cross-section at the initial height may migrate to the cross-section at other heights during the sintering experiment. Not all of the microstructure features can be traced in Figure 3a–c because of the thermal expansion and traction among the particles. However, the particles can be observed in the 3D images.

4. Discussion

Interaction mechanisms, such as thermal effects and non-thermal effects between mass and microwave fields, remarkably influence microstructure evolution. Morphological parameters, such as sintering neck growth rate and particle surface area, were obtained and quantitatively analyzed to examine kinetic mechanisms.

Sintering neck formation and growth were quantitatively evaluated. In traditional sintering theories, the dynamics of stable neck growth summarized by Kuczynski [13] is shown as follows. This theory was applied to quantify the sintering neck growth rate during microwave sintering and to provide a reference on the corresponding mass diffusion mechanisms.

$$\left(\frac{x}{a}\right)^n = \frac{F(t)}{a^m}t \qquad (1)$$

where x/a is the ratio of inter-particle neck radius to the particle radius; $F(t)/a^m$ is a constant that involves particle size, temperature, and geometric and material terms; t is the sintering time; and n is the sintering neck growth exponent. Equation (1) indicates that the plot of $\log(x/a)$–$\log(t)$ shows a linear relationship with a slope equal to $1/n$. According to the exponential criterion, different n represents different major diffusion mechanisms. Watershed algorithm was used to reveal the sintering-neck kinetic mechanisms and to obtain the morphological parameters. Sintering neck extraction is illustrated in Figure 4. The sintering neck between two particles can be distinguished and the size of sintering neck can be counted [14]. A total of 100 cross-sections of the same region at different sintering times identified by the marking points and microstructure features were selected and subjected to statistical analyses. The same watershed operation was applied to these cross-sections, and the size of the sintering neck was the average value of the selected region.

The line $\log(x/a)$–$\log(t)$ is shown in Figure 4. For comparisons, the line of aluminum microwave sintering in the present work [11] is also shown in Figure 5b.

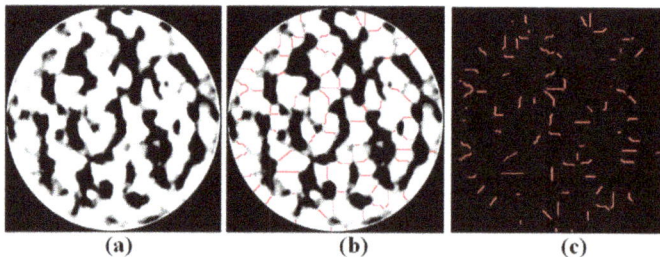

Figure 4. Sintering neck extraction by watershed method: The cross section image (**a**), the result of watershed algorithm (**b**), and the extracted sintering necks (**c**).

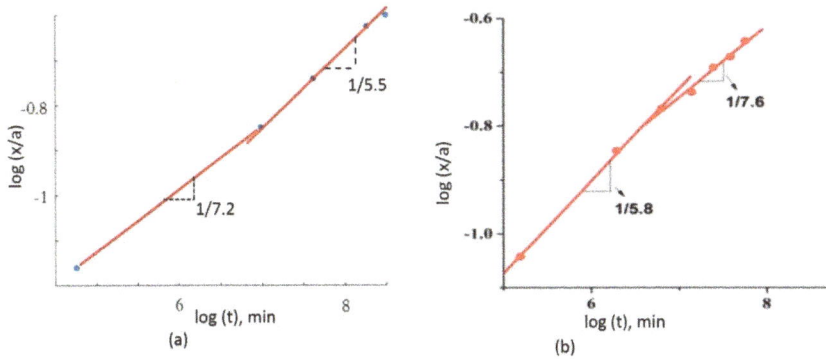

Figure 5. Sintering neck growth curves at $\log(x/a)$–$\log(t)$ of Ti (**a**) and Al (**b**).

Figure 5 shows that the scatter diagram of $\log(x/a)$–$\log(t)$ for both titanium and aluminum satisfies the linear relationship, but the differences are also significant. In the early period of microwave sintering, n indicated that the sintering neck growth rate of titanium was slower than that of aluminum, and the dominant diffusion mechanisms of titanium and aluminum were surface diffusion and volume diffusion, respectively. As sintering proceeded, the sintering neck growth became quite different from the intermediate period. The sintering neck growth rate of titanium significantly accelerated in the intermediate period; by contrast, the sintering neck growth rate of aluminum decelerated. The dominant diffusion mechanism was also different, and n indicated that volume diffusion and surface diffusion were the dominant mechanisms of titanium and aluminum, respectively. These results revealed that the sintering processes of the two metals behaved differently. The reasons for these phenomena were analyzed on the basis of the interaction mechanisms between mass and microwave fields.

We discussed the difference in the sintering neck growth rate in the early period. The major heating mechanism of metal powder in the microwave field is eddy current loss on the particle surface [15]. The power loss of metal particle is in accordance with the following equation [16,17]:

$$P = \frac{1}{2} \int \vec{E} \cdot \vec{J_s} dV = \frac{1}{2}\sigma \int \left|\vec{J_s}\right|^2 dV \tag{2}$$

where J_s is the surface current calculated as $\vec{J_s} = \vec{n} \times \vec{H_t}$; $\vec{H_t}$ is the tangential magnetic vector; and σ is the conductance. The eddy current loss of aluminum was much higher than that of titanium because the electrical conductivity of aluminum (approximately 35.5×10^6 S/m) is higher than that of titanium (approximately 2.6×10^6 S/m). Therefore, aluminum yields a faster sintering neck growth rate than titanium in the early period, as shown in the sintering neck growth curve.

The sintering neck growth of titanium accelerated, but the sintering neck growth of aluminum decelerated when $\log(x/a)$ reached approximately -0.8. In the previous work, alumina possibly covers the grain surface and hinders mass diffusion to slow down sintering neck growth. A further analysis was conducted to clarify the acceleration of the sintering neck growth rate of titanium from the intermediate period.

The eddy current was located on a very thin layer of the particle surface because of the restriction of the skin depth of microwave (approximately 7 μm for titanium at 2.45 GHz, less than the size of one particle). Therefore, the total free surface area of particles significantly influences the eddy current loss [12]. Figure 6 reveals the statistical results of the total particle free surface area as the average values of the same several layers. Figure 6 also shows that the surface area reduced rapidly in the early period. However, the surface area remarkably decreased as the sintering process progressed.

Figure 5a indicates that the eddy current loss was reduced in the intermediate period of the sintering process; by contrast, the sintering neck growth rate accelerated. In addition to eddy current loss, other heating mechanisms caused by thermal and non-thermal effects occurred, and these mechanisms also contributed to the acceleration of sintering neck growth. Other possible reasons for the sintering neck growth acceleration were proposed.

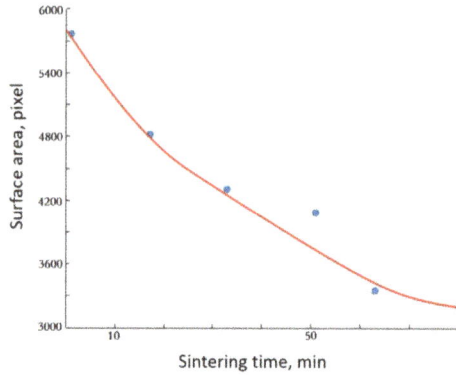

Figure 6. Variation in the total particle free surface area.

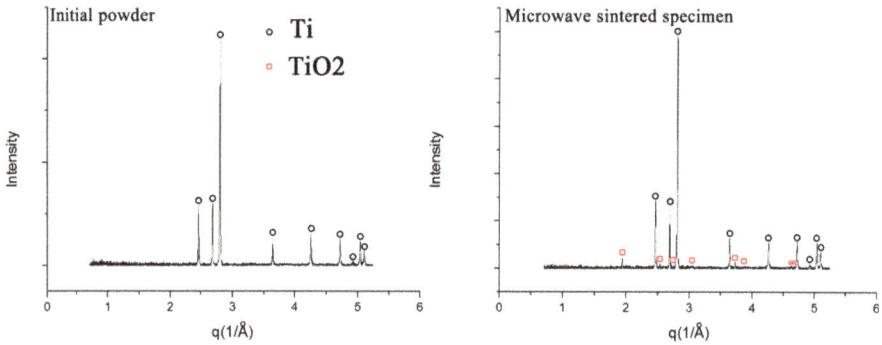

Figure 7. X-ray diffraction patterns of the powder and specimen.

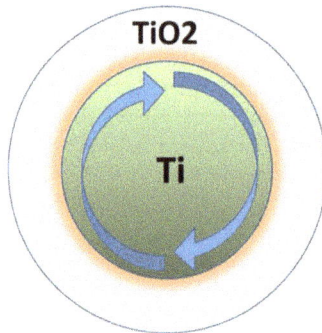

Figure 8. Schematic of particle loss.

(1)   Oxidation may occur to a certain extent because of the constraints in experimental conditions, such as open furnace chamber and slow protective atmosphere flow rate, although sintering proceeded in an argon-protected atmosphere. Titanium dioxide formed and covered the grains as sintering proceeded because of the slow protective gas flow rate. The X-ray diffraction patterns of the powder before and after microwave sintering are represented in Figure 7. Oxide was not detected in the initial powder before sintering. By contrast, the diffraction peaks of titanium dioxide appeared, and this finding indicated the formation of oxide after microwave sintering was completed. The dominant microwave heating mechanism of the pure metal powder was eddy current loss. When titanium dioxide covered the metal particles (Figure 8 showing the thickness of titanium dioxide as a representation, not the actual thickness), the inner metal was heated by the eddy current and the outer titanium dioxide was heated by dipolar polarization loss. Therefore, the total power loss of metal oxides may be higher than that of pure metal particles; as a result, the sintering neck of titanium grew rapidly from the intermediate period. In addition, the sintering neck growth of aluminum decelerated in the present work because of the poor microwave-absorbing property of alumina.

(2)   When titanium dioxide was formed, polarization charges accumulated at the interface of metal and oxide because of the heterogeneity between titanium and titanium dioxide. Therefore, the interfacial polarization effect cannot be ignored as another important heating mechanism. The heating effect caused by interfacial polarization probably accelerated the sintering neck growth. For the microwave sintering of aluminum, the difference between the interface of aluminum-alumina and titanium-titanium dioxide resulted in different strengths of interfacial polarization. Therefore, the sintering neck growth of aluminum decelerated in the present work.

(3)   Local plasma-activated sintering [18] promoted the sintering neck growth. The local electromagnetic field may be strong in some regions of the sintering neck, pores, and burrs of the rough surface because of the focusing influence of the microstructure; thus, the protective atmosphere of argon was ionized at the local regions. Evaporation-condensation mechanisms may be the local dominant diffusion mechanism because of plasma-generated ultra-high temperature; this mechanism likely accelerates the sintering neck growth and densification. Moreover, the atomic radius of most air components is smaller than that of argon; as such, the ionization energy of air is higher than that of argon and the ionization of air becomes more difficult than that of argon. Therefore, the microwave sintering of aluminum was not accelerated.

5. Conclusions

SR-CT was applied to investigate titanium powder microwave sintering *in situ*. The cross-sections of the same microstructure features at different sintering times were extracted on the basis of the reconstructed images. The sintering phenomena differ from conventional sintering, such as particle smoothing and rounding and contacted particles isolation. These unique phenomena may be attributed to the thermal and non-thermal effects of microwave field, including eddy current, interfacial polarization effect, and micro-focusing effect. The microstructure morphological parameters, including sintering neck size and particle surface area, were obtained and quantitatively analyzed to examine the different sintering mechanisms of the two kinds of aluminum and titanium metals. Our results revealed that the sintering neck growth rate accelerated from the intermediate period. Considering that the particle-free surface area associated with the eddy current loss was reduced, we can infer that other heating mechanisms, in addition to eddy current loss, occurred. Possible reasons, such as mixing heating of eddy current loss, dielectric loss, interfacial polarization, and local plasma-activated sintering, were proposed. The microwave sintering of titanium was enhanced because of these interaction mechanisms, including thermal and non-thermal effects.

Acknowledgments: This research was supported by the National Natural Science Foundation of China (No. 11272305, No. 11172290, No. 11472265, No. 11402160, No. 10902108) and the National Basic Research Program of China (973 Program, No. 2012CB937504) and Anhui Provincial Natural Science Foundation (No. 1508085MA17).

Author Contributions: Yu Xiao, Feng Xu and Yongcun Li analyzed the experimental data and wrote the paper; Feng Xu and Xiaofang Hu conceived and designed the experiment; Yu Xiao, Wenchao Liu and Bo Dong prepared the sample and performed the experiment.

Conflicts of Interest: The authors declare no conflict of interest.

References

1. Roy, R.; Agrawal, D.; Cheng, J.; Gedevanishvili, S. Full sintering of powdered-metal bodies in a microwave field. *Nature* **1999**, *399*, 668–670.
2. Janney, M.A.; Kimrey, H.D.; Allen, W.R.; Kiggans, J.O. Enhanced diffusion in sapphire during microwave heating. *J. Mater. Sci.* **1997**, *32*, 1347–1355. [CrossRef]
3. Saitou, K. Microwave sintering of iron, cobalt, nickel, copper and stainless steel powders. *Scr. Mater.* **2006**, *54*, 875–879. [CrossRef]
4. Chhillar, P.; Agrawal, D.; Adair, J.H. Sintering of molybdenum metal powder using microwave energy. *Powder Metall.* **2008**, *51*, 182–187. [CrossRef]
5. Ma, J.; Diehl, J.F.; Johnson, E.J.; Martin, K.R.; Miskovsky, N.M.; Smith, C.T.; Weisel, G.J.; Weiss, B.L.; Zimmerman, D.T. Systematic study of microwave absorption, heating, and microstructure evolution of porous copper powder metal compacts. *J. Appl. Phys.* **2007**. [CrossRef]
6. Ahn, J.H.; Lee, J.N.; Kim, Y.C.; Ahn, B.T. Microwave-induced low-temperature crystallization of amorphous Si thin films. *Curr. Appl. Phys.* **2002**, *2*, 135–139. [CrossRef]
7. Birnboim, A.; Calame, J.P.; Carmel, Y. Microfocusing and polarization effects in spherical neck ceramic microstructures during microwave processing. *J. Appl. Phys.* **1999**, *85*, 478–482. [CrossRef]
8. Chen, G.; Liss, K.D.; Cao, P. An *in situ* Study of NiTi Powder Sintering Using Neutron Diffraction. *Metals* **2015**, *5*, 530–546. [CrossRef]
9. Grupp, R.; Nöthe, M.; Kieback, B.; Banhart, J. Cooperative material transport during the early stage of sintering. *Nat. Commun.* **2011**. [CrossRef] [PubMed]
10. Bale, H.A.; Haboub, A.; MacDowell, A.A.; Nasiatka, J.R.; Parkinson, D.Y.; Cox, B.N.; Marshall, D.B.; Ritchie, R.O. Real-time quantitative imaging of failure events in materials under load at temperatures above 1600 °C. *Nat. Mater.* **2013**, *12*, 40–46. [CrossRef] [PubMed]
11. Xu, F.; Li, Y.; Hu, X.; Niu, Y.; Zhao, J.; Zhang, Z. In situ investigation of metal's microwave sintering. *Mater. Lett.* **2012**, *67*, 162–164.
12. Li, Y.C.; Xu, F.; Hu, X.F.; Kang, D.; Xiao, T.Q.; Wu, X.P. In situ investigation on the mixed-interaction mechanisms in the metal-ceramic system's microwave sintering. *Acta Mater.* **2014**, *66*, 293–301.
13. Kuczynski, G.C. Self-diffusion in sintering of metallic particles. *AIME Trans.* **1949**, *185*, 169–178.
14. Gonzalez, R.C.; Woods, R.E.; Eddins, S.L. *Digital Image Processing Using Matlab*; Publishing House of Electronics Industry: Beijing, China, 2005; pp. 315–317.
15. Cheng, J.; Roy, R.; Agrawal, D. Experimental proof of major role of magnetic field losses in microwave heating of metal and metallic composites. *J. Mater. Sci. Lett.* **2001**, *20*, 1561–1563. [CrossRef]
16. Huang, Y. *Electromagnetic Field and Microwave Technology*, 1st ed.; Posts & Telecom Press: Beijing, China, 2007; pp. 77–101.
17. Feng, C. *Electromagnetic Field*, 2nd ed.; Higher Education Press: Beijing, China, 1983; pp. 322–362.
18. Tracy, M.; Groza, J.R.; Yamazaki, K.; Sudarshan, T.S. Preliminary studies on the densification of fine tungsten powders by plasma activated sintering (PAS) process. In Proceedings of the 2nd International Conference on Tungsten and Refractory Metals, McLean, VA, USA, 17–19 October 1994; pp. 291–297.

![metals logo] *metals*

Article

Study on Sintering Mechanism of Stainless Steel Fiber Felts by X-ray Computed Tomography

Jun Ma *, Aijun Li and Huiping Tang *

State Key Laboratory of Porous Metal Materials, Northwest Institute for Nonferrous Metal Research,
Xi'an 710016, China; ajli@alum.imr.ac.cn
* Correspondence: majun.2008@stu.xjtu.edu.cn (J.M.); hptang@c-nin.com (H.T.); Tel.: +86-29-8623-1095 (J.M.);
  Fax: +86-29-8626-4926 (J.M.)

Academic Editor: Klaus-Dieter Liss
Received: 1 October 2015; Accepted: 4 January 2016; Published: 13 January 2016

Abstract: The microstructure evolution of Fe-17 wt. % Cr-12 wt. % Ni-2 wt. % Mo stainless steel fiber felts during the fast sintering process was investigated by the synchrotron radiation X-ray computed tomography technique. The equation of dynamics of stable inter-fiber neck growth was established for the first time based on the geometry model of sintering joints of two fibers and Kucsynski's two-sphere model. The specific evolutions of different kinds of sintering joints were observed in the three-dimensional images. The sintering mechanisms during sintering were proposed as plastic flow and grain boundary diffusion, the former leading to a quick growth of sintering joints.

Keywords: metal fiber; sintering mechanism; X-ray tomography; three-dimensional structure

---

1. Introductions

Fe-17 wt. % Cr-12 wt. % Ni-2 wt. % Mo stainless steel fiber felts (316L SSFF) are super-light porous materials with an open net structure and a porosity ranging from 70% to 95%. They are applied to gas/liquid/solid filtration, and have recently showed great potential in the fields of impact energy absorption, fuel cell, heat transfer, *etc.* [1]. Typically, these materials are made by bundle-drawn 316L stainless steel fibers with diameters less than 50 μm. Two key steps are included in the manufacturing process of the materials: firstly, the fibers are mixed and sediments set to form green felts (unsintered felts) through an air-laid process, and then the green felts are sintered in vacuum at a high temperature to form metallurgical bonds between fibers. At the meso-level, the fibers distribute in a random way within the sediment layers, and the inter-fiber sintering joints bond all the fibers as a whole. The strength of the sintering joints greatly affects the mechanical property of 316L SSFF. Thus, the formation process of sintering joints is worth studying. However, the sintering mechanism of metal fiber felts has not received adequate attention. Kostomov [2] studied the relationship between sintering temperature, holding time and the macro-shrinkage of metal fiber felts and revealed that the dominant mechanism was viscous flow during the sintering of porous metal fiber felts. Using the viscous flow mode, which is initiated by surface tension, Kostomov also obtained a far higher apparent viscidity of metal fibers than that of metal powders [2]. However, since the macro-shrinkage of metal fibers is controlled not only by the formation of inter-fiber joints, Kostomov's conclusion is doubtful. In fact, Balshin found that the elasticity energy of the metal fibers greatly enhanced the shrinkage rate of metal fiber felts during sintering [3]. Pranatis studied the sintering of parallel metal fibers using the two-sphere model of Kuczynski [4] and suggested that the dominant sintering mechanism was surface diffusion combined with volume diffusion, while under the same sintering conditions the dominant sintering mechanism of metal powders was only surface diffusion [5]. G. Matsumura studied the sintering process of parallel Fe fibers with a diameter of 200 μm and suggested that surface diffusion was the dominant sintering mechanism at 895 °C while volume diffusion was the dominant sintering mechanism at 1300 °C [6]. These studies of parallel fibers provided an analysis of the sintering mechanisms of fibers at the microcosmic level; however, they could not give a convincible explanation of the sintering mechanisms of metal fiber felts because the sintering of porous metal fiber felts is performed among the adjacent fibers with random angles in the felts. The three-dimensional (3D) structure of the sintering joints of fibers in the porous metal fiber felts is far more asymmetric than that of the sintering joints of powders with a sphere shape, so the structure evolutions of inter-fiber joints during sintering are hard to observe and study. The adjacent joints along the fibers could impact each other during the sintering process. Furthermore, the bundle-drawn 316L stainless steel fibers experience large deformation processes during the manufacturing process and contain a large density of dislocations which could affect the formation process of the sintering joints [7]. There is no ready micro-model for the sintering of inter-fiber joints. To our best knowledge, no study has been done on the sintering mechanism of 316L SSFF.

The synchrotron radiation X-ray computed tomography (SR-CT) technique is a kind of non-destructive testing technology which can realize the observation of the 3D microstructure evolution of 316L SSFF. In this study, the microstructure of 316L SSFF was reconstructed by SR-CT, and then the radii of the sintering necks and the angles between the sintered fibers were measured in the 3D image of 316L SSFF. The specific evolutions of fiber-joints during sintering were observed. Furthermore, the sintering model of inter-fiber joints was established based on the two-sphere model and the sintering mechanisms at different temperatures were analyzed.

2. Experimental Procedures

Three columns of 316L SSFF with dimension of ø1.2 mm × 2 mm, fiber diameter of 28 µm and porosity of 83% were sealed in vacuum quartz tubes with an inner diameter of 1.2 mm, respectively. All the samples were heated in vacuum drying oven at 120 °C for 0.5 h to eliminate the water vapor before being sealed in quartz tubes and the vacuum degree in the sealed quartz tubes was 10^{-4} Pa. Then the vacuum-sealed columns were continually CT-scanned and heated. The dislocations in the fibers may accelerate the formation of inter-fiber joints; however the dislocations may annihilate during the normal heating-up process (heating rate at 10 °C/min). Therefore, to maximize the effect of dislocations on the sintering process, a fast heating process was applied to the samples. All the samples were loaded into the muffle furnace only when the furnace had reached the preset temperature, and after certain isothermal hold the samples were withdrawn from the furnace and cooled subsequently to room temperature in air. The heating rate in this sintering process was estimated to be 300–500 °C/min. Another advantage of the fast heating process is that it can also prevent overgrowing of grains in the fibers, so it can improve the strength of fibers and in turn raise the strength of the 316L SSFF. The experimental procedure is given in Table 1:

Table 1. Specific experimental procedure.

Sintering Temperature(°C)	First Holding Time (min)	Test	Second Holding Time (min)	Test	Third Holding Time (min)	Test
1000	10		10		20	
1100	5	CT scan	5	CT scan	10	CT Scan
1200	5		5		10	

The CT scans were carried out on the BL13W1 beam line at Shanghai Synchrotron Radiation Facility (SSRF, Shanghai, China). The spatial resolution of SR-CT was 0.74 µm and the size of vision field was 1.4 mm × 1.4 mm. During the CT scan each sample spinned over 180° and 900 projection images were taken at an increment of 0.2°, each image with an exposure time of 6 s and a beam energy of 35 KeV. Then the projection images were transformed to slice images, and the 3D structure was reconstructed based on the slices. The neck radii of the sintering joints and the angles between the sintered fibers were measured in the 3D structure and the measuring process is shown in Figure 1.

The black broken line in Figure 1b represents a slicing plane which is vertical to the two sintered fibers and equally divides the obtuse angle between the two fibers. This slicing plane was used to cut the sintering joint (in the red broken circle) in half to obtain a representative section of the sintering joint in the transverse direction. After the cutting, as shown in Figure 1c, the sintering joint was rotated over 90° to make the representative section envisaged and the size of inter-fiber neck radius was measured, as shown in Figure 1d.

Figure 1. Measurement of neck radius of sintering joint from the 3D structure of 316L SSFF, (**a**) Selection of joint, (**b**) Selection of slicing plane, (**c**) Cutting of joint, (**d**) Measurement of neck radius.

3. Results and Discussions

3.1. Dynamics Equation of Stable Neck Growth of Metal Fibers

The evolution of the sintering joint between two fibers is too hard to chase and characterize due to the asymmetric 3D structure; thus, a simplified two-dimensional (2D) geometry model was established to depict the evolution of the sintering joint. The key parameter to determine the shape of inter-fiber joints is the acute angle θ between the two fibers. The value of θ ranges from 0° to 90° to depict all the angle states between fibers. A plane which was vertical to the two fibers and equally divided the obtuse angle between the two fibers was selected to slice the sintering joint in half to obtain a representative section of the sintering joint in the transverse direction. The top view of the sintering joint and the transverse section of the sintering joint are shown in Figure 2, in which x represents the inter-fiber neck radius in the section and can be used as an indication of the extent of the sintering joint's growth, and a is the radius of the fibers. When θ = 0°, the shape of the section is identical to Kuczynski's two-sphere model. When θ ≠ 0°, the shape of the section turns into two symmetrical ellipses sintered together, as shown in Figure 2b. The 3D shape of the inter-fiber joints with θ ≠ 0° is not as symmetrical as the two-sphere model, so the curvatures of the neck surfaces in the adjacent sections of the sintering joints are different from each other and this may cause the migration of atoms between the neck surfaces in the adjacent sections. To predigest the deduction of sintering equations, it is assumed that the migration of atoms between the adjacent neck surfaces can be ignored. Through the comparison of magnitude of the migration rate of atoms between the adjacent neck surfaces at the joint with that between the neck surface and the fiber surface within one section (as shown in the Appendix), the assumption is verified. Thus, the growth of the sintering neck in the section shown in Figure 2b can be regarded as controlled only by migration of atoms within the section. Based on this model, the equations of dynamics of stable neck growth can be established in a way similar to that used by Kuczynski in the two-sphere model. According to the geometry relations in Figure 2, the value of b can be expressed as follows:

$$b = a/\cos(\theta/2) \tag{1}$$

The neck surface curvature radius ρ can be expressed as follows:

$$\rho = x^2 \times \cos^2(\theta/2)/2a \tag{2}$$

The equations of dynamics of stable inter-fiber neck growth are given based on the above model as follows:

Plastic flow mechanism:

$$x^2/a^2 = 4bwn_0 L \exp(-E/kT) t/\cos^2(\theta/2) \tag{3}$$

b: bergers vector; w: coefficient of frequency; n_0: volume density of atoms; E: activation energy of dislocation; L: average sliding distance of dislocations; k: Boltzmann constant; t: isothermal holding time; T: absolute sintering temperature [8].

Surface diffusion mechanism:

$$x^7/a^3 = 56D_s\gamma\delta^4/kT\cos^6(\theta/2) \times t \tag{4}$$

D_s: coefficient of surface diffusion; γ: surface energy; δ: constant of crystal lattice.

Grain boundary diffusion mechanism:

$$x^6/a^2 = 96\delta_g D_g\gamma\delta^3/kT\cos^4(\theta/2) \times t \tag{5}$$

δ_g: thickness of grain boundary; D_g: coefficient of grain boundary diffusion [9].

The letter n is used to depict the exponent of x in Equations (3)–(5). The certain value of n corresponds to the certain sintering mechanism. The value of n is also equivalent to the inverse slope of the $\ln(x/a)$-$\ln(t)$ curve of inter-fiber joints and the curve can be obtained from experiments.

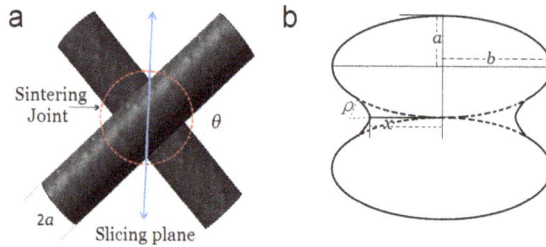

Figure 2. The simplified model of sintering neck of fibers; (**a**) Top view of sintering joint; (**b**) Section of sintering joint.

3.2. Specific Evolutions of Sintering Joints

Figure 3 shows the evolutions of four typical sintering joints holding at 1200 °C for 5 min, 10 min and 20 min. The sintering joint shown in Figure 3a had a fiber angle of 90°, and achieved a large value of neck radius when the holding time was 5 min. Its neck growth was not evident in the subsequent isothermal holding. The sintering joint shown in Figure 3b had a small initial value of neck radius when the holding time was 5 min and showed a high neck growth rate during the subsequent isothermal holding. Notably, Figure 3b also exhibits a special evolution process of sintering joints. The right-hand surface of the inter-fiber neck in Figure 3b had a higher initial curvature than the left-hand surface had, and the upper fiber approached the bottom fiber in the right-hand during the subsequent sintering process, causing the change of the relative angle between the couple fibers. It can be explained as follows: the asymmetric curvatures at the surfaces of the sintering neck could give rise to asymmetric massive transportation along the neck, eventually resulting in the angle change between the couple fibers. Detailed analysis was given by Exner [10,11]. The sintering joint shown in Figure 3c had a fiber angle of 60° and also had a small initial value of neck radius when the holding time was 5 min, and showed a high neck growth rate in the following process. An anomalous evolution process was observed in Figure 3d: the neck radius reduced at first, then a quick growth followed. A variety of mechanics could lead to this phenomenon, such as the couple fibers' relative "rolling" caused by grain boundary sliding or the tension by adjacent sintering joints [12]. A similar phenomenon was also observed in the *in-situ* CT observation of microwave sintering of Al powders [13].

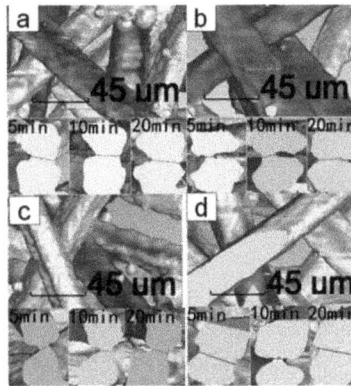

Figure 3. The evolutions of four sintering joints sintered at 1200 °C for 5 min, 10 min and 20 min; (**a**) Sintering joint of 90° experienced a slow neck growth, (**b**) Sintering joint of 90° experienced a fast neck growth, (**c**) Sintering joint of 60° experienced a fast neck growth, (**d**) Sintering joint of 90° experienced a anomalous evolution process.

3.3. Determination of Sintering Mechanism of 316L SSFF

The change of values of sintering neck radius x during the isothermal holding of 316L SSFF was measured in the CT images. Figure 4 presents the $\ln(x/a)$-$\ln(t)$ curves of inter-fiber joints with a fiber angle of 90°. The values of $\ln(x/a)$ used here were the averages of 10–20 joints in each sample. According to the values of n, which was determined by the inverse slopes of $\ln(x/a)$-$\ln(t)$ curves, the sintering mechanisms of the 316L SSFF at different temperatures could be identified.

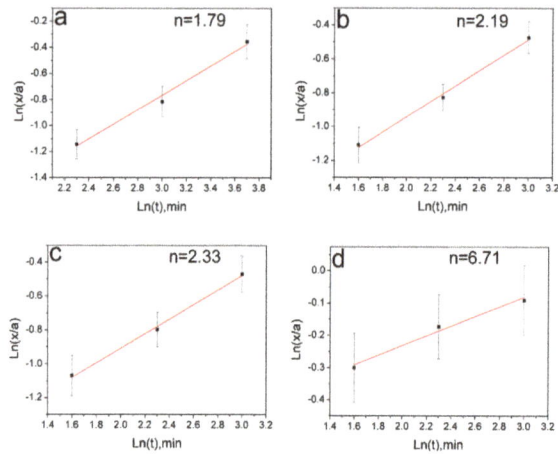

Figure 4. The ln(x/a)-ln(t) curves of sintering joints with a fiber angle of 90° during isothermal holding at (**a**) 1000 °C; (**b**) 1100 °C; (**c**) 1200 °C (n = 2.33); and (**d**) 1200 °C (n = 6.71).

It is shown in Figure 4a,b that the values of n are close to 2 for the samples sintered at 1000 °C or 1100 °C. Thus, according to Equation (3), the dominant sintering mechanism should be plastic flow. With the sintering temperature increasing to 1200 °C, the plastic flow (n = 2.33) and the grain boundary diffusion (n = 6.71) were both observed, as shown in Figure 4c,d. It indicates that a transition of sintering mechanisms emerged around 1200 °C. At this temperature, some sintering joints had entered the later sintering stage in which the grain boundary diffusion controlled the neck growth process.

In general, the starting-up condition of plastic flow is the larger Laplace stress at the sintering neck surface rather than the critical initial shear stress of dislocation. Normally, the Laplace stress at the surface of metals is not big enough to make the dislocations slide or initiate. However, it was reported that the values of critical initial shear stress of dislocation in metals reduce quickly with the rise of heating temperature; for example, when the temperature was close to melting point, the critical initial shear stress of dislocation reduced to one-tenth of its original value [14]. Thus, it is possible that, at this high sintering temperature of 1000 °C–1200 °C, the Laplace stress at the neck surface can initiate dislocations and make them slide. Initiation of dislocations near the neck during the high temperature sintering process of metals has been reported, for example in Scatt's investigation on the sintering between single crystal Cu powder with a diameter of 250 μm and a Cu plate, a large number of dislocations were found to initiate near the sintering neck when $x/a \approx 0.2$, which was regarded as a sign of plastic flow [15]. Another supporting evidence is the phenomenon that, during holding at 1200 °C, there was a distinct transition between the plastic flow and grain boundary diffusion when the x/a reached a critical value (0.6–0.7). This can be explained as follows: the Laplace stress at the neck is inverse to the neck surface curvature radius ρ. The ρ becomes larger with the increase of the relative neck radius x/a according to the Equation (2), which leads to the decrease of the Laplace stress. When the Laplace stress decreases to a value lower than that of the critical initial shear stress of dislocation in the metal, plastic flow near the sintering neck would stop. In this case, the critical value of x/a is supposed to be 0.6–0.7. It is also worth noting that the growth of the neck radius under the plastic flow is much quicker than that under other mechanisms such as surface diffusion or grain boundary diffusion, which is beneficial to the formation of high-strength sintering-joints.

4. Conclusions

The 3D structure of 316L SSFF was reconstructed using the synchrotron radiation X-ray computed tomography (SR-CT) technique and the characteristic sizes of sintering joints were measured in the 3D structure. The equations of dynamics of stable inter-fiber neck growth were established based on the geometry model of sintering joint of fibers and Kucsynski's two-sphere model. Two kinds of anomalous sintering processes of fiber joints were observed. The sintering mechanisms of 316L SSFF were analyzed based on the $\ln(x/a)$-$\ln(t)$ relations of the inter-fiber joints. When sintering temperatures are relative low (1000 °C or 1100 °C), the growth of inter-fiber joints is controlled by the plastic flow which leads to a quick growth of the sintering neck. When the temperature rises to 1200 °C, the sintering mechanism changes toward grain boundary diffusion.

Acknowledgments: This work was supported by the National Nature Science Foundation of China (No. 51134003), Key Scientific and Technological Innovation Team Project of Shaanxi Province (No. 15KCT-11) and the Opening Project of State Key Laboratory of Explosion Science and Technology (Beijing Institute of Technology) (No. KFJJ15-02M).

Author Contributions: Jun Ma and Huiping Tang conceived and designed the experiments; Jun Ma performed the experiments; Jun Ma and Aijun Li analyzed the data.

Conflicts of Interest: The authors declare no conflict of interest.

Appendix

To verify the assumption that atom migration between adjacent neck surfaces in the joint can be ignored, the magnitude of the migration rate of atoms between the surfaces of adjacent necks in the joint was compared with that between the neck surface and the fiber surface within one section. Two specific sections in the joint with a fiber angle of 90° were chosen to make the calculation simplified, as shown in Figure A1. The Sections 1# and 2# are shown in Figure A1b,c and they were obtained by cutting the joint with slicing planes 1# and 2#, respectively. The direction of slicing plane 1# was the same as that of the slicing plane shown in Figure 2a, while slicing plane 2# was chosen to be vertical to the axis of the upper fiber and include the axis of the bottom fiber. The curvature of the neck surface in Section 2# is the highest of that in all the sections of the joints according to the 3D shape of the joint, consequently higher than that in Section 1#. The atom migration between the neck surfaces in Sections 1# and 2# surely exists, and it may affect the neck growth in Section 1#. To determine the extent of this effect, the atom migration rate between the neck surfaces in Sections 1# and 2# was compared to that between the neck surface and the fiber surface within Section 1#. According to the geometry relations in Sections 1# and 2#, the values of curvature of the two neck surfaces were given as follows:

$$1/\rho_1 = 2.83a/x^2 \tag{A1}$$

$$1/\rho_2 = 4a/x^2 \tag{A2}$$

ρ_1: curvature radius of the neck surface in Section 1#; ρ_2: curvature radius of the neck surface in Section 2#; a: radius of the fibers; x: radius of the necks (same value for both necks).

The vacancy concentration gradient between the two neck surfaces can be expressed as follows:

$$\nabla C_{v1} = 4C_v{}^\circ \gamma \Omega (1/\rho_1 - 1/\rho_2)/kT\pi x = 4.68C_v{}^\circ \gamma \Omega a/kT\pi x^3 \tag{A3}$$

where the distance of diffusion between the two neck surfaces is $\pi x/4$;

∇C_{v1}: vacancy concentration gradient between the two neck surfaces; $C_v{}^\circ$: vacancy concentration at equilibrium; Ω: volume of single atom;

On the other hand, the vacancy concentration gradient between the neck surface and the fiber surface in Section 1# can be expressed as follows:

$$\nabla C_{v2} = 8C_v{}^\circ \gamma \Omega a^2/kTx^4 \tag{A4}$$

Metals **2016**, 6, 18

where the distance of diffusion is ρ_1;

∇C_{v2}: vacancy concentration gradient between the neck surface and the fiber surface in Section 1#. Noticing that $x \ll a$ at the early stage of sintering, then $\nabla C_{v1} \ll \nabla C_{v2}$.

The atom migration rate can be calculated based on Fick's diffusion equation:

$$J = D\prime\nabla C_v \tag{A5}$$

J: atom migration rate; $D\prime$: diffusion coefficient; ∇C_v: vacancy concentration gradient.

According to Equation (A5), the atom migration rate between the two neck surfaces is far smaller than that between the neck surface and the fiber surface in Section 1#. This is an analysis for the diffusions between neck surfaces of two special sections in the joint with a specific fiber angle. However, it can be generalized that the atom migration between neck surfaces in any adjacent sections of sintering joints can be ignored compared with that between the neck surface and fiber surface within one section. Thus, the growth of the neck radius in one section can be regarded as controlled only by migration of atoms within the section.

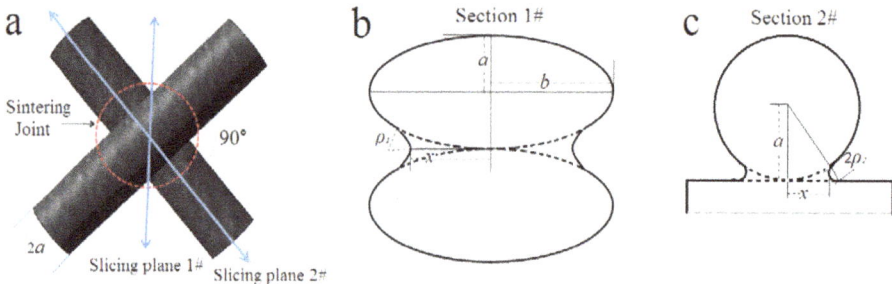

Figure A1. The sections of a sintering joint of 90° cut in different directions, (**a**) Directions of the two slicing planes, (**b**) Section 1# cut by slicing plane 1#, (**c**) Section 2# cut by slicing plane 2#.

References

1. Lefebvre, L.P.; Banhart, J.; Dunand, D.C. Porous metals and metallic foams: Current status and recent developments. *Adv. Eng. Mater.* **2008**, *10*, 775–787. [CrossRef]
2. Kostornov, A.G.; Galstyan, L.G. Sintering kinetics of porous fiber solids. *Poroshkovaya Metall.* **1984**, *254*, 41–45. [CrossRef]
3. Balshin, M.Y.; Rybalchenko, M.K. Some problems of fiber metallurgy. *Poroshkovaya Metall.* **1964**, *3*, 16–22. [CrossRef]
4. Kuczynski, G.C. Self-diffusion in sintering of metallic particles. *Trans. Am. Inst. Min. Metall. Eng.* **1949**, *185*, 169–178.
5. Swinkels, F.B.; Ashby, M.F. A second report on sintering diagrams. *Acta Metall.* **1981**, *29*, 259–281. [CrossRef]
6. Matsumara, G. Sintering of Iron Wires. *Acta Metall.* **1971**, *19*, 851–855. [CrossRef]
7. Shyr, T.W.; Shie, J.W.; Huang, S.J.; Shun, T.Y.; Weng, S.H. Phase transformation of 316L stainless steel from wire to fiber. *Mater. Chem. Phys.* **2010**, *122*, 273–277. [CrossRef]
8. Guo, S.J. *Powder Sintering Theory*; Metallurgical Industry Press: Beijing, China, 1998; pp. 185–187.
9. Coble, R.L. Initial sintering of alumina and hematite. *J. Am. Ceram. Soc.* **1958**, *41*, 55–62. [CrossRef]
10. Exner, H.E.; Bross, P. Material transport rate and stress distribution during grain boundary diffusion driven by surface tension. *Acta Metall.* **1979**, *27*, 1007–1012. [CrossRef]
11. Exner, H.E.; Muller, C. Particle rearrangement and pore space coarsening during solid-state sintering. *J. Am. Ceram. Soc.* **2009**, *92*, 1384–1390. [CrossRef]
12. Grupp, R.; Nothe, M.; Kieback, B.; Banhart, J. Cooperative material transport during the early stage of sintering. *Nat. Commun.* **2011**. [CrossRef] [PubMed]

13. Xu, F.; Li, Y.; Hu, X.; Niu, Y.; Zhao, J.; Zhang, Z. *In situ* investigation of metal's microwave sintering. *Mater. Lett.* **2012**, *67*, 162–164. [CrossRef]

14. Guo, S.J. *Powder Sintering Theory*; Metallurgical Industry Press: Beijing, China, 1998; p. 189.

15. Schatt, W.; Friedrich, E.; Joensson, D. Spannungsverteilung und versetzungsvervielfachung in der sinterkontaktregion. *Acta Metall.* **1983**, *3*, 121–128. [CrossRef]

metals

MDPI

Article

Monitoring of Bainite Transformation Using *in Situ* Neutron Scattering

Hikari Nishijima [1,†], Yo Tomota [2,*,‡], Yuhua Su [3], Wu Gong [3] and Jun-ichi Suzuki [4]

[1] Department of Applied Beam Science, Graduate school of Science and Enginieering, Ibaraki University, 4-12-1 Naka-narusawa, Hitachi 316-8511, Japan; nishijima.hikari@suzuki-metal.co.jp

[2] Graduate school of Science and Engineering, Ibaraki University, 4-12-1 Naka-narusawa, Hitachi 316-8511, Japan

[3] Japan Atomic Energy Agency, 2-4 Shirane Shirakata Tokai, Ibaraki 319-1195, Japan; yuhua.su@j-parc.jp (Y.S.); gong.wu@jaea.go.jp (W.G.)

[4] Comprehensive Research Organization for Science and Society, 162-1 Shirakata, Tokai, Ibaraki 319-1106, Japan; j_suzuki@cross.or.jp

*   Correspondence: yo.tomota.22@vc.ibaraki.ac.jp; Tel./Fax: +81-297-37-6710

†   Current address: Nippon Steel & Sumikin SG Wire Co. Ltd., 7-5-1 Higashinarashino, Narashino, Chiba 275-8511, Japan.

‡   Current address: National Institute of Materials Science, 1-2-1 Sengen, Tsukuba, Ibaraki 305-0047, Japan.

Academic Editor: Klaus-Dieter Liss
Received: 22 November 2015; Accepted: 6 January 2016; Published: 9 January 2016

Abstract: Bainite transformation behavior was monitored using simultaneous measurements of dilatometry and small angle neutron scattering (SANS). The volume fraction of bainitic ferrite was estimated from the SANS intensity, showing good agreement with the results of the dilatometry measurements. We propose a more advanced monitoring technique combining dilatometry, SANS and neutron diffraction.

Keywords: small angle neutron scattering; dilatometry; bainite transformation; *in situ* measurement; neutron diffraction

1. Introduction

Multi-phase steels containing carbon-enriched austenite have been extensively studied because of their excellent combination of strength and ductility/toughness. To obtain multi-phase structures, various material processings have been developed such as intercritical annealing followed by quenching to produce dual-phase steels [1–3], isothermal holding to yield carbon-enriched retained austenite for transformation-induced plasticity (TRIP steels) [4,5], isothermal holding at a low temperature to realize ultra-fine lamellar structure (nano-bainite steel) [6–10], and the partial quenching followed by up-heated isothermal holding (Q&P steels) [11–15]. In particular, nano-bainite steels consisting of nano-scale lamellae of bainitic ferrite and carbon-enriched austenite have attractively exhibited tensile strength greater than 2 GPa and fracture toughness of approximately 30 MPa m$^{1/2}$ [6,7]. The nano-bainite formed by isothermal holding at 300~400 °C shows an extremely slow transformation rate [6,16,17]. This heat treatment is favorable for producing large mechanical components with small residual stresses. However, the acceleration of the transformation must enlarge the application of nano-bainite steels. We have found that a small amount of low temperature ausforming (e.g., at 300 °C) [18,19] or partial quenching below Ms temperature [20] is effective to accelerate nano-bainite transformation. The dislocation structure introduced in austenite at low temperatures is found to assist bainite transformation with strong variant selection where partial dislocations introduced by ausforming play an important role for bainite transformation [19]. In these advanced steels, the processing is complicated enough that it is very important to monitor microstructure evolution quantitatively.

For studying microstructure evolution, *in situ* observations during processing using synchrotron X-ray [21,22] or neutron diffraction (ND) [19,20] have successfully been employed so far. Diffraction profiles provide the insights on volume fractions of constituents, which show good agreement with the results obtained by dilatometry [17], carbon contents in austenite [23], texture [19], and dislocation density [20]. Line broadening analysis for the ND profile provides "coherently diffracting mosaic size" probably related to dislocation cell size in engineering steels. The sizes larger than 1.0 μm such as austenite grain size or ferrite lath size cannot be evaluated by diffraction but hopefully by SANS or Bragg edge (BE) measurements. *In situ* BE measurement during bainite transformation was examined by Huang *et al.* [24]. They reported the changes in austenite volume fraction and carbon concentration with progress of bainite transformation but gave us no data on ferrite lath size. The two populations of austenite with different carbon contents have also been demonstrated by diffraction [19,22], but no information concerning the size of bainite lath has been reported so far. If we employ *in situ* SANS, the insights on the shape and size of the bainite lath would be obtained. In this study, we introduce a dilatometer into SANS-J-II at JRR-3/JAEA. The volume fraction of nano-bainite estimated from SANS data is compared with the results of conventional dilatometry. Then, we measure dilatometry, SANS and ND simultaneously to understand the mechanism of microstructure evolution during heat processing using an industrial neuron diffractometer, iMATERIA, at MLF/J-PARC. The traditional dilatometry provides only the phase fraction estimated from the amount of expansion or contraction of a specimen, whereas neutron experiments provide details in crystallography, chemical compositions, internal stresses, *etc.* This paper reports trials of such a monitoring system combined with complementary multi-methods.

2. Experimental Procedures

2.1. Specimen Preparation

The chemical compositions of the steel used in this study were 0.79C–1.98Mn–1.51Si–0.98 Cr–0.24Mo–1.06Al–1.58Co–balanced Fe (mass %). The steel was prepared by vacuum induction melting [8,9]. The ingot was homogenized at 1200 °C for 14.4 ks, followed by hot-rolling in the temperature range 1200–1000 °C to reduce the thickness from 40 mm to 10 mm through 10 successive rolling passes. Plate specimens with $15 \times 15 \times 1$ mm^3 were prepared for SANS measurements at SANS-J-II (JRR-3/Japan Atomic Energy Agency, Tokai, Japan) and iMATERIA (Japan Proton Accelerator Research Complex (J-PARC), Tokai, Japan).

2.2. Small Angle Neutron Scattering Methods

In situ SANS measurements were performed using the SANS-J-II small angle neutron scattering instrument installed at the cold neutron beam line in the JRR-3 research reactor of the Japan Atomic Energy Agency (Tokai, Japan). For the SANS measurement, two two-dimensional (2D) detectors were used to detect neutrons scattered in the 0.005 to 0.199 nm^{-1} scattering vector q-range ($q = (4\pi/\lambda)\sin\theta$, where a half of scattering angle θ, and neutron wavelength λ = 0.656 nm), covering a real microstructure size of 3 to 1000 nm. The detector was positioned 10 m away from the specimen to measure the SANS profiles in the q-range of 0.005 to 0.237 nm^{-1}. Experimental set up is shown in Figure 1. As seen, a dilatometer and a 1.0 T magnet were installed for temperature control and separation of nuclear and magnetic scattering, respectively. A thermo-couple was spot-welded on the specimen surface to control temperature of the specimen. The specimen was heated up to 900 °C with a heating speed of 2 °C/s, held there to obtain an austenite single phase microstructure for starting, and then cooled down to 300 °C with a cooling rate of 5 °C/s, followed by isothermal holding at 300 °C in vacuum under a magnetic field of 1.0 T. The time interval for data acquisition was set to be 10 min (600 s) during the isothermal holding. Using the data in the parallel direction or vertical with respect to the magnetic field direction summing azimuthal sector within 30 degrees, SANS profile, *i.e.*, scattering intensity, was counted *versus q* to obtain profiles of nuclear and magnetic scattering components (for more details see Figure 3).

Concerning the iMATERIA at MLF/J-PARC, by which simultaneous measurements of dilatometry, ND and SANS were examined, the detailed explanation was omitted here because of a preliminary experiment. Brief explanation will be given in Section 3.3 together with some experimental results.

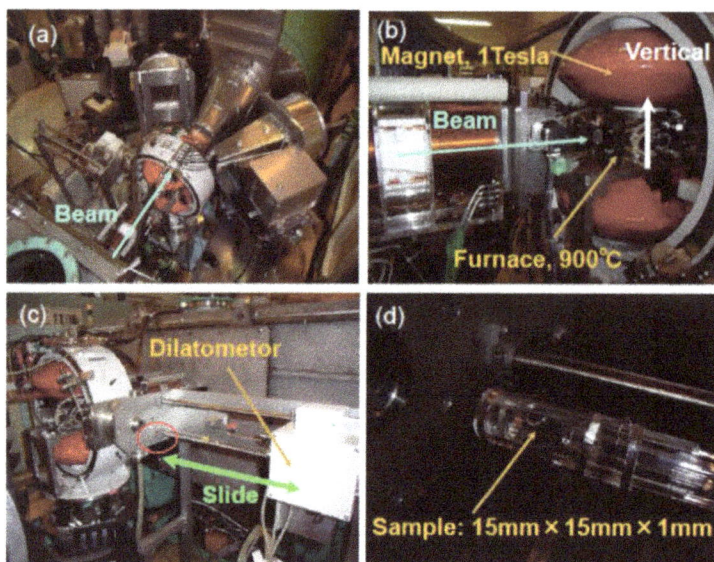

Figure 1. Experiment view of SANS with a 1 T magnet and a dilatometer to monitor the kinetics of bainite transformation at SANS-J-II/JRR-3 (JAEA): (**a**) overall top view; (**b**) magnet; (**c**) dilatometer and (**d**) specimen holder.

2.3. Data Analysis on Small Angle Neutron Scattering

The SANS intensities (*I*) obtained were plotted as a function of *q*, where lower *q*-values correspond to larger size of grain or particle. From the $\ln I(q)$ *versus* $\ln q$ plots, several microstructural parameters can be determined at different *q* regions, *i.e.*, the radius of gyration (R_g) representing "the effective size of the scattering particle" can be determined by the Guinier plot in the lower *q* region [25–29].

The *q*-range for the Guinier approximation depends on the particle size. The particle shape can be recognized from the *q*-dependence of the scattering intensity, *i.e.*, a slope of "−1" indicates cylinder shape, "−2" disc, and "−4" sphere. On the other hand, the Porod law holds in the high-*q* region and can be used to calculate other structural parameters. A slope of "−4" suggests that the interface of the particle is smooth and thereby the scattering intensity is proportional to the total interface area.

3. Results and Discussion

3.1. Monitoring of Bainite Transformation by Dilatometry

Figure 2 shows the temperature history of a specimen measured with a thermo-couple and the change in length obtained by dilatometry (DL). As is usually performed, the data obtained by the traditional dilatometry indicate, apparently, dilatation caused by bainite transformation at 300 °C. Though the change in specimen length can be converted to the ferrite volume fraction, *i.e.*, kinetics of bainite transformation, the insights on chemical, crystallographic and microstructural features cannot be found. Therefore, as was described above, *in situ* neutron scattering/diffraction has been performed; *in situ* ND gave us the information on not only the change in the ferrite volume fraction but also the formation of two populations of austenite (the higher carbon concentration region and the lower one) [19,23], texture evolution [19], and dislocation density [20], as was mentioned above. Here, the size of the transformed product is expected to be monitored by SANS measurement.

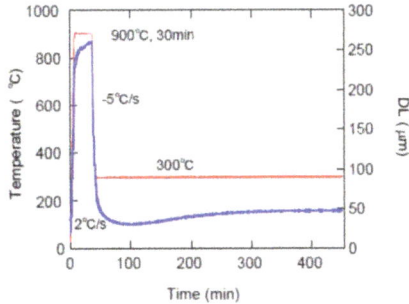

Figure 2. Change in temperature and the length of a specimen (DL) during heat treatment.

3.2. Monitoring of Bainite Transformation with in Situ SANS

Two-dimensional SANS patterns at different holding times are presented in Figure 3. It is found that the scattering intensity increases with the progress of bainite transformation both in the nuclear and magnetic components. These intensities were collected within 30° (see Figure 3c) in the parallel direction or vertical with respect to the magnetic field direction indicated in Figure 3a.

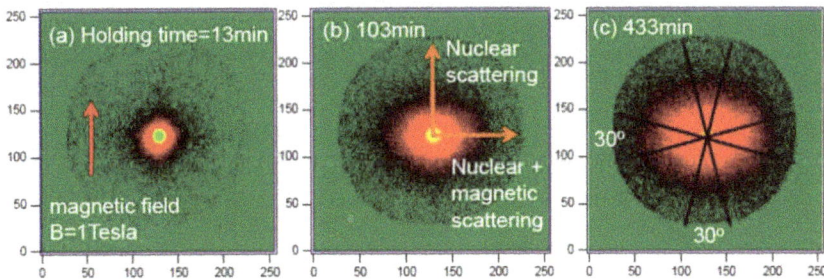

Figure 3. Change in the two-dimensional scattering pattern with holding time at 300 °C: (a) 13 min, (b) 103 min and (c) 433 min.

The nuclear component of the SANS profile obtained with a time interval of 10 min was plotted in Figure 4 as a function of the q-value. After the onset of bainite transformation, the SANS intensity of the nuclear component increased, apparently, with the holding time, *i.e.*, progress of bainite transformation. Here, two interesting features are noticed; one is an increase in scattering intensity in a high q region, the so-called "Porod region" with a slope of -4, and the other is the slope change in a low q region, the so-called "Guinier region". In the Porod region, the scattering intensity is proportional to the total area of the interface of scattering inhomogeneity, the bainitic ferrite lath in the present case, so that the intensity increase means the progress of transformation.

The scattering intensities at lines A and B in Figure 4 were plotted as a function of holding time in Figure 5a,b, respectively. The curves are similar to the dilatation curve shown in Figure 2 and the change in volume fraction determined by *in situ* neutron diffraction in the previous study [18] (the results by dilatometry, SANS and ND obtained by iMATERIA are presented together in Figure 8). As is presented in Figure 6 as an example, the morphology of bainitic ferrite lath would be assumed by a disc shape with a different radius but the same thickness. Hence, the total area of the ferrite/austenite interface is postulated to be proportional to the ferrite volume fraction. This indicates that the scattering intensity in the Porod region is proportional to the ferrite volume fraction showing the bainite transformation kinetics. Because the ferrite phase is magnetic while the austenite phase is non-magnetic at 300 °C, the magnetic scattering component also increases with the increase of the ferrite volume fraction (compare curves labeled N and N + M), suggesting that the ferrite volume fraction could be determined not only by the nuclear scattering component but also by the magnetic scattering component. This means that a magnet is not needed for the evaluation of bainite transformation.

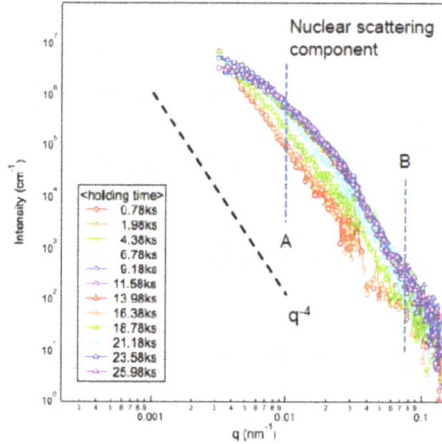

Figure 4. Change in nuclear component of SANS profile with progress in bainite transformation.

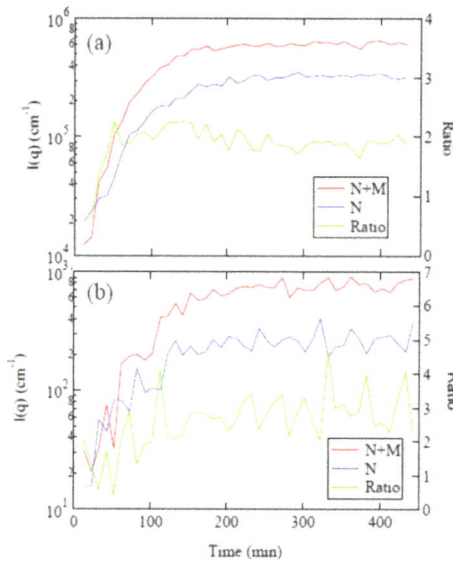

Figure 5. SANS intensity at the Porod region as a function of holding time at 300 °C: (**a**) at $q = 0.01$ nm^{-1} (line A in Figure 4) and (**b**) $q = 0.073$ nm^{-1} (line B in Figure 4).

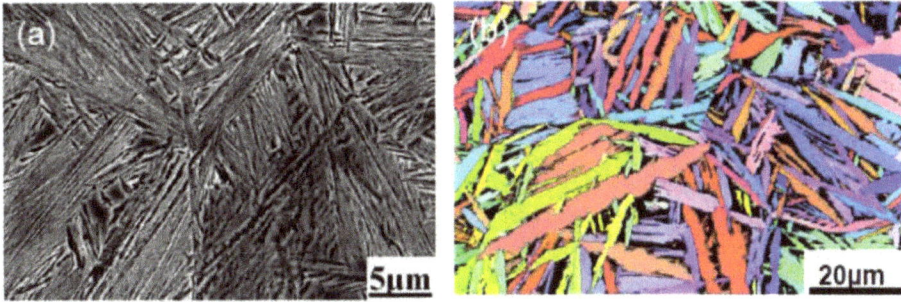

Figure 6. Microstructure of bainite formed at 300 °C: (**a**) SEM image and (**b**) EBSD/IPF map.

In the Guinier region, the slope of the I–q curve would be -2 (disc shape) if it appeared completely. However, the size of the bainite lath is too large to be detected in the present q-range. Hence, as was done previously for non-metallic inclusions in steels [30], we need to expand the measuring q-range to cover smaller values. The influence of multiple scattering [31] is also suspected and, hence, we cannot follow the change in the size of the bainitic lath in this experiment. From microstructure observations, it is very likely that the first lath is large in a scale of prior austenite grain size and that the later-formed laths must be shortened because the pre-formed laths inhibit further growth, although the thickness is nearly constant. Therefore, it is believed that the SANS intensity at the Porod region is proportional to the bainite volume fraction.

3.3. In Situ Measurements of Dilatometry, Small Angle Scattering and Diffraction at iMATERIA

In previous studies, we have successfully used *in situ* neutron diffraction to elucidate the transformation mechanism, particularly the effects of ausforming [18] and partial quenching [19]. The results in the previous Section 3.2 suggest that SANS is of use to monitor the transformation product. Hence, the combined measurements using the conventional dilatometry, ND and SANS were aimed at performing by introducing a new dilatometer into the engineering neutron diffractometer, iMATERIA at MLF/J-PARC, by which back-scatter ND and SANS can be measured simultaneously. A trial was performed using the same steel and some tentative results are presented here to show how to effectively do such an experiment.

Figure 7 shows the results of dilatometry and neutron diffraction obtained at the iMATERIA. As seen, the dilatometry result in Figure 7a is quite similar to that in Figure 2. Changes in austenite 111 and ferrite 110 diffraction profiles are presented in Figure 7b, in which the appearance of two populations of austenite with different amounts of carbon concentration found in the previous studies [19,20] is well confirmed because of the higher resolution of the back-scatter detector at the iMATERIA. The ferrite volume fraction (V_α) was calculated using Equation (1) [32,33] from the hkl diffraction intensities determined with the Z-Rietveld software (J-PARC, Tokai, Japan) [34].

$$V\alpha = \frac{\frac{1}{m}\sum^{m}\frac{I^{\alpha}_{hkl}}{R^{\alpha}_{hkl}}}{\frac{1}{m}\sum^{m}\frac{I^{\alpha}_{hkl}}{R^{\alpha}_{hkl}} + \frac{1}{n}\sum^{n}\frac{I^{\gamma}_{hkl}}{R^{\gamma}_{hkl}}} \tag{1}$$

where I^{γ}_{hkl}, I^{α}_{hkl}, n and m refer to the measured integrated intensities of austenite, those of ferrite, the number of ferrite peaks and those of austenite, respectively, while R^{α}_{hkl} and R^{γ}_{hkl} stand for theoretical values for texture-free material. The obtained results are plotted in Figure 7c showing a good coincidence with the dilatometry result in Figure 7a.

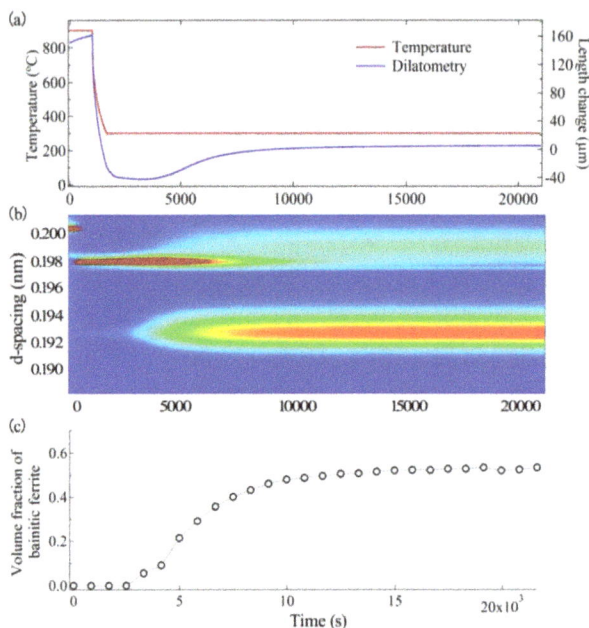

Figure 7. Results of preliminary *in situ* measurements during bainite transformation performed at iMATERIA: (**a**) temperature (red line) and dilatometric change (blue line); (**b**) changes in diffraction profile of austenite 111 and ferrite 110 peaks; and (**c**) volume fraction of ferrite determined from diffraction profiles.

The data analysis for SANS measurements at the iMATERIA is now in progress, so that the scattering intensity at q = 0.4–0.42 nm^{-1} was tentatively counted. The scattering intensity obtained was plotted in Figure 8 as a function of holding time together with the results obtained by ND and dilatometry. As can be observed, these three results are in good agreement. This was the first trial to employ SANS for monitoring bainite transformation at the iMATERIA, and, hence, the detailed analysis has not been made yet; nuclear and magnetic components were not separated because a magnetic field was not applied to the specimen in this experiment.

Three methods of neutron scattering and diffraction are applicable to monitor bainite transformation, *i.e.*, ND, SANS and transmission BE [21] measurements. All of these methods can evaluate the transformation kinetics and have the following different possible features:

(1) ND: crystal structure and volume fraction of constituents, texture, carbon concentration, elastic strain, and dislocation density: bulk average or three-dimensional (3D) distribution by scanning technique, although it requires scan time.
(2) SANS: volume fraction, shape and size of the second phase: bulk average.
(3) BE: phase volume fraction and carbon concentration of austenite, hopefully simultaneous 2D mapping of volume fraction, grain size, elastic strain and texture.

Hence, the combination of these methods would give us more fruitful information to understand microstructural evolution during processing for advanced steels.

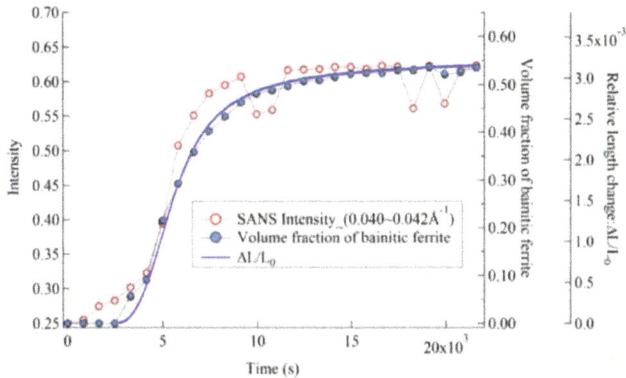

Figure 8. SANS intensity and volume fractions of ferrite determined by neutron diffraction and dilatometry as a function of holding time at 300 °C.

4. Conclusions

Bainite transformation was monitored by *in situ* measurements of conventional dilatometry as well as SANS and ND. The volume fraction of bainitic ferrite was estimated from the SANS intensity in the Porod region, showing good agreement with the results obtained by dilatometry and ND. A more advanced monitoring technique combining dilatometry, SANS, ND and BE was proposed.

Acknowledgments: We would like to thank T. Ishigaki, S. Koizumi, A. Hoshikawa and K. Iwase of Ibaraki University (BL20 (iMATERIA) instrument scientists) for their help to install a dilatometer. The SANS experiments at JRR-3 was performed through the proposal #2010-A-72 and a preliminary experiment at MLF/J-PARC through proposal #2012PM0003. This study was financially supported by a Grant-in-Aid for Scientific Research A (# 21246106). The authors sincerely thank H. Beladi of Deaken University for supplying the present steel and valuable discussion during his stay at NIMS.

Author Contributions: H.N. performed this research as her master degree thesis at Ibaraki University, Y.T. coordinated this research and wrote the manuscript with help from the other authors, Y. H. and W. G. helped with experiments and data analyses, and J.S. instructed the measurement method and analysis of SANS as an instrumental scientist.

Conflicts of Interest: The authors declare no conflict of interest.

References

1. Furukawa, T. Structure-property relationships of dual phase steels. *J. Jpn. Inst. Met.* **1980**, *19*, 439–446. [CrossRef]
2. Tomota, Y.; Tamura, I. Strength and deformation behavior of two-ductile- phase alloys. *J. Jpn. Inst. Met.* **1985**, *14*, 657–664. [CrossRef]
3. Tomota, Y.; Tamura, I. Mechanical Behavior of Steels Consisting of Two Ductile Phases. *Trans. ISIJ* **1982**, *22*, 665–677. [CrossRef]
4. Matsumura, O.; Sakuma, Y.; Takechi, H. Enhancement of elongation by retained austenite in intercritical annealed 0.4C–1.5Si–0.8Mn steel. *Trans. ISIJ* **1987**, *27*, 570–579. [CrossRef]
5. Tomota, Y.; Tokuda, H.; Adachi, Y.; Wakita, M.; Minakawa, N.; Moriai, A.; Morii, Y. Tensile Behavior of TRIP-Aided Multi-Phase Steels Studied by *in situ* Neutron Diffraction. *Acta Mater.* **2004**, *52*, 5737–5745. [CrossRef]
6. Caballero, F.G.; Bhadeshia, H.K.D.H.; Mawella, J.A.; Jones, D.G.; Brown, P. Very strong low temperature bainite. *Mater. Sci. Technol.* **2002**, *18*, 279–284. [CrossRef]
7. Garcia-Mateo, C.; Caballero, F.G.; Bhadeshia, H.K.D.H. Development of Hard Bainite. *ISIJ Int.* **2003**, *43*, 1238–1243. [CrossRef]

8.   Beladi, H.; Adachi, Y.; Timokhina, I.; Hodgson, P.D. Crystallographic analysis of nanobainitic steels. *Scr. Mater.* **2009**, *60*, 455–458. [CrossRef]
9.   Timokhina, I.; Beladi, H.; Xiong, X.Y.; Adachi, Y.; Hodgson, P.D. Nanoscale microstructural characterization of a nanobainitic steel. *Acta Mater.* **2011**, *59*, 5511–5522. [CrossRef]
10.   Carcia-Mateo, C.; Jimenz, J.A.; Yen, H.-W.; Miller, M.K.; Morales-Rivas, L.; Kuntz, M.; Ringer, S.P.; Yang, J.-R.; Caballero, F.G. Low temperature bainitic ferrite: Evidence of carbon super-saturation and tetragonality. *Acta Mater.* **2015**, *91*, 162–173. [CrossRef]
11.   Speer, J.G.; Streicher, A.M.; Matlock, D.K.; Rizzo, F.C.; Krauss, G. *Austenite Formation and Decomposition*; Damm, E.B., Merwinm, M., Eds.; TMS/ISS: Warrendale, PA, USA, 2003; pp. 502–522.
12.   Speer, J.; Matlock, D.; de Cooman, B.; Schroth, J. Carbon partitioning into austenite after martensite transformation. *Acta Mater.* **2003**, *51*, 2611–2622. [CrossRef]
13.   Edmonds, D.; He, K.; Rizzo, F.; de Cooman, B.; Matlock, D.; Speer, J. Quenching and partitioning martensite—A novel steel heat treatment. *Mater. Sci. Eng. A* **2006**, *438*, 25–34. [CrossRef]
14.   Santofimia, M.J.; Zhao, L.; Sietsma, J. Overview of mechanism involved during the quenching and partitioning process in steels. *Metall. Mater. Trans. A* **2011**, *42A*, 3620–3626. [CrossRef]
15.   Yuan, L.; Ponge, D.; Wittig, J.; Choi, P.; Jimenez, J.A.; Raabe, D. Nanoscale austenite reversion through partitioning, segregation and kinetic freezing: Example of a ductile 2 GPa Fe–Cr–C steel. *Acta Mater.* **2012**, *60*, 2790–2804. [CrossRef]
16.   Caballero, F.G.; Bhadeshia, H.K.D.H. Very Strong Bainite. *Curr. Opin. Solid State Mater. Sci.* **2004**, *8*, 251–257. [CrossRef]
17.   Koo, M.S.; Xu, P.; Suzuki, H.; Tomota, Y. Bainitic transformation behavior studied by simultaneous neutron diffraction and dilatometric measurement. *Scr. Mater.* **2009**, *61*, 797–800. [CrossRef]
18.   Gong, W.; Tomota, Y.; Koo, M.S.; Adachi, Y. Effect of ausforming on nanobainite steel. *Scri. Mater.* **2010**, *63*, 819–822. [CrossRef]
19.   Gong, W.; Tomota, Y.; Adachi, Y.; Paradowska, A.M.; Kelleher, J.F.; Zhang, S.Y. Effects of ausforming temperature on bainite transformation, microstructure and variant selection in nanobainite steel. *Acta Mater.* **2013**, *61*, 4142–4154. [CrossRef]
20.   Gong, W.; Tomota, Y.; Harjo, S.; Su, Y.H.; Aizawa, K. Effect of prior martensite on bainite transformation in nanobainite steel. *Acta Mater.* **2015**, *85*, 243–249. [CrossRef]
21.   Babu, S.S.; Specht, E.D.; David, S.A.; Karapetrova, E.; Zachack, P.; Peer, M.; Bhadeshia, H.K.D.H. *In situ* Observations of Lattice Parameter Fluctuations in Austenite and Transformation to Bainite. *Metall. Mater. Trans. A* **2005**, *36A*, 3281–3289. [CrossRef]
22.   Stone, H.J.; Peet, M.J.; Bhadeshia, H.K.D.H.; Withers, P.J.; Babu, S.S.; Specht, E.D. Synchrotron X-ray studies of austenite and bainitic ferrite. *Proc. Roy. Soc. A* **2008**, *464*, 1009–1027. [CrossRef]
23.   Gong, W. Transformation Kinetics and Crystallography of Nano-bainite Steel. Ph.D. Thesis, Ibaraki University, Hitachi, Japan, March 2012.
24.   Huang, J.; Vogel, S.C.; Poole, W.J.; Militzer, M.; Jacques, P. The study of low-temperature austenite decomposition in a Fe–C–Mn–Si steel using the neutron Bragg edge transmission technique. *Acta Mater.* **2007**, *55*, 2683–2691. [CrossRef]
25.   Muller, G.; Uhlemann, M.; Ulbricht, A.; Bohmert, J. Influence of hydrogen on the toughness of irradiated reactor pressure vessel steels. *J. Nucl. Mater.* **2006**, *359*, 114–121. [CrossRef]
26.   Ohnuma, M.; Suzuki, J.; Wei, F.G.; Tsuzaki, K. Direct observation of hydrogen trapped by NbC in steel using small-angle neutron scattering. *Scr. Mater.* **2008**, *58*, 142–145. [CrossRef]
27.   Buckley, C.E.; Birnbaum, H.K.; Bellmann, D.; Staron, P. Calculation of the radial distribution function of bubbles in the aluminum hydrogen system. *J. Alloy. Compd.* **1999**, *293–295*, 231–236. [CrossRef]
28.   Yasuhara, H.; Sato, K.; Toji, Y.; Ohnuma, M.; Suzuki, J.; Tomota, Y. Size Analysis of Nanometer Titanium Carbide in Steel by Using Small-Angle Neutron Scattering. *Tetsu-to-Hagane* **2010**, *96*, 545–549. [CrossRef]
29.   Su, Y.H.; Morooka, S.; Ohnuma, M.; Suzuki, J.; Tomota, Y. Quantitative Analyses on Cementite Spheroidization in Pearlite Structure by Small-Angle Neutron Scattering. *Metall. Mater. Trans. A* **2015**, *46A*, 1731–1740. [CrossRef]
30.   Oba, Y.; Koppoju, S.; Ohnuma, M.; Kinjo, Y.; Morooka, S.; Tomota, Y.; Suzuki, J.; Yamaguchi, D.; Koizumi, S.; Sato, M.; *et al.* Quantitative Analysis of Inclusions in Low Carbon Free Cutting Steel Using Small-Angle X-ray and Neutron Scattering. *ISIJ Int.* **2012**, *52*, 458–464. [CrossRef]

31. Schelten, J.; Schmatz, W. Multipe-Scattering Treatment for Small Angle Scattering Problem. *J. Appl. Cryst.* **1980**, *13*, 385–390. [CrossRef]

32. Järvinen, M. Texture Effect in X-ray Analysis of Retained Austenite in Steels. *Textures Microstruct.* **1996**, *26–27*, 93–101. [CrossRef]

33. Gnäuperl-Herold, T.; Creuziger, A. Diffraction study of the retained austenite content in TRIP steels. *Mater. Sc. Eng. A* **2011**, *528*, 3594–3600. [CrossRef]

34. Oishi, T.; Yonemura, M.; Morishima, T.; Oshikawa, A.; Torii, S.; Ishigaki, T.; Kamiyama, T. Application of matrix decomposition algorithms for singular matrices to Pawley method in Z-Rietveld. *J. Appl. Cryst.* **2012**, *45*, 299–308. [CrossRef]

metals

MDPI

Review

In Situ Characterization Techniques Based on Synchrotron Radiation and Neutrons Applied for the Development of an Engineering Intermetallic Titanium Aluminide Alloy

Petra Erdely [1,*], Thomas Schmoelzer [2,†], Emanuel Schwaighofer [2,‡], Helmut Clemens [2], Peter Staron [3], Andreas Stark [3], Klaus-Dieter Liss [4] and Svea Mayer [2]

[1] Recipient of a DOC Fellowship of the Austrian Academy of Sciences at the Department of Physical Metallurgy and Materials Testing, Montanuniversität Leoben, Roseggerstraße 12, 8700 Leoben, Austria
[2] Department of Physical Metallurgy and Materials Testing, Montanuniversität Leoben, Roseggerstraße 12, 8700 Leoben, Austria; thomas.schmoelzer@ch.abb.com (T.S.); emanuel.schwaighofer@alumni.unileoben.ac.at (E.S.); helmut.clemens@unileoben.ac.at (H.C.); svea.mayer@unileoben.ac.at (S.M.)
[3] Institute of Materials Research, Helmholtz-Zentrum Geesthacht, Max Planck-Straße 1, 21502 Geesthacht, Germany; peter.staron@hzg.de (P.S.); andreas.stark@hzg.de (A.S.)
[4] The Bragg Institute, Australian Nuclear Science and Technology Organisation, New Illawarra Road, Lucas Heights, NSW 2234, Australia; kdl@ansto.gov.au
* Correspondence: petra.erdely@unileoben.ac.at; Tel.: +43-3842-402-4213; Fax: +43-3842-402-4202
† Current address: ABB Corporate Research Center, Segelhofstrasse 1K, 5405 Baden-Dättwil, Switzerland.
‡ Current address: Böhler Edelstahl GmbH & Co KG, Mariazellerstraße 25, 8605 Kapfenberg, Austria.

Academic Editor: Hugo F. Lopez
Received: 30 November 2015; Accepted: 23 December 2015; Published: 4 January 2016

Abstract: Challenging issues concerning energy efficiency and environmental politics require novel approaches to materials design. A recent example with regard to structural materials is the emergence of lightweight intermetallic TiAl alloys. Their excellent high-temperature mechanical properties, low density and high stiffness constitute a profile perfectly suitable for their application as advanced aero-engine turbine blades or as turbocharger turbine wheels in next-generation automotive engines. As the properties of TiAl alloys during processing as well as during service are dependent on the phases occurring, detailed knowledge of their volume fractions and distribution within the microstructure is of paramount importance. Furthermore, the behavior of the individual phases during hot deformation and subsequent heat treatments is of interest to define reliable and cost-effective industrial production processes. *In situ* high-energy X-ray diffraction methods allow tracing the evolution of phase fractions over a large temperature range. Neutron diffraction unveils information on order-disorder transformations in TiAl alloys. Small-angle scattering experiments offer insights into the materials' precipitation behavior. This review attempts to shine a light on selected *in situ* diffraction and scattering techniques and the ways in which they promoted the development of an advanced engineering TiAl alloy.

Keywords: titanium aluminides based on γ-TiAl; high-energy X-ray diffraction; phase transformations; neutron diffraction; order/disorder transformations; small-angle scattering; thermo-mechanical processing; heat treatments; microstructure evolution

---

1. Introduction

Presently, jet engines are the prevalent form of propulsion systems in commercial airplanes. Increasing economic and environmental pressure, however, has triggered an ever-rising awareness of the issues associated with the consumption of fossil fuels. As a result, a growing demand for environmentally-friendly engine options has emerged. Substantial improvements in fuel efficiency, noise and greenhouse gas emissions can be effected by reducing the weight of selected components. This end can be achieved either by adapting the overall design or by implementing advanced structural materials that exhibit mechanical properties equal to conventional materials while having lower densities. The latter approach has been realized successfully, for instance, by installing intermetallic titanium aluminide turbine blades in the last stages of the low-pressure turbine of the GEnx™ engine by General Electrics [1]. In this way, the twice as heavy nickel-base superalloy blades could be replaced. Pratt & Whitney recently introduced TiAl turbine blades in their so-called Geared Turbofan™ engine, which will be used to power the Airbus A320neo family besides other aircrafts [2,3].

Intermetallic titanium aluminides based on the ordered γ-TiAl phase provide a set of properties well suited for structural high-temperature applications. At a density of about 4 g·cm^{-3}, they combine a high melting point with a high elastic modulus. Strength and creep properties, as well as oxidation and burn resistance are good, even at high temperatures [4–9]. Capable of withstanding extreme conditions, titanium aluminides have found applications in the automotive and aircraft engine industry, as reported in [10,11]. As engine valves in sports and racing cars, as turbocharger turbine wheels and as low-pressure turbine blades in jet engines, they are expected to replace conventional high-temperature materials of high density in a temperature range up to 750 °C [4,6–8,10–19]. During the last few decades, considerable efforts have been made to develop γ-TiAl-based alloys suitable for service in these demanding areas, while being economically competitive at the same time. Various processing routes and alloy compositions have been studied intensively [9,12,20,21].

Among the recently-developed third-generation titanium aluminide alloys, the β-solidifying TNM alloys in particular strive to overcome processing-related difficulties inherent to brittle intermetallic materials [11,20]. With a nominal composition of Ti–43.5Al–4Nb–1Mo–0.1B (at. %), their distinct characteristic is an adjustable β/β_o phase fraction, which is stabilized by the alloying elements Nb and Mo (hence the name TNM) [20,22]. At elevated temperatures, a substantial volume fraction of the disordered β-Ti(Al) phase (A2 structure) is obtained [23]. The presence of this phase improves the material's hot workability by providing a sufficient number of independent slip systems [7,24–27]. Furthermore, it allows near-conventional hot-die forging due to an enlarged processing window [28,29]. At service temperatures, however, the ordered β_o-TiAl phase (B2 structure) reduces the material's creep strength significantly [11,25]. Heat treatments must be applied after forging to reduce the β_o phase fraction to a minimum. By simultaneously adjusting the material's microstructure, attractive mechanical properties can be tailored.

Depending on the temperature profile of the heat treatment applied, as well as on the material's chemical composition, different microstructural features can be adjusted in TiAl alloys, ranging from near gamma (NG) and duplex (D) microstructures over nearly-lamellar (NL) to fully-lamellar (FL) microstructures [6,7,10,30]. FL microstructures exhibit the best creep properties at the expense of a diminished room-temperature ductility [31,32]. NL microstructures offer the most balanced properties, providing strong creep resistance and moderate ductility at ambient temperature [33]. In Figure 1a, the NL microstructure of a heat-treated TNM alloy is shown. The image was obtained by means of scanning electron microscopy (SEM). In the back-scatter electron (BSE) mode, the three main phases α_2, β_o and γ can be distinguished by their differing contrast: β_o appears brightest, γ darkest and α_2 in an intermediate contrast. The insert in the right upper corner represents a high-resolution transmission electron microscopy (HRTEM) image of the lamellae within an α_2/γ colony.

A further outstanding characteristic of TNM alloys is their solidification pathway via the β phase [28]. Due to an intrinsic grain refinement, the solidified microstructure appears equiaxed and free from a significant casting texture [34]. Additionally, as opposed to the microstructure after peritectic solidification, the material is nearly segregation-free. In spite of the given advantages, cast microstructures are frequently rather coarse. Optimized heat treatments are necessary to adjust the material's microstructure [21,35,36] and properties [37].

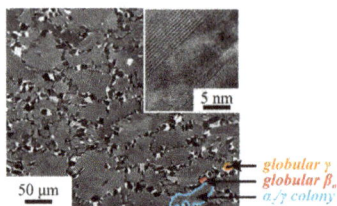

Figure 1. Nearly lamellar microstructure of a nominal TNM alloy (SEM-back-scatter electron (BSE) image) The microstructure, which was arranged during a two-step heat treatment, consists of lamellar α_2/γ colonies surrounded by globular γ and β_0 grains (dark and bright contrast, respectively). The insert (HRTEM image) features a lamellar colony. (Reproduced with permissions from Reference [11]. Copyright © Wiley-VCH Verlag GmbH & Co. KGaA, Weinheim, Germany, 2013).

Currently, so-called TNM[+] alloys are attracting notice, as the creep properties, as well as the microstructural stability can be increased as compared to regular TNM alloys. To this end, harmonized additions of C and Si are accommodated in the alloy composition [38,39]. Carbon and Si are known to act as solid solution strengtheners [40,41]. In addition to this, precipitates (cubic Ti_3AlC, hexagonal Ti_2AlC or Ti_5Si_3) tend to form during high-temperature creep experiments [42–44]. Detailed knowledge of the associated precipitation processes, however, is still needed for their deliberate application to improve a material's high-temperature properties.

The basis for addressing such research questions is provided by a profound understanding of the thermodynamics and kinetics of the phase transformations occurring. The resultant control over the composition and arrangement of the phases prevalent in a material allows the targeted adjustment of its properties. In the case of TiAl alloys, conventional *ex situ* characterization methods, although generally available and rather cost effective, encounter certain difficulties. First, the alloy system itself is rendered complex by a multitude of phases and phase transformations, whose occurrence and exact transformation temperatures are strongly dependent on the material's chemical composition [28]. Thermodynamic calculations cannot, at present, be consulted, except for trends, since commercially-available Ti–Al databases have been found to be unreliable in describing phase diagrams containing high amounts of β/β_0-stabilizing elements [11]. Secondly, ordering reactions exhibit extremely fast kinetics [45]. Thus, the disordered high-temperature state cannot be preserved for study at room temperature even by rapid quenching. Phase fractions might change as well, introducing errors in the attempted investigation of high-temperature states in heat-treated and water-quenched specimens [46]. For detailed information on the advantages and limitations of the method of quantitative phase evaluation through microscopic images, the reader is referred to [47]. Thirdly, SEM and X-ray diffraction using laboratory X-ray sources usually suffer from the restriction to the surface or to near-surface areas of the specimens. These areas might in some cases not yield sufficient grain statistics or, following extended heat treatments, be affected by the formation of an α case, an Al-depleted near-surface region of several μm in thickness [48–50]. Transmission electron microscopy (TEM) is also restricted to small sample volumes.

Diffraction and scattering techniques using high-energy X-rays and neutrons have been shown to overcome most of the problems mentioned above [48,49,51,52]. Having been used as a tool for material characterization since the early beginnings of research activities on TiAl alloys [46,53–59], they have successfully promoted the development of the alloy in various aspects (Figure 2). In this review, selected *in situ* characterization techniques shall be presented through a set of recent examples, highlighting the ways in which they contributed to the development of TNM alloys. The topics addressed range from fundamental research questions, such as the establishment of phase diagrams or the investigation of transformation and precipitation kinetics, to processing- or application-related problems, such as the hot deformation behavior of TNM alloys or the adjustment of optimized microstructures during heat treatments (Figure 2).

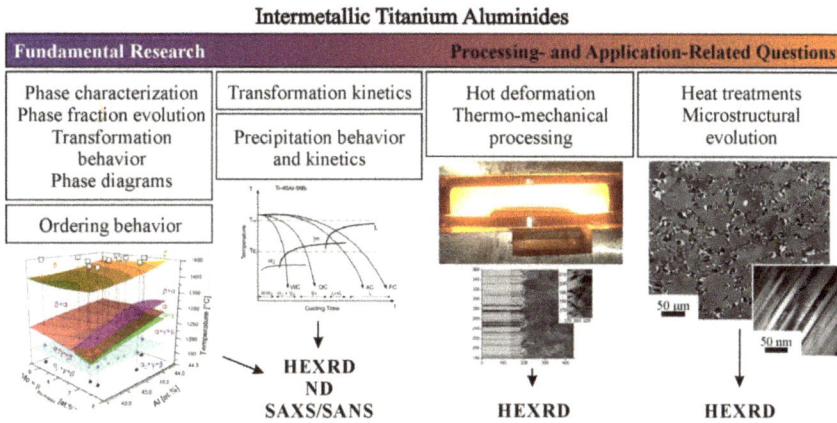

Figure 2. Scheme of the application range of *in situ* characterization techniques based on synchrotron radiation and neutrons in the development of intermetallic titanium aluminides. Neutron diffraction (ND), high-energy X-ray diffraction (HEXRD), and small-angle scattering methods (small-angle X-ray scattering (SAXS)/small-angle neutron scattering (SANS)) address a wide variety of aspects, which cover fundamental research questions, as well as processing- or application-related problems. All aspects, which are illustrated by representative images, are discussed in the present review with a focus on β-solidifying TNM alloys.

2. Peculiarities in the Investigation of Titanium Aluminide Alloys Using X-rays and Neutrons

Their different nature notwithstanding, X-rays and neutrons are both suited for diffraction and scattering experiments in materials science [60]. While neutrons, which possess a mass, are described as particle waves, massless X-rays (photons) represent electromagnetic radiation. Consequently, their modes of interaction with the material under investigation differ. Neutrons interact predominantly with the nuclei of the atoms. X-rays, on the other hand, are scattered by their electrons. This leads to different cross-sections and atomic form factors, while the mathematics derived to describe the scattering of X-rays and neutrons from matter are basically the same.

The description of the scattering of X-rays and neutrons from condensed matter includes the contributions of all atoms in the form of a discrete sum. Thereby, the phase difference due to different positions of the atoms within the ensemble is embraced. If several elements are distributed randomly on a crystal lattice, a virtual atom with a mean scattering length is conceived for the calculation of the structure factor $F_{hkl}(\mathbf{G})$. $\mathbf{G}$ is the scattering vector with Miller indices hkl. Superstructure reflections forbidden by the structure factor do not appear in disordered crystals. The disordered state results in phase factors of equal magnitude, but of opposite sign, causing the corresponding reflections to vanish. For ordered structures, each atom is included in the calculation of $F_{hkl}(\mathbf{G})$ regarding its specific position in the crystal. If different elements possess different scattering lengths, the structure factor for the formerly forbidden reflections hence attains a finite value. This case defines so-called superstructure reflections, which can be observed in the diffraction pattern.

In this regard, TiAl alloys feature an additional, yet useful peculiarity. Since the neutron scattering lengths of Ti (-3.438 fm) and Al (3.449 fm) [61] happen to be almost equal in magnitude but of opposite sign, disordered phases, in which the ratio Ti:Al is close to one, yield only very weak diffraction peaks, because the average scattering length approaches zero. In ordered Ti–Al crystals, the calculated $F_{hkl}(\mathbf{G})$ are close to zero if scattering lengths add up, *i.e.*, for fundamental reflections. The calculation of $F_{hkl}(\mathbf{G})$

for superstructure reflections, however, results in large structure factors, because they contain the difference of scattering lengths of each site. High intensities are obtained, emphasizing the presence of ordered structures (Figure 3a). When X-rays are used to probe the material (Figure 3b), an inverse situation emerges, as the X-ray scattering lengths of Ti and Al are of equal sign, but, due to their different atomic numbers (Z_{Ti} = 22, Z_{Al} = 13), of different magnitude. Superstructure reflections exhibit low intensities, whereas fundamental reflections appear strong. Since these statements are valid for all prevalent phases, neutron and X-ray diffraction are in fact most complementary techniques if applied to intermetallic TiAl alloys. While neutron diffraction offers distinct advantages for the investigation of order/disorder transitions, X-ray diffraction can be employed to study phase transformations and phase fraction evolutions, in which the fundamental reflections play a major role.

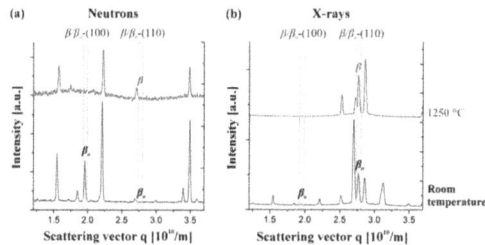

Figure 3. Diffraction patterns of a TNM alloy obtained by (**a**) neutron and (**b**) X-ray diffraction at elevated temperatures (top), as well as at room temperature (bottom) [62]. Differences between the patterns of the ordered β_o phase at room temperature and of its disordered form β at 1250 °C are exemplarily accentuated. In the case of neutron scattering, the β-(100) superstructure peak appears strong, whereas the fundamental β-(110) peak is hardly discernible from the background. In the case of X-rays, the fundamental peak exhibits a high intensity, while the superstructure peak at room temperature shows only a small intensity.

3. *In Situ* Studies of Phase Evolutions

3.1. HEXRD: State-of-the-Art Instrumental Setup

For *in situ* investigations of the phase transformation behavior of titanium aluminides, high-energy X-ray diffraction (HEXRD) has evolved into a powerful and versatile tool [48,50,51,63–66]. At synchrotron radiation sources, a white X-ray beam is produced by the deflection of charged particles, either electrons or positrons, by means of a bending magnet or an insertion device, *i.e.*, a wiggler or an undulator, as shown in Figure 4 [67]. A monochromator is used to select a narrow energy range from the broad spectrum. For titanium aluminides, energies in the range of 87 to 120 keV, corresponding to wavelengths of 0.143 to 0.103 Å, and wave numbers of 44.1 to 60.1 Å^{-1} have been shown to yield good diffraction results for specimens of 5 mm in thickness [51].

Specimens are investigated in transmission geometry (Figure 4), as the high-energy X-rays can penetrate relatively large sample volumes. Due to their small wavelengths, they yield narrow Debye-Scherrer cones according to Bragg's law. Thus, large parts of the diffraction patterns, *i.e.*, the diffraction rings, can be recorded at once with the aid of flat-panel area detectors. The latter provide a high temporal resolution, enabling the performance of real-time *in situ* experiments. Poly- and mono-crystalline materials can be investigated by means of HEXRD, as well as texturized or strained materials. As ideally whole diffraction rings are covered, even single diffraction spots from monocrystalline samples can be detected. For conventional powder diffraction purposes, the rings are azimuthally integrated, while anisotropic, directionally-dependent properties can be studied by limiting the integration to segments of the azimuthal angle [68]. Basic aspects and common sources of error in the powder diffraction approach are discussed in [69]. For detailed information on features specific to data collection from neutron and synchrotron radiation sources, the reader is referred to [70].

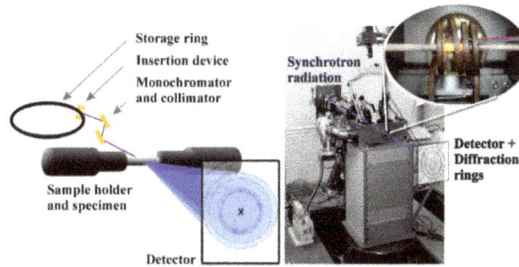

Figure 4. Typical experimental setup for HEXRD. The scheme to the left illustrates the pathway of high-energy X-rays (blue) from their generation at an insertion device in the storage ring, up to their diffraction from the specimen in the sample holder. An area detector at the front captures sets of Debye-Scherrer rings [52]. The photograph on the right shows the quenching and deformation dilatometer [64,66] as an example of a versatile sample environment installed at the PETRA III High Energy Materials Science (HEMS) beamline. In the insert at the top, a heated specimen is glowing during an *in situ* experiment. (Reproduced with permissions from Reference [52]. Copyright © Wiley-VCH Verlag GmbH & Co. KGaA, Weinheim, Germany, 2011).

For *in situ* diffraction experiments, specific conditions concerning temperature, pressure and atmosphere must be adjusted at the sample position. For this purpose, a variety of sample holders and furnaces is currently available. For a large part of the HEXRD experiments presented in this review, a quenching and deformation dilatometer supplied by Bähr Thermoanalyse GmbH, Germany (now TA Instruments), with built-in windows for the X-ray beam, was used (see the photograph in Figure 4) [64]. In this setup, thermal expansion curves and diffraction patterns can be measured simultaneously. Furthermore, thermo-mechanical processing and heat treatments on specimens can be investigated in an *in situ* manner. The dilatometer setup is currently available for high-energy X-rays at the HZG-operated High Energy Materials Science (HEMS) beamline (P07) at PETRA III at DESY in Hamburg, Germany [66]. In the near future, a dilatometer will also be available for neutron scattering at the Heinz Maier-Leibnitz Zentrum (MLZ) in Garching, Germany.

3.2. Studies of Phase Evolutions in TNM Alloys

Knowledge of phase fraction evolutions as a function of temperature and time has adopted a key role in the design and further development of advanced materials. It can be applied to evaluate an alloy's potential regarding future processing routes and applications. The alloy's chemical composition can be optimized as its influence on phase fractions, transition temperatures and kinetics is analyzed. The accuracy of predicted phase diagrams can be verified. If necessary, thermodynamic databases can be modified on the basis of experiments and thermodynamic models improved. Process- and application-related questions can be addressed, treating, for example, the alloy's deformation behavior or its response to heat treatments.

In situ diffraction techniques provide a valuable tool to collect pieces of this fundamental information. Shull *et al.* [46] conducted high-temperature X-ray diffraction experiments on TiAl alloys for phase identification at elevated temperatures. The equipment used comprised a θ-2θ diffractometer and a linear position-sensitive proportional counter. Specimens were heated in a furnace chamber. The *in situ* experiments proved the presence of the eutectoid reaction for a Ti–45Al (at. %) alloy and allowed the investigation of the α + γ phase field region. Novoselova *et al.* [48] applied synchrotron X-ray diffraction to a Ti–46Al–1.9Cr–3Nb (at. %) alloy using a two-circle high-resolution powder diffractometer and an X-ray energy of approximately 9.55 keV. Samples were heated in steps to temperatures ranging from 20 to 1450 °C. Diffraction patterns were acquired under isothermal conditions. In this way, phase fraction evolutions as a function of temperature could be derived in an *in situ* manner for every phase in a titanium aluminide alloy. Nowadays, phase fraction evolutions are often determined by means of *in situ* HEXRD (Section 3.1) while working on a heating ramp. All experimental approaches offer distinctive advantages, but also possess inherent limitations and sources of error. General considerations on data collection and evaluation are provided in [69–71]. For further details on the respective methods, the reader is referred to the original publications cited throughout this work.

In terms of TNM alloys, *in situ* HEXRD studies on phase fraction evolutions have played an important role in the alloy design and the establishment of the phase diagram. Calculated phase diagrams were found to indicate trends, but to be unreliable in two main aspects [72]. First, the level of the eutectoid temperature T_{eu} was underestimated, and the position of the eutectoid point was shifted concerning its Al concentration. Secondly, in contrast to thermodynamic calculations, experimental data proved the existence of a single α phase field in TNM alloys containing more than 42 at. % Al. On this account, the phase diagram has been verified by Boeck [73] on the basis of heat treatment studies. These studies, in combination with differential scanning calorimetry (DSC), conventional XRD and HEXRD experiments, provided data for the eventual modification of the phase diagram [37,72,74].

In Figure 5a, the experimentally-verified quasi-binary section through the TNM alloying system is given [37]. An exemplary phase fraction evolution of a TNM alloy is shown in Figure 5b [37]. The investigated alloy possesses an exact chemical composition of Ti–43.67Al–4.08Nb–1.02Mo–0.1B (at. %). The *in situ* experiments for the determination of the phase fraction evolution were conducted at the HZG-operated beamline HARWI II at DESY. Due to a low diffusivity at low temperatures, which entails retarded kinetics, as well as due to poor grain statistics related to excessive grain growth close to single phase field regions at high temperatures, the temperature range for the quantitative determination of phase diagrams by means of HEXRD is restricted. HEXRD experiments were, thus, conducted covering temperatures ranging from 1100 °C to 1350 °C. Phase fraction evolutions above 1350 °C were estimated as qualitative trends by means of thermodynamic calculations. The temperature range below 1100 °C was studied by means of long-term heat treatments and *ex situ* XRD measurements. From the evolution of experimentally-determined phase fractions, transition temperatures could be extracted with high accuracy. Measurements on several model alloys of differing chemical compositions provided the basis for the establishment of a phase diagram.

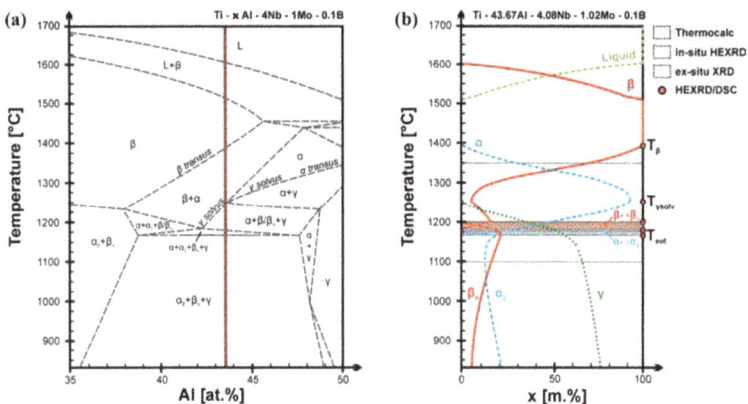

Figure 5. Quasi-binary section through the TNM alloying system (**a**) and experimentally-determined evolution of phase fractions of a Ti–43.67Al–4.08Nb–1.02Mo–0.1B (at. %) alloy (**b**). (Reprinted from Reference [37], Copyright (2014), with permission from Elsevier).

To investigate the influence of different amounts of β-stabilizing alloying elements, *in situ* and *ex situ* HEXRD experiments were conducted on three TNM alloys, two of them containing elevated amounts of Nb or Mo, respectively [74]. Phase fraction evolutions were evaluated by means of quantitative Rietveld analysis. The diffraction experiments were performed at the beamline ID-15B at the European Synchrotron Radiation Facility (ESRF) in Grenoble, France. The samples were slowly heated in a temperature range of 1100 to 1400 °C. DSC measurements and quantitative metallography were performed as complementary characterization techniques. The evolution of the experimentally-acquired phase fractions clearly shows that increased amounts of Nb and Mo shift the β/β_0 minimum at elevated temperatures, which is a characteristic of TNM alloys (Figure 5), to larger phase fractions [74]. A comparison with *ex situ* experiments, however, indicates that the quantitative amount of the β phase at its minimum might be overestimated by the results of the HEXRD experiments on the two modified TNM alloy variants. The heating rate during the *in situ* experiments was chosen as a compromise between limited beam time, avoidance of grain coarsening and time for the achievement of thermodynamic equilibrium conditions. It might have been too high to fully reach thermodynamic equilibrium, especially as the β phase, stabilized by slowly diffusing elements Nb and Mo, exhibits a prolonged dissolution behavior. Nevertheless, Schloffer *et al.* [75] showed that phase fractions retrieved from light-optical microscopy, conventional XRD and *in situ* HEXRD are in good agreement, at least up to the dissolution temperature of the γ phase ($T_{\gamma,solv}$).

3.3. Neutron Diffraction Studies on Order/Disorder Transformations in TNM Alloys

Ordering phenomena strongly affect the mechanical behavior of the phases present in a material and, thus, the mechanical properties of the material as a whole. In TiAl alloys, for instance, the disordered β phase is known to improve hot workability at elevated temperatures [26,76,77], while the ordered β_0 phase has a negative impact on ductility at ambient temperature [78], as well as on creep strength at service temperature [29]. Consequently, the experimental determination of ordering temperatures is of great interest. Using laboratory X-ray sources, *ex situ* studies have been conducted to investigate the site occupancy within different ordered phases [54,79,80]. Complementary investigations of the ordering behavior were conducted by means of TEM [45]. HEXRD experiments were performed to investigate, for example, the site occupancy within the α/α_2 phase in a Ti–45Al–7.5Nb–0.5C (at. %) alloy [81]. A sudden rearrangement of the atomic site occupancies provided evidence for the disordering temperature of the α_2 phase.

Due to short diffusion pathways, ordering reactions are characterized by extremely fast kinetics. The preservation of the disordered states of the α and β phase to room temperature is, thus, impeded, even if the specimens are water-quenched [45,82]. Consequently, the ordering behavior of TiAl alloys can only be investigated at those temperatures, at which the ordering reactions occur, *i.e.*, in an *in situ* manner.

As explained in Section 2, neutron diffraction offers particular benefits regarding the investigation of ordering transitions in TiAl alloys. Hence, the $\alpha \leftrightarrow \alpha_2$ and $\beta \leftrightarrow \beta_0$ ordering/disordering temperatures in a TNM alloy (termed T_{eu} and $T_{\beta,ord}$ in the Ti–Al system) were examined at the structure powder diffractometer SPODI [83,84] at the neutron source FRM II at the MLZ [85]. Specimens were heated in steps ranging from 1000 °C to 1450 °C. Isothermal segments of 3 h allowed the acquisition of diffraction patterns at selected temperatures (Figure 6). Due to the system's peculiarity regarding structure factor considerations, all peaks in the patterns represent superstructure reflections. The Nb-(110) peak, which is indicated in grey, stems from the sample holder.

At temperatures ranging from room temperature to 1100 °C, superstructure reflections of all three phases (α_2, β_0 and γ) are visible. All phases are present in their ordered state. At 1200 °C, the α_2 reflections have vanished, the α_2 phase has disordered. The transition temperature T_{eu} must therefore lie in the temperature range between 1100 and 1200 °C. Between 1200 and 1250 °C, the $\beta_0 \leftrightarrow \beta$ ordering reaction takes place, the β_0 reflections vanish. Only γ superstructure peaks can be observed at 1250 °C. At 1275 °C, also the γ reflections have disappeared. As the γ phase does not disorder, the disappearance of its superstructure reflections implies its dissolution at $T_{\gamma,solv}$. Thus, at 1275 °C, the material consists of the disordered phases α and β, as earlier work has proven for temperatures above 1300 °C [23].

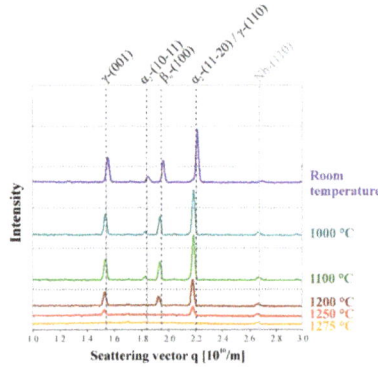

Figure 6. Neutron diffraction patterns of a TNM alloy obtained during isothermal segments at selected temperatures [85]. Between 1100 and 1200 °C, the α_2 reflections vanish, while the β_o phase disorders between 1200 and 1250 °C. The γ phase, which shows no disordering, dissolves in the temperature range between 1250 and 1275 °C.

Due to the extensive pattern acquisition times at SPODI, only a few datasets could be generated. These sets could be used for a rough determination of the temperature ranges in which the order transitions of the α/α_2 and β/β_o phases occur. For a more accurate determination of the ordering/disordering temperatures, neutron diffraction experiments were performed at WOMBAT at the OPAL reactor of the Australian Nuclear Science and Technology Organization (ANSTO) in Lucas Heights, Australia [74,86]. The facility hosts a two-dimensional position-sensitive detector, which is optimized for high-intensity rather than high-resolution measurements and, thus, permits fast data acquisition rates. At heating/cooling rates of 0.166 K·s^{-1}, the temperatures at which order transitions occur could be determined accurately [74]. An extract of the results obtained in the course of the heating experiments is given in Figure 7. Normalized intensities of the α_2-(10-11) and β_o-(100) superstructure peaks are presented as a function of temperature. The sharp decrease in intensity indicates the loss of order at relatively fast disordering kinetics. For nominal TNM alloys, T_{eu} was determined to lie at 1174 °C, while $T_{\beta,ord}$ was located at 1225 °C [74]. These findings are in perfect agreement with the results obtained previously at SPODI [85].

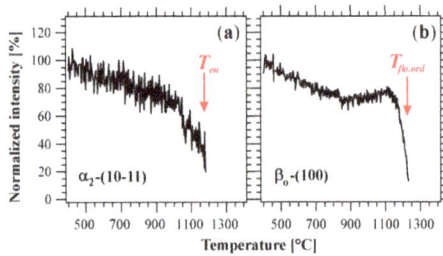

Figure 7. Intensity evolutions of the α_2-(10-11) (a) and β_o-(100) (b) superstructure peaks during heating The loss of order is indicated through the sharp decrease in intensity. The intensities, which are plotted as a function of temperature, are normalized to the intensity observed at 400 °C and given in %. (Reproduced with permissions from Reference [52]. Copyright © Wiley-VCH Verlag GmbH & Co. KGaA, Weinheim, Germany, 2011).

For the sake of completeness, it should be noted that with an increasing brilliance of today's synchrotron radiation sources, ordering and disordering reactions in TiAl alloys can also be studied by means of *in situ* HEXRD experiments. Due to the structure factor, however, the intensities of superstructure reflections are generally very weak [87].

3.4. Recent Advancements in the Alloy Development: The Introduction of Carbon to TNM Alloys

The increasing demand for lightweight high-temperature construction materials, higher efficiency and better performance has sparked the need for even more creep-resistant γ-TiAl-based alloys. To meet the requirements, advanced TNM alloys are developed that exhibit good creep resistance and a stable microstructure well above the usual application limit of about 750 °C [38,39].

An increase in creep resistance has often been linked primarily to improvements in the chemical composition of an alloy [10–12,21]. For γ-TiAl-based alloys, a positive effect on creep properties has been reported for the addition of β-stabilizing substitutional solid solution hardening elements (e.g., Nb, Mo, Ta, W), as well as for the addition of interstitial hardening elements (e.g., C, N), which can also form carbides and nitrides. In high Nb- and Mo-containing alloys, solid solution hardening has been shown to be a dominating hardening effect in the γ-TiAl phase. Furthermore, the alloying elements Nb and Mo have been shown to reduce diffusivity and retard thermally-activated dislocation climb during creep [88]. Below their solubility limit, a solid solution hardening effect is also attributed to additions of C and N, albeit these elements are solved interstitially [40,41,89]. Carbon, in particular, has been suggested to impede the growth of α grains due to segregation and to decrease the rate of diffusion in the α phase in cast γ-TiAl-based alloys [21]. Concerning mechanical properties, additions of C and N were found to increase the yield and fracture stresses by the mechanisms of grain refinement, solid solution and precipitation hardening [90]. Especially the precipitation of fine carbides has been shown to considerably increase the high-temperature strength of TiAl alloys [91]. Cubic perovskite Ti_3AlC (*p*-type) carbides have been reported to enhance strength and creep performance for a variety of γ-TiAl-based alloys [10,12,44,91–94]. Hexagonal Ti_2AlC (*h*-type) carbides, on the contrary, are thermodynamically stable at higher temperatures, but are known to be less efficient regarding the impediment of dislocation mobility [10,91]. However, they can act as efficient grain refiners during solidification, yielding a nearly texture-free solidification microstructure, as reported in [38].

Generally, an increase in creep resistance is linked to a decrease in dislocation mobility. However, at room temperature, the ductility of a material may be lost. Therefore, in the course of alloy development, ways of adjusting balanced mechanical properties have to be established. Furthermore, the interplay between different alloying effects has to be considered. Carbon acts as a strong stabilizer of the hexagonally-close-packed α phase and significantly increases the eutectoid temperature T_{eu}. To preserve the solidification pathway via the β phase, the amount of β-stabilizing elements, such as Nb and Mo, has to be corrected, *i.e.*, slightly increased [38]. Thus, grain refinement, minimum segregation and almost texture-free solidification microstructures can still be obtained in C-containing TNM alloys. In this regard, *in situ* diffraction techniques offer a valuable tool to investigate the influence of alloying additions on the solidification behavior of a material, the phase fractions present, as well as their evolution as a function of temperature and time, as reported in [38].

For the investigation of the influence of C additions on the phase evolution in TNM alloys, TNM variants containing 0, 0.25, 0.5, 0.75 and 1 at. % C, respectively, were produced by melting, hot isostatic pressing (HIP), near-conventional forging and subsequent heat treatments [38]. *In situ* HEXRD experiments were conducted at the HZG beamline HARWI II at DORIS III in Hamburg, Germany. Specimens of 5 mm in diameter and 10 mm in length were heated in a modified quenching and deformation dilatometer (Section 3.1; Figure 4). First, the specimens were heated rapidly to 1100 °C. An isothermal segment of 10 min allowed the approach of conditions near thermodynamic equilibrium. Afterwards, the specimens were heated to 1300 °C at a slow heating rate of 2 K·min⁻¹. From 1300 °C to 1425 °C, they were heated at higher rates again to avoid excessive grain growth, which would result in poor grain statistics and the risk of thermocouple tear-off.

The data on the phase fractions as a function of temperature, combined with the results from SEM, were used to derive the quasi-binary phase diagram given in Figure 8 [38]. TNM alloys with a C content below roughly 0.5 at. % still solidify via the β phase (*i.e.*, L → L + β → β → β + α → ...), as is known from standard TNM alloys. Thus, intrinsic grain refinement occurs, which stems from the β → α phase transformation according to the Burgers orientation relationship. Increasing C content entails a change of the solidification pathway from β to peritectic solidification (*i.e.*, L → L + β → L + β + α → β + α → ...), as can be observed in the TNM alloys with C contents of 0.5 at. % and above. The peritectic solidification leads to a coarse-grained columnar microstructure with a sharp solidification texture. In the TNM-1C alloy, however, a grain-refining effect was reported to occur due to the formation of *h*-type carbides, which lead to enhanced heterogeneous nucleation during subsequent solid-state phase transformations [38]. Although peritectically solidified, TNM-1C alloys exhibit a fine-grained and texture-free microstructure.

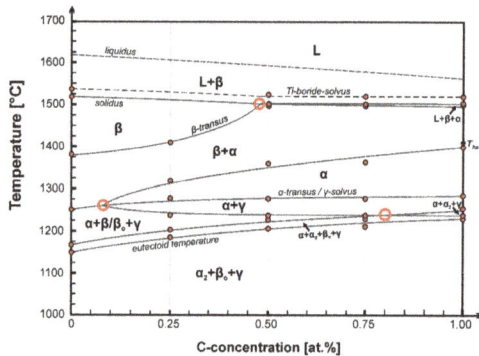

Figure 8. Experimentally-determined quasi-binary section of the TNM-(0–1)C (at. %) system. Evidently, C acts as a strong α stabilizer [38]. (Reprinted from Reference [38], Copyright (2014), with permission from Elsevier).

3.5. Small-Angle Scattering Experiments on Carbide Precipitation in TNM Alloys

In a forged and heat-treated TNM alloy containing 1 at. % C (in the following termed TNM-1C), Ti_2AlC *h*-type carbides were found (see the previous section), whereas no Ti_3AlC *p*-type carbides could be detected. TEM investigations confirmed the absence of both *h*- and *p*-type carbides within cast/HIP, as well as forged and heat-treated alloy variants containing lower amounts of C. In these alloys, the solubility limit of C may not have been exceeded. In this regard, it has to be noted that the overall composition of the alloys may impact the results concerning the presence and type of carbides. Small-angle neutron scattering (SANS) studies proved the presence of *p*-type precipitates in a Ti–48.5Al–0.4C (at. %) alloy after a solution treatment for 4 h at 1250 °C followed by ageing for 24 h at 750 °C and quenching [42]. As a result of the precipitation-hardening treatment, the material's high-temperature strength was improved. Further SANS experiments, which targeted the effect of an addition of 0.5 at. % C to Ti–45Al–5Nb (at. %) alloys, showed that in contrast to the Nb-free alloys, the Nb-containing alloys exhibit no strengthening precipitates, but rather a homogeneous C distribution with a few C-enriched regions [43]. Investigations on a powder-metallurgically produced TNB alloy, on the other hand, which differs in its Al and Nb content from the TNM specimens and is moreover Mo-free, evidenced the presence of *p*-type carbides at 0.75 at. % C [95]. Recently, Wang *et al.* [96] systematically investigated the nucleation, growth and coarsening of carbides in C-containing Ti–45Al–5Nb (at. %) alloys by means of HEXRD and TEM studies.

Generally, not only the chemical composition, but also the heat treatment history determines if, which type and to what extent carbides are present in a material. In a follow-up study on a TNM-1C alloy, the precipitation behavior and thermal stability of the Ti_3AlC p-type carbides during isothermal annealing and subsequent re-heating to 1200 °C were quantified [44]. HEXRD experiments were conducted to study the phase transformation and precipitation behavior of the carbides from a supersaturated matrix. Thereby, the matrix was formed by ultrafine γ lamellae, which precipitated themselves during the heat treatment in supersaturated α_2 grains, building up hierarchical structures on the nanoscale. *In situ* small-angle X-ray scattering (SAXS) experiments using synchrotron radiation were conducted to gain a deeper insight into the size distribution, precipitation kinetics and thermal stability of the carbide precipitates.

The SAXS experiments were conducted at the HEMS beamline at PETRA III [44]. In Figure 9, the experimental setup is schematically illustrated. In principle, it resembles closely the HEXRD setup described in Section 3.1. A major difference, however, lies in the fact that the sample-to-detector distance must be larger for SAXS experiments, as the X-rays scattered at small angles are studied with high resolution. As indicated in Figure 9, a second area detector might be positioned in the optical path of the diffracted X-rays to record the HEXRD pattern at times or to record part of the diffraction rings simultaneously to the *in situ* SAXS experiment.

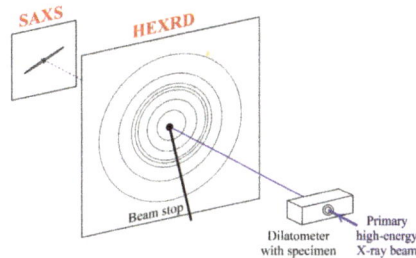

Figure 9. Schematic of the experimental setup for SAXS experiments conducted on C-containing TNM alloy variants [44]. The primary X-ray beam impinges on the specimen within the dilatometer (see Figure 4) and is subsequently scattered/diffracted. Two-dimensional detectors record the SAXS (and if desired, the HEXRD) signal.

For the *in situ* investigation of the carbide precipitation behavior in the TNM-1C alloy, cylindrical specimens of 5 mm in diameter and 10 mm in length were inductively heated in an adapted quenching and deformation dilatometer (Section 3.1; Figure 4). The recorded diffraction patterns were azimuthally integrated using the software program Fit2D [97]. The derived scattering curves were fit within SANSFit based on a least-squares method iteration, as conducted in [42]. The p-type carbides were modelled as elongated ellipsoids of revolution. A random orientation of particles was suggested, obeying a lognormal size distribution.

Figure 10 illustrates the time-temperature profile of the conducted heat treatment. The first annealing step, which consisted of an isothermal segment of 1 h in the $(\alpha + \beta)$ phase field region at 1415 °C followed by quenching, was performed *ex situ* in a high-temperature furnace to avoid thermocouple tear-off. The resulting microstructure at room temperature consisted primarily of supersaturated α_2 grains, as well as a small amount of ordered β_o phase that additionally contained martensitic α_2'. The subsequent isothermal stabilization or precipitation annealing and the ensuing re-heating were conducted in the dilatometer in the optical path of the synchrotron beam, while the SAXS and HEXRD signals were continuously recorded. Regarding the microstructure, the isothermal annealing at 750 °C for 4 h provoked the transformation of the supersaturated α_2 grains to lamellar α_2/γ colonies as γ lamellae precipitated obeying the Blackburn orientation relationship [98]. Furthermore, lens-shaped γ particles precipitated within β_o according to a modified Kurdjumov-Sachs orientation relationship. P-type Ti_3AlC carbides occurred in the γ phase according to the orientation relationship reported in [91]. The precipitation of p-type carbides was attributed to the low solubility of C in the γ phase, which was found to be six-times lower than in the α$_2$ phase, while the β$_o$ phase exhibits in general almost no solubility for C [38,41]. The re-heating of the specimens from 750 to 1200 °C yielded a fine-grained α$_2$/γ phase arrangement due to the occurrence of discontinuous coarsening (cellular reaction) [99]. As the thermal activation of diffusion processes is enhanced with increasing temperature, the microstructural instability becomes increasingly pronounced. The *p*-type carbides dissolve in γ as the C solubility increases.

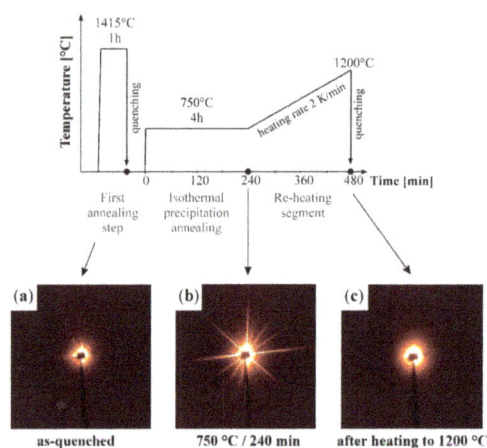

Figure 10. Time-temperature profile of the heat treatment conducted as described in [44]. Dots at certain stages of the profile correspond to the SAXS detector images displayed below (**a,b,c**). See the text for details.

In Figure 10, SAXS detector images recorded at selected stages during the *in situ* experiment (indicated as dots on the time scale of the heat treatment plot) are shown. In all cases, a pronounced small-angle scattering signal is visible around the beam stop. Figure 10a corresponds to the quenched and supersaturated state after the first annealing step at 1415 °C. Small streaks in the SAXS signal are attributed to the martensitic structure of α$_2'$ in β$_o$, as described in [44]. During the initial stage of the annealing segment at 750 °C, these streaks vanish. After an incubation time of roughly 30 min, however, new streaks can be observed. Appearing more pronounced, they are assigned to smaller structural features within the hierarchical structure. Furthermore, the incubation time for nucleation indicates that the streaking does not stem from ultrafine γ lamellae, which form directly after thermal activation at 720 °C, but rather from the *p*-type carbides precipitating in γ-laths. Figure 10b shows the SAXS signal at the end of the isothermal precipitation segment. Heating to 1200 °C leads to the disappearance of the streaks (Figure 10c), which can be attributed to the dissolution of the *p*-type carbides within the γ-lamellae.

The SAXS detector images were evaluated regarding the carbide precipitation kinetics as well as their thermal stability. In Figure 11, selected averaged scattering curves during isothermal annealing and subsequent heating are shown. The time specifications are given in accordance with the time scale set in Figure 10, *i.e.*, 0 min refers to the beginning of the isothermal segment. The magnitude of the scattering vector *q* inversely corresponds to the characteristic dimensions (e.g., radii) of the precipitated particles, while the macroscopic scattering cross-section $d\Sigma/d\Omega$ is directly proportional to the fraction and number of *p*-type carbide precipitations. After a distinct incubation time of roughly 30 min, the scattering curves increase in $d\Sigma/d\Omega$, indicating an increase of the precipitate volume fractions. After 333 min, the specimen has already entered the heating segment (at a temperature of 936 °C), and an inverse trend begins to manifest itself. This means that the *p*-type carbides begin to dissolve, provoked by the increasing temperature (see the time-temperature profile in Figure 10). From the scattering curves, the number density, volume fraction and characteristic dimensions of the precipitates can be estimated [44].

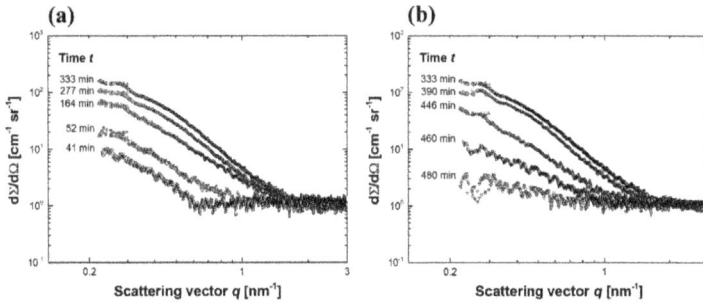

Figure 11. Selected averaged scattering curves during isothermal annealing and subsequent heating. Until 333 min (936 °C), p-type carbides precipitate in the γ laths of the α_2/γ colonies (**a**). Afterwards, their volume fraction decreases again (**b**). The indicated time specifications refer to the time scale set in the time-temperature profile depicted in Figure 10. (Reprinted from Reference [44], Copyright (2014), with permission from Elsevier).

4. TNM Alloys during Processing at Elevated Temperatures

4.1. General Considerations on the Mapping of Deformation, Recovery and Recrystallization

For the investigation of the structure and structural changes in crystalline materials, a multitude of methods based on X-ray diffraction exists. Conventional powder diffraction methods that average over many crystallites offer information on lattice parameters, phase fractions, texture and internal stresses. If monochromatic radiation is used, a fine-grained and homogeneous material will yield a diffraction pattern related to the intersection of the detector (point, line or plane) with the created Debye-Scherrer cones. In transmission geometry, Debye-Scherrer rings will, for example, appear on an area detector. If a monochromatic beam impinges only on a small number of crystallites, be it due to a small beam cross-section or due to a coarse-grained microstructure, information on individual crystallites can be extracted from the resulting reflection spots, which sit on the according positions on the Debye-Scherrer rings [77]. In this regard, three-dimensional X-ray diffraction microscopy (3DXRD) has emerged as a powerful technique enabling the tracking of reflections of individual grains over time [100–103]. Depending on the selected setup, nucleation and growth studies, grain-by-grain volumetric mapping or orientation imaging microscopy can be performed to study undeformed or deformed material regarding recrystallization, grain growth and grain orientations.

Alternatively, if only a few grains are illuminated, white-beam Laue diffraction can be applied. The result is a Laue pattern characteristic of the crystal lattice and its orientation with respect to the incident beam. A typical field of application of the white-beam Laue technique is the adjustment of the orientation of TiAl single crystals as performed by Inui *et al.* [104]. The advantage of this method lies in the fact that reflections of the lattice planes stay visible in the Laue pattern if the crystal is rotated about an arbitrary axis. Using a monochromatic beam, only lattice planes in the diffraction condition are visible [101]. Consequently, the Laue technique is well suited for the investigation of the deformation and recrystallization behavior of single crystals [105,106]. However, it is less advantageous for the investigation of fine-grained polycrystalline samples. In these samples a large number of grains is illuminated at once, resulting in complex Laue patterns that are difficult to evaluate.

In these cases, it may be favorable to conduct HEXRD experiments using monochromatic X-rays (Section 3.1; Figure 4). To study recovery, recrystallization and grain growth processes, the intensity distribution of one or more Debye-Scherrer rings can be evaluated as a function of time. If a large number of grains is illuminated by the incident beam, information concerning texture [107] and strain can be derived from the inhomogeneity along the Debye-Scherrer rings. If reflections from single grains can be distinguished, the patterns can be evaluated regarding a variety of additional parameters, such as the number of spots, their morphology, their arrangement and their evolution with time [50]. Information regarding grain sizes, grain refinement or coarsening, mosaic spread and grain orientations can be retrieved. In terms of heat treatments, for instance, the rearrangement of differently-oriented grains in a massively-transformed TiAl sample has been studied upon heating [82,108]. Careful analysis of single diffraction patterns even allowed tracing the phase transformations of single grains, as diffuse streaking between the parent and the newly-formed grains was observed in a TiAl alloy, as reported in [109]. Individual reflection spots might also transform characteristically during an experiment emulating, for example, hot deformation. A simplified model of the development of reflection spots during hot working is proposed in [52]. The way in which the diffraction pattern is affected depends on the processes dominating in the material, e.g., whether a material or phase predominantly shows recovery or rather recrystallization under the conditions applied. The evaluation of diffraction patterns can help to distinguish these processes and to characterize the evolution of polycrystalline materials undergoing thermo-mechanical processing.

An intuitive approach to visualize the evolution of reflection spots pertaining to one crystallographic plane is to plot the diffracted intensity along the azimuthal angle of a specific ring as a function of time (azimuthal angle *vs.* time (AT) plot) [50]. For this plot, an integration over a small Bragg-angle range has to be performed, which includes only the Debye–Scherrer ring of interest. Yan *et al.* [110] used this form of data representation to illustrate the formation of texture during the deformation of an initially coarse-grained Cu sample. In a similar experimental setup, recovery and recrystallization processes were studied in a Zr alloy during hot compression [111]. Mosaicity was derived from the azimuthal angle spread of single reflection spots. The formation of subgrains was studied as single spots of large angular spread split into many sharp reflections separated by small angles. Recrystallization was identified at a later stage of the experiment by the rapid appearance or disappearance of reflection spots.

4.2. HEXRD Studies on the Hot Deformation Behavior of TNM Alloys

The approach of creating AT plots (Section 4.1) was adopted to study the hot deformation behavior of a TNM alloy [112]. HEXRD experiments were conducted at the ID15 beamline at ESRF. Cylindrical specimens of 4 mm in diameter and 8 mm in length were heated resistively to 1300 °C between the anvils of an electro-thermo-mechanical tester supplied by Instron, Norwood, USA, and subsequently compressed. In Figure 12, the AT plots of two selected Debye–Scherrer rings are given. Figure 12a illustrates the evolution of the α_2/α-(20-21) ring, Figure 12b the one of β_0/β-(002). In Figure 12c, the temperature profile, as well as true stress and true strain curves are given as a function of time.

At the beginning of the experiment, the evolution of the individual reflection spots over time results in distinct lines in the AT plots, which are in the following referred to as timelines. For both phases, α_2/α and β_0/β, continuous timelines are observed during heating. While the specimen is held at 1300 °C within the ($\alpha + \beta$) phase field region, the α reflections remain steady, while the β phase reflections fluctuate. Upon the onset of deformation, the intensity distribution immediately changes for both phases. Concerning the α phase, diverting, yet continuous timelines are discernible, which exhibit a blurred appearance. The β phase reflections, on the other hand, broaden partly and become sharp spots. After 215 s, the sample was unloaded for 30 s to investigate the material's response. Upon this, the α phase timelines remain again continuous, but blurred, while the β phase reflections fluctuate fast and yield sharp timelines.

The inserts in the respective figures highlight the material's behavior from the start of deformation up to the end of the interim unloading period. Diverted timelines stem from the rotation of a crystal lattice of a grain about an axis parallel to the incident beam. In contrast to this, the broadening of a reflection indicates an increasing defect density within the grain [113]. Sharp dots, which relate to one original reflection and occur within a small angular range, are caused by subgrains stemming from a dynamic recovery process [77,111]. As peak broadening followed by the formation of new reflection spots can be observed in the insert of Figure 12b, fast dynamic recovery processes are suggested to dominate the β phase. In the α phase, however, the reflections are diverted around small azimuthal angles and appear blurred, indicating the deformation by dislocation slip and the presence of high defect densities. The absence of reflections with low angular spread points to slow recovery processes instead of discontinuous recrystallization processes [77]. For the validation of the results of the presented *in situ* HEXRD experiment, electron-backscatter diffraction (EBSD) measurements were performed on the deformed specimens. These investigations, which are elaborately described in [52,112], have confirmed the results derived from the *in situ* experiments. No indication of appreciable dynamic recrystallization was found for both phases α/α_2 and β/β_o.

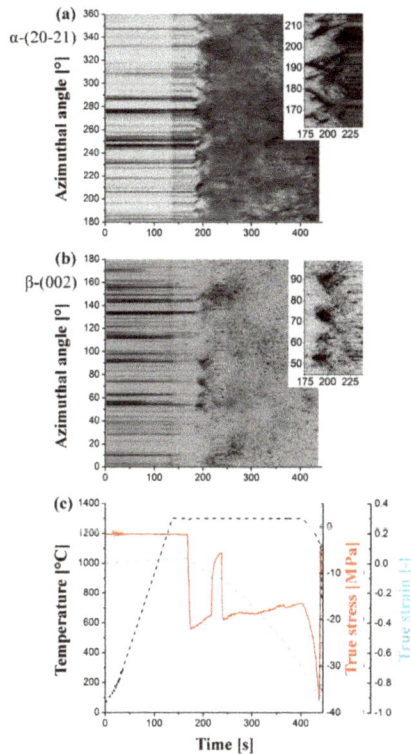

Figure 12. Azimuthal angle *vs.* time (AT) plots of the α_2/α-(20-21) (a) and the β_o/β-(002) reflections (b) illustrating the hot deformation behavior of a TNM alloy. The temperature profile, true stress and true strain are given as a function of time (c). The *in situ* deformation experiment was conducted within the ($\alpha + \beta$) phase field region [112]. (Reproduced with permissions from Reference [52]. Copyright © Wiley-VCH Verlag GmbH & Co. KGaA, Weinheim, Germany, 2011).

In situ HEXRD experiments during deformation were also conducted on TNM alloys containing minor additions of C and Si [114]. For this purpose, the deformation unit of the quenching and deformation dilatometer (Section 3.1) was used at the HZG beamline HARWI II at DESY. Phase fraction and texture evolutions, as well as temperature uncertainties arising during compression, were studied. The information retrieved complemented uniaxial compressive hot-deformation tests performed on a Gleeble 3500 simulator at temperatures ranging from 1150 to 1300 °C and strains rates between 0.005 and 0.5 s^{-1}.

5. Heat Treatments for Balanced Mechanical Properties

5.1. Microstructural Evolution: The Formation of Lamellar α_2/γ Colonies in a TNM Alloy

For the adjustment of optimized properties, the control over the phases present in a material, specifically in terms of composition, volume fraction and arrangement, is indispensable. Processes taking place during phase transformations, such as the precipitation or dissolution of phases, offer opportunities for targeted manipulation. Knowledge of these transformations is, thus, of great technological importance. Comprehensive insight into the prevailing processes can be gained, for example, by taking a two-fold approach. While *in situ* diffraction and scattering techniques offer information on the evolution of phase fractions and kinetic aspects, complementary *in situ* or *ex situ* studies of the microstructure (e.g., by means of SEM or TEM) unveil the arrangement of the various phases in real space.

In TiAl alloys, mechanical properties strongly depend on the occurrence and structure of lamellar α_2/γ colonies [31,32,115,116]. For this reason, the $\alpha \rightarrow \alpha_2/\gamma$ transformation has been the subject of thorough investigation, both from technological and scientific points of view [4,48,52,109,117–119]. In TiAl alloys, the γ phase dissolves upon heating at $T_{\gamma,solv}$. Depending on the chemical composition, the alloy then enters the ($\alpha + \beta$) or the single α phase field region (Figure 5). If the specimen is again cooled down from this phase field region, γ lamellae tend to form within supersaturating α grains [5,118,120]. The cooling rate has been proven to be a decisive parameter for the lamellar spacing within the forming α_2/γ colonies [118,121]. However, if the alloy is cooled at a sufficiently high rate, the microstructure present at elevated temperatures is roughly preserved, and the formation of the γ phase is suppressed. Ordering reactions take place, but the α_2 grains remain in a supersaturated state. In the course of a subsequent heat treatment, (ultra)fine γ lamellae can precipitate in the supersaturated α_2 grains [28,109,122,123]. In this case, the lamellar spacing strongly depends on the temperature profile during the heat treatment, *i.e.*, the heating rate and maximum temperature during continuous heating or the holding temperature in isothermal heat treatments [116,124]. The formation of fine γ lamellae improves the mechanical properties of a material in terms of strength, plastic fracture strain, fracture toughness and creep resistance at service temperature [5,10,11,53,116,120,122,125].

To study the onset of the γ lamellae precipitation in the course of continuous heating in a Ti–45Al–7.5Nb (at. %) alloy, *in situ* experiments were conducted at the beamline ID15A/B at the ESRF [109]. A two-dimensional detector enabled the investigation of diffuse and inter-Bragg peak scattering. The reflections recorded in reciprocal space were observed to form streaks, which correlated with the precipitation and formation of ultrafine γ lamellae. High temperature laser scanning confocal microscopy was used as a complementary technique to verify the obtained results [109]. *In situ* TEM studies of the initial stages of lamellae formation in a Ti–45Al–7.5Nb (at. %) alloy were performed by Cha *et al.* [126]. During heating, γ laths were found to precipitate at roughly 750 °C according to the Blackburn relationship [98], following the formation and motion of dislocations next to interfaces at 730 °C. As *in situ* TEM, however, comprises many experimental difficulties, the majority of reported TEM studies is restricted to *ex situ* characterizations. Nevertheless, valuable data on lamellae thickness and spacing can be obtained, which are generally not accessible by means of other characterization methods.

To investigate the precipitation of γ lamellae in TNM alloys, specimens were subjected to a two-step heat treatment (Figure 13) while being simultaneously probed by means of HEXRD [52]. The first step in the heat treatment was designed to create supersaturated α_2 grains through rapid quenching from 1230 °C. The second heat treatment step consisted of a continuous heating ramp, during which fine γ lamellae precipitated in the α_2 grains [28,72]. In Figure 13a, the temperature profile of the heat treatment is shown superimposed by the intensity evolution of the γ-(200) peak, which was evaluated by fitting a Gaussian to the relevant *q*-range of the azimuthally-integrated diffraction patterns. The rapid heating of the specimen during the first heat treatment step results in a decreasing peak intensity, indicating a strong decrease in the amount of γ phase with increasing temperature. During the isothermal segment at 1230 °C, the γ phase fraction further decreases as the phase approaches thermodynamic equilibrium. After 300 s, the specimen is quenched to room temperature. Upon this, an increase in peak intensity is observed, indicating a small increase in the amount of γ phase. The precipitation and growth of the γ phase has not completely been suppressed. However, the hindrance was sufficient to create supersaturated α_2 grains, which contain only a few γ lamellae (Figure 13b, top). After quenching, the specimen is continuously heated at a low heating rate. Until about 765 °C, no major changes are observed in the peak intensity evolution. At temperatures well above 865 °C, however, the γ phase fraction increases rapidly. This increase is attributed to the precipitation of fine γ lamellae (Figure 13b, bottom). *Ex situ* TEM studies performed by Cha *et al.* [115,116] document that an average interface spacing in the range of 5 to 40 nm can be adjusted in heat-treated TiAl alloys.

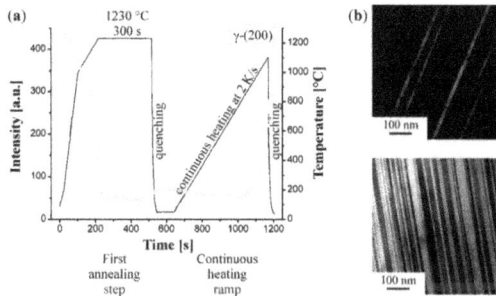

Figure 13. Temperature profile applied during the heat treatment of a TNM alloy, shown superimposed by the intensity evolution of the γ-(200) peak (blue) (**a**). The heat treatment was conducted in two steps and resulted in the precipitation of fine γ lamellae. Exemplary TEM images show the presence of only few γ lamellae in the supersaturated α_2 grains after quenching (**b**, top). After annealing, the number and volume fraction of fine γ lamellae has significantly increased (**b**, bottom). (Reproduced with permissions from Reference [52]. Copyright © Wiley-VCH Verlag GmbH & Co. KGaA, Weinheim, Germany, 2011).

5.2. Studies Enabling the Targeted Optimization of Heat Treatment Steps

In TNM alloys, heat treatments must be applied after hot deformation as high amounts of the ordered β_o phase would deteriorate the material's properties at room and service temperature [11,20,25,29,127]. Adapted post-forging heat treatments have been reported in [20,28,29,115,116,128–130]. These heat treatments typically comprise three steps [131]. In the first heat treatment step, the material is homogenized as reported in [129]. Subsequently, a high-temperature annealing step is conducted either in the (α + β + γ) or in the (α + β) phase field region. The aim is the minimization of the β phase fraction, the setting of the volume fraction of globular γ grains and the adjustment of the size of α grains, which act as precursors of the lamellar α_2/γ colonies. The third annealing step represents a stabilization treatment near 900 °C and is followed by furnace cooling. According to the Blackburn orientation relationship [98], γ lamellae precipitate in supersaturated α_2 grains or α_2/γ colonies (Section 5.1), causing an increase in hardness. The lamellar spacing within the colonies is usually attempted to be kept small to gain a substantial increase in strength through a modified Hall-Petch relationship [115,116]. For detailed information regarding the mechanical properties of TNM alloys, the reader is referred to [11,28,37,114]. In ultrafine lamellae, however, the driving force for discontinuous coarsening (cellular reaction) is greatly increased, which has a negative impact on the material's creep properties [39,99,132]. Processing parameters, such as annealing temperatures or heating and cooling rates, have to be carefully selected to obtain balanced properties [115,133,134].

As the mechanical properties of TNM alloys are determined by the precipitation of γ lamellae during the third annealing step, the adjustment of a favorable starting condition during the second heat treatment step, *i.e.*, supersaturated α_2 grains for a maximum driving force in γ precipitation, is of great importance. The microstructural evolution upon cooling has been thoroughly studied in several γ-TiAl-based alloys [109,118,133,135–139]. The influence of varying cooling rates on the grain size and the appearance of lamellar structures in particular was studied and linked to the resulting mechanical properties. The applied methods included short- and long-term heat treatments combined with metallography, dilatometry, differential scanning calorimetry, micro-hardness testing and TEM. Diffraction methods, such as HEXRD, have been used for the investigation of phase fraction evolutions as a function of temperature and time [50,51,65,77,82,128,130].

To investigate the influence of technologically relevant cooling rates on the evolution of the γ phase during the second heat treatment step, a TNM alloy with increased amounts of β-stabilizing alloying elements Nb and Mo was studied by means of *in situ* HEXRD [87]. The diffraction experiments were conducted partly at HARWI II and partly at PETRA III. In a dilatometer setup in the synchrotron beam (Section 3.1; Figure 4), forged and homogenized specimens were annealed for 10 min at 1235 °C in the (α + β + γ) phase field region below the γ-solvus temperature $T_{\gamma,solv}$ and subsequently subjected to linear cooling rates ranging from 35 to 1200 K· min^{-1}. After reaching 600 °C, the specimens were quenched at the highest rate possible. The correlation between the linear cooling rates, the phase fraction evolutions and the resulting microstructures was investigated.

In accordance with [118,133,138], the resulting α/α_2 phase fractions upon cooling were found to increase significantly with increasing cooling rate. At high cooling rates, the time available for diffusion-controlled processes is shorter, and a larger fraction of the untransformed, supersaturated α_2 phase remains. With the aid of the *in situ* HEXRD experiments, the phase fractions could be accurately quantified. Furthermore, the increase in the amount of α/α_2 phase was observed to occur primarily at the expense of the γ phase, whose phase fractions decreased with increasing cooling rate. From these results, a plot analogous to a continuous-cooling-transformation (CCT) diagram was derived (Figure 14), including all investigated cooling rates. In this plot, broad black lines confine regions in which the indicated γ phase fraction is exceeded. Experimental points representing conditions close to thermodynamic equilibrium were obtained from preliminary diffraction experiments [87].

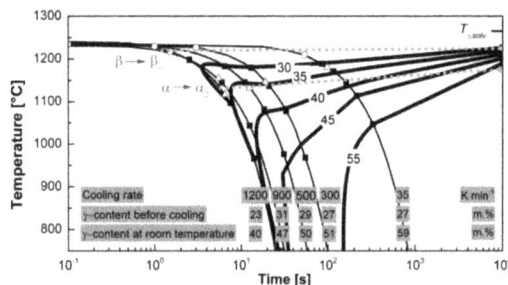

Figure 14. Continuous-cooling-transformation (CCT) diagram for the γ phase precipitation during the heat treatment of a TNM alloy. Broad black lines confine regions in which the given γ phase fraction (in m %) is certainly exceeded. The cooling paths plotted as narrow black lines result from the linear cooling rates of 35, 300, 500, 900 and 1200 K· min^{-1}, respectively. Thus, the influence of technologically relevant cooling rates on the quantitative gain in the γ phase can be derived [87]. Actual γ phase fractions (in m %) before and after cooling are indicated in the boxes at the bottom. (Reprinted from Reference [87], Copyright (2015), with permission from Elsevier).

The temperatures at which specific amounts of γ phase are present decrease in accordance with [135] with increasing cooling rate, while the phenomenon is enhanced for higher cooling rates. The γ phase fractions determined at the end of the isothermal annealing segment, *i.e.*, before cooling and after cooling to room temperature, are given in Figure 14. Comparing the four specimens that were cooled at constant cooling rates in the range of 35 to 900 K·min^{-1}, the quantitative gain in γ phase upon cooling generally decreases as the cooling rate increases. The specimen cooled at 1200 K·min^{-1} exhibits at room temperature, as expected, the lowest γ phase fraction (40 m %). The quantitative gain in γ phase, however, is slightly elevated when compared to the specimen cooled at 900 K·min^{-1} (from 31 to 47 m %). This behavior is attributed to the fact that this specimen exhibited with 23 m % the lowest γ phase fraction at the end of the isothermal annealing segment, corresponding to a higher actual annealing temperature. Due to differences in the actual starting temperatures, the gain in γ phase as a quantity is not suitable as an absolute means of comparison. Comparison of the γ phase fractions at room temperature is more reliable.

As all specimens were cooled from the (α + β + γ) phase field region, a certain amount of the γ phase is present in the specimens from the beginning. From Figure 14, it can be concluded that even at cooling rates as high as 1200 K·min^{-1} further gain in γ phase cannot be suppressed. During cooling, the overall amount of the γ phase increases. At high cooling rates, however, a sufficient amount of supersaturated α/α$_2$ grains can be obtained, rendering possible the formation of fine lamellar structures in an ensuing annealing step, as described at the beginning of this section. Investigation of the resulting microstructures by means of SEM and TEM offers additional information for the design of a suitable time-temperature profile in the second heat treatment step typical of TNM alloys.

6. Summary

In situ diffraction and scattering techniques based on synchrotron radiation and neutrons offer unique opportunities for the development of advanced structural materials. They enable the exploration of a multitude of aspects, which are usually not accessible with the aid of conventional methods. The present review attempts to shine a light on selected *in situ* diffraction and scattering techniques, highlighting the ways in which they promoted the development of intermetallic TNM alloys based on the γ-TiAl phase.

In situ studies of phase evolutions, as discussed in Section 3, provide answers to fundamental research questions. They reveal phase fractions and transition temperatures and allow to study influencing parameters under controlled experimental conditions. Measurements on several model alloys of differing chemical compositions provide the basis for the establishment of phase diagrams [37]. Furthermore, the specific influence of certain alloying elements can be studied. For instance, it was found that the amount and type of β-stabilizing alloying elements significantly influences the phase evolution character of the β phase, which plays a decisive role in post-forging heat treatments [65]. Diffraction techniques also promoted the development of TNM alloys containing harmonized additions of C and Si, as they enabled the investigation of the influence of the alloying additions on the solidification behavior and phase fraction evolution within the material [38]. SAXS and SANS studies offered insights into the carbide precipitation behavior of these advanced alloys [44]. For the investigation of order/disorder transitions, typically, neutron diffraction experiments are conducted. Due to peculiarities in the structure factors of titanium aluminide alloys (Section 2), they offer distinct advantages. *In situ* studies are again indispensable, as ordering reactions are characterized by extremely fast kinetics. Consequently, the ordering behavior of TiAl alloys can only be investigated at those temperatures at which the ordering reactions occur [65,85].

The second part of the present review exemplarily introduces approaches to processing- and application-related questions. For the investigation of the hot deformation behavior, for example, AT plots can be applied (Section 4). With this technique, an understanding of the processes prevailing in TNM alloys during hot working could be established [112]. After hot deformation, heat treatments are applied to TNM alloys to adjust balanced mechanical properties (Section 5). For this purpose, control over the phases present in the material, specifically in terms of composition, volume fraction and arrangement, is indispensable. To extend the necessary knowledge bases, the formation of lamellar α_2/γ colonies was investigated from a diffraction point of view and complemented by the results from conventional characterization techniques [52]. Finally, *in situ* HEXRD techniques were used as a tool for the targeted optimization of heat treatments [87].

Continuous progress in hardware development, especially detector technology, as well as the construction of dedicated diffraction beamlines of increasing brilliance provide the necessary base for *in situ* experiments at high frame rates. Simultaneously to this development, new techniques are emerging that comprise valuable tools for a multitude of applications in materials science. Besides HEXRD, neutron diffraction and SAXS/SANS, great potential for future investigations is seen, for example, in nano-beam experiments and in the 3DXRD technique. Diffraction and scattering methods have significantly contributed to the characterization and development of advanced multi-phase γ-TiAl-based alloys during the last 25 years. Undoubtedly, also in the future, characterization techniques based on synchrotron radiation and neutrons will promote the understanding of this class of innovative, lightweight, high-temperature materials.

Acknowledgments: Some of the research studies presented in this review were conducted at the HZG beamlines HARWI II and HEMS at DESY. The support of the DESY management, user office and HZG beamline staff is gratefully acknowledged. In particular, the authors would like to thank Thomas Lippmann, Norbert Schell, Lars Lottermoser, Torben Fischer and René Kirchhof for their assistance at the beamlines. Some studies are based on experiments performed at the ESRF and the MLZ. Access and support of the respective managements, user offices and beamline staff are appreciated. The contribution of Kun Yan to the successful realization of the neutron scattering experiments at ANSTO is gratefully acknowledged. Parts of the research pursued by the authors affiliated with the Montanuniversität Leoben received support by the European Commission under the 7th Framework Programme through the "Research Infrastructures" action of the "Capacities" Programme (Contract No. CP-CSA_INFRA-2008-1.1.1 Number 226507-NMI3), by the European Community's 7th Framework Programme (FP7/2007e2013) under Grant Agreement No. 226716 and by the Styrian Materials Cluster, Austria. Some of the studies were conducted within the frameworks of the German BMBF Project O3X3530A and of the Austrian FFG Project 830381 "fAusT" of the Österreichisches Luftfahrtprogramm TAKE OFF.

Author Contributions: All authors contributed equally to this work. P.E. wrote the bulk of the manuscript in close collaboration with the other authors. The contribution of the authors to the experimental work is specified in the respective original publications cited throughout this work.

Conflicts of Interest: The authors declare no conflict of interest.

References

1. Bewlay, B.P.; Weimer, M.; Kelly, T.; Suzuki, A.; Subramanian, P.R. The science, technology, and implementation of TiAl alloys in commercial aircraft engines. *MRS Proc.* **2013**, *1516*, 49–58. [CrossRef]
2. Clemens, H.; Smarsly, W.; Güther, V.; Mayer, S. Advanced intermetallic titanium aluminides. In Proceedings of the 13th World Conference on Titanium, San Diego, CA, USA, 16–20 August 2015.
3. Habel, U.; Heutling, F.; Helm, D.; Kunze, C.; Smarsly, W.; Das, G.; Clemens, H. Forged intermetallic γ-TiAl based alloy low pressure turbine blade in geared turbofan. In Proceedings of the 13th World Conference on Titanium, San Diego, CA, USA, 16–20 August 2015.
4. Yamaguchi, M.; Inui, H.; Ito, K. High-temperature structural intermetallics. *Acta Mater.* **2000**, *48*, 307–322. [CrossRef]
5. Appel, F.; Wagner, R. Microstructure and deformation of two-phase γ-titanium aluminides. *Mater. Sci. Eng. Rep.* **1998**, *R22*, 187–268. [CrossRef]
6. Kim, Y.-W.; Clemens, H.; Rosenberger, A. *Gamma Titanium Aluminides 2003*; The Minerals, Metals and Materials Society (TMS): Warrendale, PA, USA, 2003.
7. Kim, Y.-W.; Kim, S.; Dimiduk, D.M.; Woodward, C. Development of beta gamma alloys: Opening robust processing and greater application potential for TiAl-base alloys. In *Structural Aluminides for Elevated Temperature Applications*; The Minerals, Metals and Materials Society (TMS): Warrendale, PA, USA, 2008; pp. 215–216.
8. Leyens, C.; Peters, M. *Titanium and Titanium Alloys*; Wiley-VCH: Weinheim, Germany, 2003.
9. Wu, X. Review of alloy and process development of TiAl alloys. *Intermetallics* **2006**, *14*, 1114–1122. [CrossRef]

10. Appel, F.; Paul, J.D.H.; Oehring, M. *Gamma Titanium Aluminide Alloys—Science and Technology*; Wiley-VCH: Weinheim, Germany, 2011.

11. Clemens, H.; Mayer, S. Design, processing, microstructure, properties, and applications of advanced intermetallic TiAl alloys. *Adv. Eng. Mater.* **2013**, *15*, 191–215. [CrossRef]

12. Appel, F.; Oehring, M.; Wagner, R. Novel design concepts for gamma-base titanium aluminide alloys. *Intermetallics* **2000**, *8*, 1283–1312. [CrossRef]

13. Cui, W.F.; Liu, C.M.; Bauer, V.; Christ, H.-J. Thermomechanical fatigue behaviours of a third generation γ-TiAl based alloy. *Intermetallics* **2007**, *15*, 675–678. [CrossRef]

14. Baur, H.; Joos, R.; Smarsly, W.; Clemens, H. γ-TiAl for aeroengine and automotive applications. In *Intermetallics and Superalloys*; Wiley-VCH: Weinheim, Germany, 2000; pp. 384–390.

15. Tetsui, T. Effect of composition on endurance of TiAl in turbocharger applications. In *Gamma Titanium Aluminides 1999*; Kim, Y.-W., Dimiduk, D.M., Loretto, M.H., Eds.; The Minerals, Metals and Materials Society (TMS): Warrendale, PA, USA, 1999; pp. 15–23.

16. Sommer, A.; Keijzers, G. Gamma TiAl and the engine exhaust valve. In *Gamma Titanium Aluminides 2003*; The Minerals, Metals and Materials Society (TMS): Warrendale, PA, USA, 2003; pp. 3–8.

17. Lasalmonie, A. Intermetallics: Why is it so difficult to introduce them in gas turbine engines? *Intermetallics* **2006**, *14*, 1123–1129. [CrossRef]

18. Clemens, H.; Bartels, A.; Bystrzanowski, S.; Chladil, H.F.; Leitner, H.; Dehm, G.; Gerling, R.; Schimansky, F.P. Grain refinement in γ-TiAl-based alloys by solid state phase transformations. *Intermetallics* **2006**, *14*, 1380–1385. [CrossRef]

19. Clemens, H.; Smarsly, W. Light-weight intermetallic titanium aluminides—Status of research and development. *Adv. Mater. Res.* **2011**, *278*, 551–556. [CrossRef]

20. Clemens, H.; Wallgram, W.; Kremmer, S.; Güther, V.; Otto, A.; Bartels, A. Design of novel β-solidifying TiAl alloys with adjustable β/βₒ-phase fraction and excellent hot-workability. *Adv. Eng. Mater.* **2008**, *10*, 707–713. [CrossRef]

21. Imayev, R.M.; Imayev, V.M.; Khismatullin, T.G.; Oehring, M.; Appel, F. New approaches to designing of alloys based on γ-TiAl + α₂-Ti₃Al. *Rev. Adv. Mater. Sci.* **2006**, *11*, 99–108. [CrossRef]

22. Naka, S. Advanced titanium-based alloys. *Curr. Opin. Solid State Mater. Sci.* **1996**, *1*, 333–339. [CrossRef]

23. Clemens, H.; Chladil, H.F.; Wallgram, W.; Zickler, G.A.; Gerling, R.; Liss, K.-D.; Kremmer, S.; Güther, V.; Smarsly, W. In and ex situ investigations of the β-phase in a Nb and Mo containing γ-TiAl based alloy. *Intermetallics* **2008**, *16*, 827–833. [CrossRef]

24. Kainuma, R.; Fujita, Y.; Mitsui, H.; Ohnuma, I.; Ishida, K. Phase equilibria among α (hcp), β (bcc) and γ (L1₀) phases in Ti-Al base ternary alloys. *Intermetallics* **2000**, *8*, 855–867. [CrossRef]

25. Tetsui, T.; Shindo, K.; Kobayashi, S.; Takeyama, M. A newly developed hot worked TiAl alloy for blades and structural components. *Scr. Mater.* **2002**, *47*, 399–403. [CrossRef]

26. Tetsui, T.; Shindo, K.; Kaji, S.; Kobayashi, S.; Takeyama, M. Fabrication of TiAl components by means of hot forging and machining. *Intermetallics* **2005**, *13*, 971–978. [CrossRef]

27. Kremmer, S.; Chladil, H.F.; Clemens, H.; Otto, A.; Güther, V. Near conventional forging of titanium aluminides. In Proceedings of the 11th World Conference on Titanium, Kyoto, Japan, 3–7 June 2008; pp. 989–992.

28. Wallgram, W.; Schmoelzer, T.; Cha, L.; Das, G.; Güther, V.; Clemens, H. Technology and mechanical properties of advanced γ-TiAl based alloys. *Int. J. Mater. Res.* **2009**, *100*, 1021–1030. [CrossRef]

29. Huber, D.; Clemens, H.; Stockinger, M. Near conventional forging of an advanced TiAl alloy. *MRS Proc.* **2012**, *1516*, 23–28. [CrossRef]

30. Kestler, H.; Clemens, H. Production, processing and application of γ(TiAl)-based alloys. In *Titanium and Titanium Alloys*; Leyens, C., Peters, M., Eds.; Wiley-VCH: Weinheim, Germany, 2003; pp. 356–360.

31. Crofts, P.; Bowen, P.; Jones, I. The effect of lamella thickness on the creep behaviour of Ti–48Al–2Nb–2Mn. *Scr. Mater.* **1996**, *35*, 1391–1396. [CrossRef]

32. Maruyama, K.; Yamamoto, R.; Nakakuki, H.; Fujitsuna, N. Effects of lamellar spacing, volume fraction and grain size on creep strength of fully lamellar TiAl alloys. *Mater. Sci. Eng. A* **1997**, *239–240*, 419–428. [CrossRef]

33. Voisin, T.; Monchoux, J.-P.; Hantcherli, M.; Mayer, S.; Clemens, H.; Couret, A. Microstructures and mechanical properties of a multi-phase β-solidifying TiAl alloy densified by spark plasma sintering. *Acta Mater.* **2014**, *73*, 107–115. [CrossRef]

34. Jin, Y.; Wang, J.N.; Yang, J.; Wang, Y. Microstructure refinement of cast TiAl alloys by β solidification. *Scr. Mater.* **2004**, *51*, 113–117. [CrossRef]

35. Imayev, V.M.; Imayev, R.M.; Khismatullin, T.G. Mechanical properties of the cast intermetallic alloy Ti–43Al–7(Nb,Mo)–0.2B (at. %) after heat treatment. *Phys. Met. Metallogr.* **2008**, *105*, 484–490. [CrossRef]

36. Yang, J.; Wang, J.N.; Wang, Y.; Xia, Q. Refining grain size of a TiAl alloy by cyclic heat treatment through discontinuous coarsening. *Intermetallics* **2003**, *11*, 971–974. [CrossRef]

37. Schwaighofer, E.; Clemens, H.; Mayer, S.; Lindemann, J.; Klose, J.; Smarsly, W.; Güther, V. Microstructural design and mechanical properties of a cast and heat-treated intermetallic multi-phase γ-TiAl based alloy. *Intermetallics* **2014**, *44*, 128–140. [CrossRef]

38. Schwaighofer, E.; Rashkova, B.; Clemens, H.; Stark, A.; Mayer, S. Effect of carbon addition on solidification behavior, phase evolution and creep properties of an intermetallic β-stabilized γ-TiAl based alloy. *Intermetallics* **2014**, *46*, 173–184. [CrossRef]

39. Kastenhuber, M.; Rashkova, B.; Clemens, H.; Mayer, S. Enhancement of creep properties and microstructural stability of intermetallic β-solidifying γ-TiAl based alloys. *Intermetallics* **2015**, *63*, 19–26. [CrossRef]

40. Appel, F.; Lorenz, U.; Oehring, M.; Sparka, U.; Wagner, R. Thermally activated deformation mechanisms in micro-alloyed two-phase titanium aluminide alloys. *Mater. Sci. Eng. A* **1997**, *233*, 1–14. [CrossRef]

41. Scheu, C.; Stergar, E.; Schober, M.; Cha, L.; Clemens, H.; Bartels, A.; Schimansky, F.-P.; Cerezo, A. High carbon solubility in a γ-TiAl-based Ti–45Al–5Nb–0.5C alloy and its effect on hardening. *Acta Mater.* **2009**, *57*, 1504–1511. [CrossRef]

42. Staron, P.; Christoph, U.; Appel, F.; Clemens, H. SANS investigation of precipitation hardening of two-phase γ-TiAl alloys. *Appl. Phys. A.* **2002**, *75*, 1–3. [CrossRef]

43. Staron, P.; Schimansky, F.-P.; Scheu, C.; Clemens, H. SANS study of carbon addition in Ti–45Al–5Nb. *MRS Proc.* **2011**, *1295*, 195–200. [CrossRef]

44. Schwaighofer, E.; Staron, P.; Rashkova, B.; Stark, A.; Schell, N.; Clemens, H.; Mayer, S. *In situ* small-angle X-ray scattering study of the perovskite-type carbide precipitation behavior in a carbon-containing intermetallic TiAl alloy using synchrotron radiation. *Acta Mater.* **2014**, *77*, 360–369. [CrossRef]

45. Abe, E.; Kumagai, T.; Nakamura, M. New ordered structure of TiAl studied by high-resolution electron microscopy. *Intermetallics* **1996**, *4*, 327–333. [CrossRef]

46. Shull, R.D.; Cline, J.P. High temperature X-ray diffractometry of Ti-Al alloys. *High Temp. Sci.* **1988**, *26*, 95–117.

47. Schloffer, M.; Schmoelzer, T.; Mayer, S.; Schwaighofer, E.; Hawranek, G.; Schimansky, F.-P.; Pyczak, F.; Clemens, H. The characterisation of a powder metallurgically manufactured TNM titanium aluminide alloy using complementary quantitative methods. *Pract. Metallogr.* **2011**, *48*, 594–604. [CrossRef]

48. Novoselova, T.; Malinov, S.; Sha, W.; Zhecheva, A. High-temperature synchrotron X-ray diffraction study of phases in a gamma TiAl alloy. *Mater. Sci. Eng. A* **2004**, *371*, 103–112. [CrossRef]

49. Watson, I.J.; Liss, K.-D.; Clemens, H.; Wallgram, W.; Schmoelzer, T.; Hansen, T.C.; Reid, M. *In situ* characterization of a Nb and Mo containing γ-TiAl based alloy using neutron diffraction and high-temperature microscopy. *Adv. Eng. Mater.* **2009**, *11*, 932–937. [CrossRef]

50. Liss, K.-D.; Yan, K. Thermo-mechanical processing in a synchrotron beam. *Mater. Sci. Eng. A* **2010**, *528*, 11–27. [CrossRef]

51. Liss, K.-D.; Bartels, A.; Schreyer, A.; Clemens, H. High-energy X-rays: A tool for advanced bulk investigations in materials science and physics. *Textures Microstruct.* **2003**, *35*, 219–252. [CrossRef]

52. Schmoelzer, T.; Liss, K.-D.; Staron, P.; Mayer, S.; Clemens, H. The contribution of high-energy X-rays and neutrons to characterization and development of intermetallic titanium aluminides. *Adv. Eng. Mater.* **2011**, *13*, 685–699. [CrossRef]

53. Kim, Y.-W. Effects of microstructure on the deformation and fracture of γ-TiAl alloys. *Mater. Sci. Eng. A* **1995**, *192*, 519–533. [CrossRef]

54. Swaminathan, S. Debye-Waller factors in off-stoichiometric γ-TiAl: Effect of ordering of excess Al atoms on Ti sites. *Philos. Mag. Lett.* **1996**, *73*, 319–330. [CrossRef]

55. Kim, Y.-W. Microstructural evolution and mechanical properties of a forged gamma titanium aluminide alloy. *Acta Metall. Mater.* **1992**, *40*, 1121–1134. [CrossRef]

56. Grinfeld, M.A.; Hezzledine, P.M.; Shoykhet, B.; Dimiduk, D.M. Coherency stresses in lamellar Ti–Al. *Metall. Mater. Trans. A* **1998**, *29A*, 937–942.
57. Hashimoto, K.; Kimura, M.; Mizuhara, Y. Alloy design of gamma titanium aluminides based on phase diagrams. *Intermetallics* **1998**, *6*, 667–672. [CrossRef]
58. Kimura, M.; Hashimoto, K. High-temperature phase equilibria in Ti–Al–Mo system. *J. Phase Equilibria* **1999**, *20*, 224–230. [CrossRef]
59. Qin, G.W.; Oikawa, K.; Sun, Z.M.; Sumi, S.; Ikeshoji, T.; Wang, J.J.; Guo, S.W.; Hao, S.M. Discontinuous coarsening of the lamellar structure of γ-TiAl-based intermetallic alloys and its control. *Metall. Mater. Trans. A* **2001**, *32*, 1927–1938. [CrossRef]
60. Reimers, W.; Pyzalla, A.R.; Schreyer, A.; Clemens, H. *Neutrons and Synchrotron Radiation in Engineering Materials Science—From Fundamentals to Application*, 2nd ed.; Wiley-VCH: Weinheim, Germany, 2008.
61. Sears, V.F. Neutron scattering lengths and cross sections. *Neutron News* **1992**, *3*, 29–37. [CrossRef]
62. Schmoelzer, T. Investigation of γ-TiAl Alloys by Means of Diffraction Methods. Ph.D. Thesis, Montanuniversität Leoben, Leoben, Austria, April 2012.
63. Liss, K.-D.; Whitfield, R.E.; Xu, W.; Buslaps, T.; Yeoh, L.A.; Wu, X.; Zhang, D.; Xia, K. *In situ* synchrotron high-energy X-ray diffraction analysis on phase transformations in Ti–Al alloys processed by equal-channel angular pressing. *J. Synchrotron Radiat.* **2009**, *16*, 825–834. [CrossRef] [PubMed]
64. Staron, P.; Fischer, T.; Lippmann, T.; Stark, A.; Daneshpour, S.; Schnubel, D.; Uhlmann, E.; Gerstenberger, R.; Camin, B.; Reimers, W.; *et al.* *In situ* experiments with synchrotron high-energy X-rays and neutrons. *Adv. Eng. Mater.* **2011**, *13*, 658–663. [CrossRef]
65. Schmoelzer, T.; Mayer, S.; Sailer, C.; Haupt, F.; Güther, V.; Staron, P.; Liss, K.-D.; Clemens, H. *In situ* diffraction experiments for the investigation of phase fractions and ordering temperatures in Ti–44 at. % Al–(3–7) at. % Mo alloys. *Adv. Eng. Mater.* **2011**, *13*, 306–311. [CrossRef]
66. Erdely, P.; Stark, A.; Clemens, H.; Mayer, S. *In situ* high-energy X-ray diffraction on an intermetallic β-stabilised γ-TiAl based alloy. *BHM Berg-und Hüttenmännische Monatshefte* **2015**, *160*, 221–225. [CrossRef]
67. Treusch, R. Radiation sources: Production and properties of synchrotron radiation. In *Neutrons and Synchrotron Radiation in Engineering Materials Science*; Reimers, W., Pyzalla, A., Schreyer, A., Clemens, H., Eds.; Wiley-VCH: Weinheim, Germany, 2008; pp. 97–112.
68. Daniels, J.E. Determination of directionally dependent structural and microstructural information using high-energy X-ray diffraction. *J. Appl. Crystallogr.* **2008**, *41*, 1109–1114. [CrossRef]
69. McCusker, L.B. Product characterization by X-ray powder diffraction. *Microporous Mesoporous Mater.* **1998**, *22*, 527 529. [CrossRef]
70. McCusker, L.B.; von Dreele, R.B.; Cox, D.E.; Louër, D.; Scardi, P. Rietveld refinement guidelines. *J. Appl. Crystallogr.* **1999**, *32*, 36–50. [CrossRef]
71. Spieß, L.; Teichert, G.; Schwarzer, R.; Behnken, H.; Genzel, C. *Moderne Röntgenbeugung*, 2nd ed.; Vieweg + Teubner: Wiesbaden, Germany, 2009.
72. Clemens, H.; Boeck, B.; Wallgram, W.; Schmoelzer, T.; Droessler, L.M.; Zickler, G.A.; Leitner, H.; Otto, A. Experimental studies and thermodynamic simulations of phase transformations in Ti–(41–45)Al–4Nb–1Mo–0.1B alloys. In Proceedings of MRS Fall Meeting, Boston, MA, USA, 1–5 December 2008; pp. 115–120.
73. Boeck, B. Untersuchung und Verifizierung des Quasibinären Zustandsdiagramms für β-erstarrende TNM-Legierungen auf Basis von Ti–(41–45)Al–4Nb–1Mo–0.1B. Diploma Thesis, Montanuniversität Leoben, Leoben, Austria, August 2008.
74. Schmoelzer, T.; Liss, K.-D.; Zickler, G.A.; Watson, I.J.; Droessler, L.M.; Wallgram, W.; Buslaps, T.; Studer, A.; Clemens, H. Phase fractions, transition and ordering temperatures in TiAl–Nb–Mo alloys: An *in* and *ex situ* study. *Intermetallics* **2010**, *18*, 1544–1552. [CrossRef]
75. Schloffer, M.; Iqbal, F.; Gabrisch, H.; Schwaighofer, E.; Schimansky, F.-P.; Mayer, S.; Stark, A.; Lippmann, T.; Göken, M.; Pyczak, F.; *et al.* Microstructure development and hardness of a powder metallurgical multi phase γ-TiAl based alloy. *Intermetallics* **2012**, *22*, 231–240. [CrossRef]
76. Grujicic, M.; Zhang, Y. Crystal plasticity analysis of the effect of dispersed β-phase on deformation and fracture of lamellar γ + α₂ titanium aluminide. *Mater. Sci. Eng. A* **1999**, *265*, 285–300. [CrossRef]
77. Liss, K.-D.; Schmoelzer, T.; Yan, K.; Reid, M.; Peel, M.; Dippenaar, R.; Clemens, H. *In situ* study of dynamic recrystallization and hot deformation behavior of a multi-phase titanium aluminide alloy. *J. Appl. Phys.* **2009**. [CrossRef]

78. Sun, F.-S.; Cao, C.-X.; Kim, S.-E.; Lee, Y.-T.; Yan, M.-G. Alloying mechanism of beta stabilizers in a TiAl alloy. *Metall. Mater. Trans. A* **2001**, *32*, 1573–1589. [CrossRef]
79. Kim, S.; Smith, G.D.W.; Roberts, S.G.; Cerezo, A. Alloying elements characterisation in γ-based titanium aluminides by APFIM. *Mater. Sci. Eng. A* **1998**, *250*, 77–82. [CrossRef]
80. Kim, S.; Smith, G.D.W. AP-FIM investigation on γ-based titanium aluminides. *Mater. Sci. Eng. A* **1997**, *239–240*, 229–234. [CrossRef]
81. Yeoh, L.A.; Liss, K.-D.; Bartels, A.; Chladil, H.F.; Avdeev, M.; Clemens, H.; Gerling, R.; Buslaps, T. In situ high-energy X-ray diffraction study and quantitative phase analysis in the α + γ phase field of titanium aluminides. *Scr. Mater.* **2007**, *57*, 1145–1148. [CrossRef]
82. Liss, K.-D.; Bartels, A.; Clemens, H.; Bystrzanowski, S.; Stark, A.; Buslaps, T.; Schimansky, F.-P.; Gerling, R.; Scheu, C.; Schreyer, A. Recrystallization and phase transitions in a γ-TiAl-based alloy as observed by ex situ and in situ high-energy X-ray diffraction. *Acta Mater.* **2006**, *54*, 3721–3735. [CrossRef]
83. Hoelzel, M.; Gilles, R.; Schlapp, M.; Boysen, H.; Fuess, H. Monte Carlo simulations of various instrument configurations of the new structure powder diffractometer (SPODI). *Phys. B Condens. Matter* **2004**, *350*, 671–673. [CrossRef]
84. Gilles, R.; Hoelzel, M.; Schlapp, M.; Elf, F.; Krimmer, B.; Boysen, H.; Fuess, H. First test measurements at the new structure powder diffractometer (SPODI) at the FRM-II. *Z. Krist.* **2006**, *1*, 183–188. [CrossRef]
85. Drössler, L.M. Characterization of β-Solidifying γ-TiAl Alloy Variants Using Advanced in and ex situ Investigation Methods. Diploma Thesis, Montanuniversität Leoben, Leoben, Austria, October 2008.
86. Studer, A.J.; Hagen, M.E.; Noakes, T.J. Wombat: The high-intensity powder diffractometer at the OPAL reactor. *Phys. B Condens. Matter* **2006**, *385–386*, 1013–1015. [CrossRef]
87. Erdely, P.; Werner, R.; Schwaighofer, E.; Clemens, H.; Mayer, S. In situ study of the time-temperature-transformation behaviour of a multi-phase intermetallic β-stabilised TiAl alloy. *Intermetallics* **2015**, *57*, 17–24. [CrossRef]
88. Park, H.S.; Nam, S.W.; Kim, N.J.; Hwang, S.K. Refinement of the lamellar structure in TiAl-based intermetallic compound by addition of carbon. *Scr. Mater.* **1999**, *41*, 1197–1203. [CrossRef]
89. Klein, T.; Schachermayer, M.; Mendez-Martin, F.; Schöberl, T.; Rashkova, B.; Clemens, H.; Mayer, S. Carbon distribution in multi-phase γ-TiAl based alloys and its influence on mechanical properties and phase formation. *Acta Mater.* **2015**, *94*, 205–213. [CrossRef]
90. Kawabata, T.; Tadanoi, M.; Izumi, O. Effect of carbon and nitrogen on mechanical properties of TiAl alloys. *ISIJ Int.* **1991**, *31*, 1161–1167. [CrossRef]
91. Tian, W.H.; Nemoto, M. Effect of carbon addition on the microstructures and mechanical properties of γ-TiAl alloys. *Intermetallics* **1997**, *5*, 237–244. [CrossRef]
92. Christoph, U.; Appel, F.; Wagner, R. Dislocation dynamics in carbon-doped titanium aluminide alloys. *Mater. Sci. Eng. A* **1997**, *239–240*, 39–45. [CrossRef]
93. Appel, F.; Paul, J.D.H.; Oehring, M.; Fröbel, U.; Lorenz, U. Creep behavior of TiAl alloys with enhanced high-temperature capability. *Metall. Mater. Trans. A* **2003**, *34*, 2149–2164. [CrossRef]
94. Appel, F.; Fischer, F.D.; Clemens, H. Precipitation twinning. *Acta Mater.* **2007**, *55*, 4915–4923. [CrossRef]
95. Gabrisch, H.; Stark, A.; Schimansky, F.-P.; Wang, L.; Schell, N.; Lorenz, U.; Pyczak, F. Investigation of carbides in Ti–45Al–5Nb–xC alloys (0 ⩽ x ⩽ 1) by transmission electron microscopy and high energy-XRD. *Intermetallics* **2013**, *33*, 44–53. [CrossRef]
96. Wang, L.; Gabrisch, H.; Lorenz, U.; Schimansky, F.-P.; Schreyer, A.; Stark, A.; Pyczak, F. Nucleation and thermal stability of carbide precipitates in high Nb containing TiAl alloys. *Intermetallics* **2015**, *66*, 111–119. [CrossRef]
97. Hammersley, A.P.; Svensson, S.O.; Hanfland, M.; Fitch, A.N.; Häusermann, D. Two-dimensional detector software: From real detector to idealised image or two-theta scan. *High Press. Res.* **1996**, *14*, 235–248. [CrossRef]
98. Blackburn, M.J. Some aspects of phase transformations in titanium alloys. In *The Science, Technology and Application of Titanium*; Jaffee, R.I., Promisel, N.E., Eds.; Plenum Press: New York, NY, USA, 1970; pp. 639–642.
99. Denquin, A.; Naka, S. Phase transformation mechanisms involved in two-phase TiAl-based alloys—II. Discontinuous coarsening and massive-type transformation. *Acta Mater.* **1996**, *44*, 353–365. [CrossRef]
100. Margulies, L.; Lorentzen, T.; Poulsen, H.F.; Leffers, T. Strain tensor development in a single grain in the bulk of a polycrystal under loading. *Acta Mater.* **2002**, *50*, 1771–1779. [CrossRef]

101. Poulsen, H.F. An introduction to three-dimensional X-ray diffraction microscopy. *J. Appl. Crystallogr.* **2012**, *45*, 1084–1097. [CrossRef]

102. Margulies, L.; Winther, G.; Poulsen, H.F. *In situ* measurement of grain rotation during deformation of polycrystals. *Science* **2001**, *291*, 2392–2394. [CrossRef] [PubMed]

103. Lauridsen, E.M.; Schmidt, S.; Suter, R.M.; Poulsen, H.F. Tracking: A method for structural characterization of grains in powders or polycrystals. *Appl. Crystallogr.* **2001**, *34*, 744–750. [CrossRef]

104. Inui, H.; Matsumuro, M.; Wu, D.-H.; Yamaguchi, M. Temperature dependence of yield stress, deformation mode and deformation structure in single crystals of TiAl (Ti–56 at. % Al). *Philos. Mag. A* **1997**, *75*, 395–423. [CrossRef]

105. Ihara, K.; Miura, Y. Dynamic recrystallization in Al-single crystals revealed by synchrotron radiation Laue technique. *Mater. Sci. Eng. A* **2004**, *387–389*, 651–654. [CrossRef]

106. Miura, Y.; Ihara, K.; Fukaura, K. Dynamic recrystallization in Al single crystals revealed by rapid X-ray Laue method. *Mater. Sci. Eng. A* **2000**, *280*, 134–138. [CrossRef]

107. Stark, A.; Rackel, M.; Tankoua, A.T.; Oehring, M.; Schell, N.; Lottermoser, L.; Schreyer, A.; Pyczak, F. *In situ* high-energy X-ray diffraction during hot-forming of a multiphase TiAl alloy. *Metals* **2015**, *5*, 2252–2265. [CrossRef]

108. Liss, K.-D.; Clemens, H.; Bartels, A.; Stark, A.; Buslaps, T. Phase transitions and recrystallization in a Ti–46 at. % Al-9 at. % Nb alloy as observed by *in situ* high-energy X-ray diffraction. *MRS Proc.* **2007**. [CrossRef]

109. Liss, K.-D.; Stark, A.; Bartels, A.; Clemens, H.; Buslaps, T.; Phelan, D.; Yeoh, L.A. Directional atomic rearrangements during transformations between the α- and γ-phases in titanium aluminides. *Adv. Eng. Mater.* **2008**, *10*, 389–392. [CrossRef]

110. Yan, K.; Liss, K.-D.; Garbe, U.; Daniels, J.; Kirstein, O.; Li, H.; Dippenaar, R. From single grains to texture. *Adv. Eng. Mater.* **2009**, *11*, 771–773. [CrossRef]

111. Liss, K.-D.; Garbe, U.; Li, H.; Schambron, T.; Almer, J.D.; Yan, K. *In situ* observation of dynamic recrystallization in the bulk of zirconium alloy. *Adv. Eng. Mater.* **2009**, *11*, 637–640. [CrossRef]

112. Schmoelzer, T.; Liss, K.-D.; Rester, M.; Yan, K.; Stark, A.; Reid, M.; Peel, M.; Clemens, H. Dynamic recovery and recrystallisation during hot-working in an advanced TiAl alloy. *Pract. Metallogr.* **2011**, *48*, 632–642. [CrossRef]

113. Sahu, P. Lattice imperfections in intermetallic Ti–Al alloys: An X-ray diffraction study of the microstructure by the Rietveld method. *Intermetallics* **2006**, *14*, 180–188. [CrossRef]

114. Schwaighofer, E.; Clemens, H.; Lindemann, J.; Stark, A.; Mayer, S. Hot-working behavior of an advanced intermetallic multi-phase γ-TiAl based alloy. *Mater. Sci. Eng. A* **2014**, *614*, 297–310. [CrossRef]

115. Cha, L.; Clemens, H.; Dehm, G. Microstructure evolution and mechanical properties of an intermetallic Ti–43.5Al–4Nb–1Mo–0.1B alloy after ageing below the eutectoid temperature. *Int. J. Mater. Res.* **2011**, *102*, 1–6. [CrossRef]

116. Cha, L.; Scheu, C.; Clemens, H.; Chladil, H.F.; Dehm, G.; Gerling, R.; Bartels, A. Nanometer-scaled lamellar microstructures in Ti–45Al–7.5Nb–(0; 0.5)C alloys and their influence on hardness. *Intermetallics* **2008**, *16*, 868–875. [CrossRef]

117. Veeraraghavan, D.; Pilchowski, U.; Natarajan, B.; Vasudevan, V.K. Phase equilibria and transformations in Ti–(25–52) at. % Al alloys studied by electrical resistivity measurements. *Acta Mater.* **1998**, *46*, 405–421. [CrossRef]

118. Beschliesser, M.; Chatterjee, A.; Lorich, A.; Knabl, W.; Kestler, H.; Dehm, G.; Clemens, H. Designed fully lamellar microstructures in a γ-TiAl based alloy: Adjustment and microstructural changes upon long-term isothermal exposure at 700 and 800 °C. *Mater. Sci. Eng. A* **2002**, *329–331*, 124–129. [CrossRef]

119. Takeyama, M.; Kumagai, T.; Nakamura, M.; Kikuchi, M. Cooling rate dependence of the α₂/γ phase transformation in titanium aluminides and its application to alloy development. In *Structural Intermetallics 1993*; Darolia, R., Lewandowski, J.J., Liu, C.T., Martin, P.L., Miracle, D.B., Nathal, M.V., Eds.; The Minerals, Metals and Materials Society (TMS): Warrendale, PA, USA, 1993; pp. 167–176.

120. Zhu, H.; Seo, D.Y.; Maruyama, K.; Au, P. Effect of microstructural stability on creep behavior of 47XD TiAl alloys with fine-grained fully lamellar structure. *Scr. Mater.* **2005**, *52*, 45–50. [CrossRef]

121. Tang, J.; Huang, B.; Zhou, K.; Liu, W.; He, Y.; Liu, Y. Factors affecting the lamellar spacing in two-phase TiAl alloys with fully lamellar microstructures. *Mater. Res. Bull.* **2001**, *36*, 1737–1742. [CrossRef]

122. Droessler, L.M.; Schmoelzer, T.; Wallgram, W.; Cha, L.; Das, G.; Clemens, H. Microstructure and tensile ductility of a Ti–43Al–4Nb–1Mo–0.1B alloy. *MRS Proc.* **2009**, *1128*, 121–126.
123. Guyon, J.; Hazotte, A.; Bouzy, E. Evolution of metastable α phase during heating of Ti48Al2Cr2Nb intermetallic alloy. *J. Alloy. Compd.* **2015**, *656*, 667–675. [CrossRef]
124. Zhu, H.; Matsuda, J.; Maruyama, K. Influence of heating rate in α + γ dual phase field on lamellar morphology and creep property of fully lamellar Ti–48Al alloy. *Mater. Sci. Eng. A* **2005**, *397*, 58–64. [CrossRef]
125. Dehm, G.; Motz, C.; Scheu, C.; Clemens, H.; Mayrhofer, P.H.; Mitterer, C. Mechanical size-effects in miniaturized and bulk materials. *Adv. Eng. Mater.* **2006**, *8*, 1033–1045. [CrossRef]
126. Cha, L.; Clemens, H.; Dehm, G.; Zhang, Z. *In situ* TEM heating study of the γ lamellae formation inside the α₂ matrix of a Ti–45Al–7.5Nb alloy. *Adv. Mater. Res.* **2011**, *146–147*, 1365–1368.
127. Clemens, H.; Mayer, S. Micro- and nanostructure evolution in intermetallic titanium aluminides. In Proceedings of the Ti-2011—12th World Conference on Titanium, Beijing, China, 19–24 June 2012; Zhou, L., Chang, H., Lu, Y., Xu, D., Eds.; Science Press: Beijing, China, 2012; pp. 395–403.
128. Schwaighofer, E.; Schloffer, M.; Schmoelzer, T.; Mayer, S.; Lindemann, J.; Güther, V.; Klose, J.; Clemens, H. Influence of heat treatments on the microstructure of a multi-phase titanium aluminide alloy. *Pract. Metallogr.* **2012**, *49*, 124–137. [CrossRef]
129. Clemens, H.; Wallgram, W.; Schloffer, M. Method for Producing a Component and Components of a Titanium-Aluminum Base Alloy. US Patent 2011,0277,891 A1, 17 November 2011.
130. Bolz, S.; Oehring, M.; Lindemann, J.; Pyczak, F.; Paul, J.; Stark, A.; Lippmann, T.; Schrüfer, S.; Roth-Fagaraseanu, D.; Schreyer, A.; *et al.* Microstructure and mechanical properties of a forged β-solidifying γ TiAl alloy in different heat treatment conditions. *Intermetallics* **2015**, *58*, 71–83. [CrossRef]
131. Huber, D.; Werner, R.; Clemens, H.; Stockinger, M. Influence of process parameter variation during thermo mechanical processing of an intermetallic β-stabilized γ-TiAl based alloy. *Mater. Charact.* **2015**, *109*, 116–121. [CrossRef]
132. Kim, Y. Ordered intermetallic alloys, part III: Gamma titanium aluminides. *JOM* **1994**, *46*, 30–39. [CrossRef]
133. Perdrix, F.; Trichet, M.F.; Bonnentien, J.L.; Cornet, M.; Bigot, J. Influence of cooling rate on microstructure and mechanical properties of a Ti–48Al alloy. *Intermetallics* **1999**, *7*, 1323–1328. [CrossRef]
134. Cao, G.; Fu, L.; Lin, J.; Zhang, Y.; Chen, C. The relationships of microstructure and properties of a fully lamellar TiAl alloy. *Intermetallics* **2000**, *8*, 647–653. [CrossRef]
135. Djanarthany, S.; Servant, C.; Penelle, R. Phase transformations in Ti₃Al and Ti₃Al + Mo aluminides. *J. Mater. Res.* **1991**, *6*, 969–986. [CrossRef]
136. Djanarthany, S.; Servant, C.; Penelle, R. Influence of an increasing content of molybdenum on phase transformations of Ti–Al–Mo aluminides—Relation with mechanical properties. *Mater. Sci. Eng. A* **1992**, *152*, 48–53. [CrossRef]
137. Schnitzer, R.; Chladil, H.F.; Scheu, C.; Clemens, H.; Bystrzanowski, S.; Bartels, A.; Kremmer, S. The production of lamellar microstructures in intermetallic TiAl alloys and their characterisation. *Pract. Metallogr.* **2007**, *44*, 430–442. [CrossRef]
138. Novoselova, T.; Malinov, S.; Sha, W. Experimental study of the effects of heat treatment on microstructure and grain size of a gamma TiAl alloy. *Intermetallics* **2003**, *11*, 491–499. [CrossRef]
139. Kenel, C.; Leinenbach, C. Influence of Nb and Mo on microstructure formation of rapidly solidified ternary TiAl–(Nb,Mo) alloys. *Intermetallics* **2016**, *69*, 82–89. [CrossRef]

![metals logo] *metals*

MDPI

Article

In Situ High-Energy X-ray Diffraction during Hot-Forming of a Multiphase TiAl Alloy

Andreas Stark [1,*], Marcus Rackel [1], Aristide Tchouaha Tankoua [2], Michael Oehring [1], Norbert Schell [1], Lars Lottermoser [1], Andreas Schreyer [1] and Florian Pyczak [1]

[1]   Institute of Materials Research, Helmholtz-Zentrum Geesthacht, Geesthacht 21502, Germany;
      marcus.rackel@hzg.de (M.R.); michael.oehring@hzg.de (M.O.); norbert.schell@hzg.de (N.S.);
      lars.lottermoser@hzg.de (L.L.); andreas.schreyer@hzg.de (A.S.); florian.pyczak@hzg.de (F.P.)
[2]   Institute of Materials Physics and Technology, Hamburg University of Technology,
      Hamburg 21073, Germany; aristidetchouaha@hotmail.com
*   Author to whom correspondence should be addressed; andreas.stark@hzg.de; Tel.: +49-4152-87-2663;
      Fax: +49-4152-87-2534.

Academic Editor: Klaus-Dieter Liss
Received: 28 September 2015; Accepted: 23 November 2015; Published: 30 November 2015

Abstract: Intermetallic γ-TiAl based alloys exhibit excellent high-temperature strength combined with low density. This makes them ideal candidates for replacing the twice as dense Ni base super-alloys, currently used in the medium temperature range (~700 °C) of industrial and aviation gas turbines. An important step towards the serial production of TiAl parts is the development of suitable hot-forming processes. Thermo-mechanical treatments often result in mechanical anisotropy due to the formation of crystallographic textures. However, with conventional texture analysis techniques, their formation can only be studied after processing. In this study, *in situ* high-energy X-ray diffraction measurements with synchrotron radiation were performed during hot-forming. Thus, it was possible to record the evolution of the phase constitution as well as the formation of crystallographic texture of different phases directly during processing. Several process temperatures (1100 °C to 1300 °C) and deformation rates were investigated. Based on these experiments, a process window can be recommended which results in the formation of an optimal reduced texture.

Keywords: crystallographic texture; X-ray diffraction; synchrotron radiation; intermetallic alloy; titanium aluminides based on γ-TiAl; hot-forming; thermo-mechanical processing; phase constitution

---

1. Introduction

The demand to reduce both fuel consumption and greenhouse gas emissions from gas turbines and combustion engines is continuously increasing. This requires the development of novel lightweight materials which can withstand extreme conditions, like high stresses at elevated temperatures. Intermetallic γ-TiAl based alloys are the most promising candidates among these materials. They have, for example, similar high temperature strength and creep resistance to the currently used Ni base superalloys but only half of their density [1]. One recent, first industrial application of TiAl alloys is as low pressure turbine blade material in civil aircraft engines at service temperatures up to about 700 °C [2]. Great efforts are made to develop a suitable hot forming processes, e.g., forging routes, for serial production of TiAl parts [3–6]. Thus, research activities are currently focused on TiAl alloys with additional amounts of β-Ti(Al) stabilizing elements like Nb or Mo because the ductile bcc high-temperature β phase improves the formability at elevated temperatures [1,7].

Thermo-mechanical treatments (TMT), such as hot-rolling or forging, are well-established processes to improve mechanical properties and to homogenize the microstructure of metals and alloys as well as for near net-shape production. However, TMT can also produce unwanted mechanical anisotropy due to the formation of crystallographic texture. Thus, the study of texture formation is of great technological interest. If texture formation is understood, it could be used to control texture evolution in the future. Hot forming of γ-TiAl based alloys often takes place in phase fields with different phase compositions to those at room or service temperatures. But with conventional texture analysis techniques the texture formed can only be studied by post process metallographic methods [8,9]. This means that the real high temperature material conditions are often masked by lower temperature phase transformations or recrystallization.

First texture measurements after hot compression of γ-TiAl based alloys were performed by Fukutomi *et al.* [10] and Hartig *et al.* [11]. They discussed either pure deformation or dynamic recrystallization (DRX) as the dominant texture forming mechanism. Computer simulations of γ-TiAl deformation texture development, based on the single crystal yield surface model of γ-TiAl according to Mecking *et al.* [12] later allowed a clear discrimination between the above mentioned texture formation mechanisms. After hot forming on an industrial scale, e.g., hot-rolled TiAl sheets, γ-TiAl textures were observed that could also be described as a mixture of deformation and DRX components [13]. However hot forming of TiAl often takes place at temperatures and in phase fields with α-Ti(Al) as the main phase. For a long time little was known about the texture formation of α-Ti(Al) and its influence on the texture of γ-TiAl, to which it transforms at lower temperatures. The α-Ti(Al) texture was measured and studied for the first time by Schillinger *et al.* [14] and Stark *et al.* [15] using oil quenched as-rolled TiAl samples.

Over the last decade, new high-energy synchrotron sources were constructed, which, in combination with advanced sample environments, provide novel tools and analysis methods for engineering and materials science [16–18]. Such synchrotron sources offer the possibility for time-resolved *in situ* studies during materials processing [19,20]. In the current paper we have used a deformation dilatometer that has been modified for working in the synchrotron beamline to perform hot-compression experiments. *In situ* high-energy X-ray diffraction (HEXRD) measurements were performed during hot-forming. This setup enables an *in situ* observation of the interaction and evolution of several microstructural parameters during processing. In particular, we can directly observe the evolution of crystallographic texture, phase fractions, or grain size during deformation and simultaneously record the process parameters, like temperature, force, and length change. Thus, we have been able to systematically analyze texture evolution of a multi-phase alloy in different phase fields, both *in situ* and time resolved.

2. Experimental Section

We studied a novel Nb rich γ-TiAl based alloy with a nominal composition of Ti-42Al-8.5Nb (in at. %). In order to start with chemically homogeneous and texture free samples the alloy was powder metallurgically produced using the EIGA technique (Electron Induction Melting Gas Atomization) [21]. Alloy powder with a particle size up to 180 μm was filled under Ar atmosphere into Ti cans, which subsequently were degassed and sealed gas tight. These cans were hot-isostatically pressed (HIPed) for 2 h at 1250 °C at 200 MPa. The HIPed material contained about 500 μg/g oxygen and 110 μg/g nitrogen. Cylindrical samples 4 and 5 mm in diameter and 10 mm length were cut by spark erosion from the HIPed powder compact.

Several heating and hot compression tests were performed using a DIL 805A/D quenching and deformation dilatometer (TA Instruments, Hüllhorst, Germany) that had been modified for working in the Helmholtz-Zentrum Geesthacht synchrotron beamline HEMS at PETRA III, DESY (Hamburg, Germany) [17,22]. This setup is displayed in Figure 1a. In order to avoid sample oxidation the experiments were performed in an Ar atmosphere. The temperature was controlled by a spot welded type S thermocouple. The phase constitution of the Ti-42Al-8.5Nb alloy was measured during heating up to 1400 °C using a heating rate of 10 °C·min^{-1}. An additional sample was heated up in 50 °C steps from 900 °C to 1300 °C. The sample was held at each temperature for at least 30 min (below 1100 °C for 1 h) to come closer to equilibrium conditions. Five temperatures between

1100 °C and 1300 °C were selected for the hot forming experiments in order to study the deformation behavior in different phase fields. A sketch with the process parameters is shown in Figure 1b. The specimens were heated up to the processing temperature at 200 °C·min^{-1} followed by isothermal holding at temperature for 5–10 min. Subsequently the specimens were deformed with deformation rates between 5×10^{-3}·s^{-1} and 3×10^{-2}·s^{-1} up to a total deformation of $\varphi = -0.5$ corresponding to about 40% height reduction. Immediately after deformation, within 1 s, the samples were quenched at a cooling rate of about 100 °C·s^{-1} by blowing with Ar gas, in order to keep the deformed microstructure.

During the experiments, high-energy X-ray diffraction (HEXRD) was performed in transmission geometry (Figure 1c). In order for a X-ray transmission through the samples a high-energy X-ray beam with an energy of 100 keV (corresponding to a wavelength of 0.124 Å) was used. The beam size was 1 mm × 1 mm. The resulting Debye-Scherrer diffraction rings were continuously recorded during deformation on a Perkin Elmer XRD 1622 (Perkin Elmer, Norwalk, CT, USA) flat panel detector with acquisition rates up to 5 Hz and an exposure time of 0.2–1 s. In order to calculate the instrumental parameters a calibration measurement was done using standardized lanthanum hexaboride (LaB$_6$) powder. The diffraction rings were azimuthally integrated using FIT2D software (ESRF, Grenoble, France) [23]. Phase fractions and crystallographic textures were determined using the MAUD program (University of Trento, Trento, Italy) [24]. The Rietveld texture analysis method implemented in the Rietveld program MAUD enables refinement of both the phase parameters and the orientation distribution functions (ODF), which were calculated by means of the E-WIMV approach. Thus the effect of texture is taken into account while refining phase parameters and fractions. Changes in the phase contents up to ±3 vol. % during deformation might be ascribed to temperature oscillations.

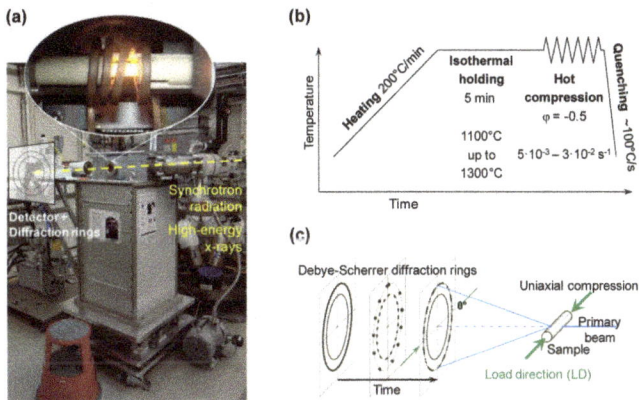

Figure 1. (**a**) The deformation dilatometer mounted in the HEMS beamline. The insert shows the interior of the measurement chamber with a heated sample just before hot forming. Schematic diagrams showing; (**b**) the compression experiments and processing parameters; and (**c**) the diffraction geometry.

Microstructural analysis was performed on deformed and quenched samples that had been cut and vibration polished. Scanning electron microscopy (SEM) was performed in the back scattered electron (BSE) mode in a Zeiss Gemini electron microscope with field emission gun (Oberkochen, Germany).

3. Results and Discussion

3.1. Development of Phase Constitution with Temperature for Ti-42Al-8.5Nb

The phase diagram of Nb rich γ-TiAl based alloys is rather complex and still under research [25,26] even though it is composed of only three elements. At temperatures up to service temperature (about 700 °C to 800 °C) γ-TiAl alloys mainly consist of the tetragonal γ-TiAl phase ($L1_0$ structure, P4/mmm) and the hexagonal α_2-Ti$_3$Al phase (DO_{19} structure, P6$_3$/mmc). The ordered α_2 phase transforms to the disordered hcp α-Ti(Al) phase (A3 structure, P6$_3$/mmc) at higher temperatures. Nb additionally stabilizes the cubic disordered bcc β-Ti(Al,Nb) phase (A2 structure, Im-3m) at high temperatures which can transform during cooling first to the ordered β_o-TiAl(Nb) phase (B2 structure, Pm-3m) and then at lower temperatures to the ω_o-Ti$_4$Al$_3$Nb phase (B8$_2$ structure, P6$_3$/mmc). Recent studies have shown that an orthorhombic phase (Cmcm) can also appear at temperatures around 600 °C [27]. More details regarding the structure and formation conditions of this orthorhombic phase will be published elsewhere.

Figure 2a shows the evolution of the diffraction patterns during heating between 900 °C and 1300 °C. The specific reflections of various phases are indicated above. From the results, the phase fractions over the temperature range were calculated by Rietveld analysis, see Figure 2b. The starting material, *i.e.*, the HIPed powder compact, consists of 51 vol. % γ-TiAl, 48 vol. % α_2-Ti$_3$Al and 1 vol. % ω_o-Ti$_4$Al$_3$Nb at room temperature. During heating, the ω_o transforms to β_o at 920 °C. The loss of the 100-α_2 and 101-α_2 superlattice reflections at 1155 °C and the 100-β_o superlattice reflection at 1170 °C indicate the $\alpha_2 \rightarrow \alpha$ and $\beta_o \rightarrow \beta$ order-disorder transformations. Above 1150 °C, the amount of γ significantly decreases until the γ solvus temperature $T_{\gamma sol}$ is reached at 1236 °C, indicating the end of the $\gamma \rightarrow \alpha$ transformation. The transformation temperatures determined from this measurement are slightly shifted to higher values, compared to equilibrium conditions, due to the continuous heating. In order to converge to the equilibrium conditions, the results were compared to a stepwise heating experiment holding the sample at each temperature for at least 30 min. The phase fractions calculated at the end of each heating step are plotted in Figure 2b. In the temperature range between 1000 °C and 1200 °C the amount of γ and β/β_o increases at the expense of α/α_2, while holding at temperature compared to continuous heating. Because the high Nb content impedes diffusion, the formation of phase equilibrium needs some time. It is interesting to note that under these conditions the β phase fraction has a local peak at about 1150 °C. The differences between both measurements at temperatures above 1200 °C might be caused by intensive grain growth, while holding at temperature.

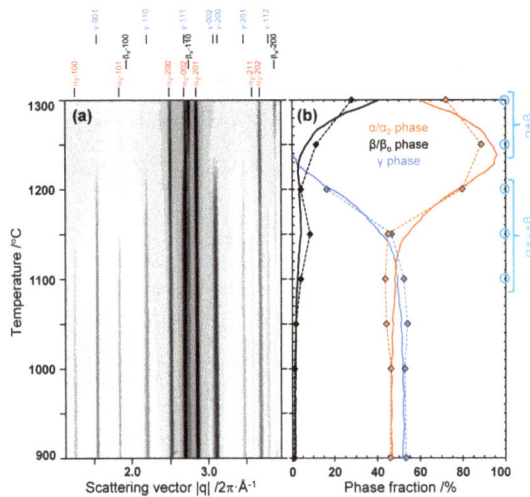

Figure 2. Phase constitution as a function of temperature for Ti-42Al-8.5Nb. (**a**) Evolution of the diffraction patterns during heating; (**b**) Phase fractions determined by Rietveld analysis. Continuous lines: heating rate of 10 °C·min^{-1}. Dashed lines: stepwise heating.

From the results we have chosen five working temperatures representing different phase fields and different phase contents for the hot forming experiments. Three experiments were performed in the (γ + α/α_2 + β/β_0) 3-phase field, one of them at 1150 °C, *i.e.*, at the relative β peak and another at 1200 °C, where the γ content is very low. Two additional temperatures were chosen in the (α + β) 2-phase field, representing a low or a high β content.

3.2. Microstructures before and after Deformation

The SEM images in Figure 3 show characteristic microstructures of the Ti-42Al-8.5Nb alloy before and after deformation. Different phases can be distinguished in the micrographs due to their different brightness in the BSE mode. The γ phase is imaged as dark grey whereas the α/α_2 phase appears light grey and the β phase and its derivates are almost white.

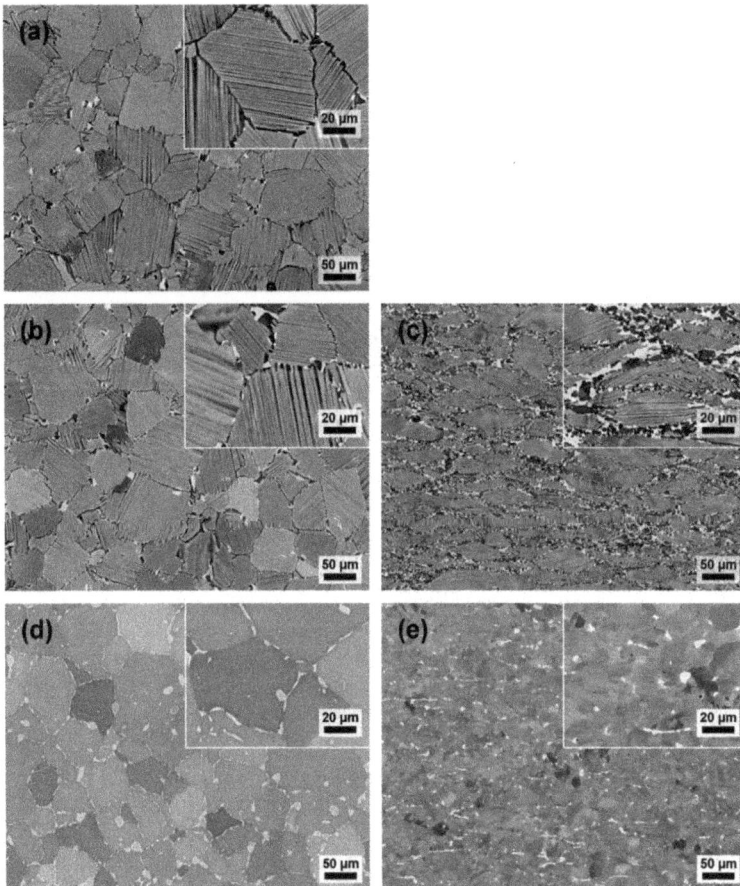

Figure 3. SEM micrographs taken in the BSE mode. (**a**) HIPed starting material at room temperature; (**b**–**e**) Microstructures of specimens from two hot forming experiments heated up to process temperatures of (**b,c**) 1150 °C and (**d,e**) 1250 °C and quenched (**b,d**) before and (**c,e**) after deformation. The γ phase appears dark grey, α_2/α as light grey, and β_0/β almost white. The load direction is vertical ↓.

The starting material (Figure 3a) consists of large ($\alpha_2 + \gamma$) lamellar colonies with a diameter of 50–100 μm. Additionally, a few small globular γ and β grains can be observed at triple points and colony boundaries. This relatively coarse grained microstructure can be attributed to HIPing at 1250 °C which is almost in a single-phase α phase field and the subsequent slow furnace cooling.

Figure 3b–e show microstructures of two representative deformation experiments, one performed in the ($\gamma + \alpha_2 + \beta_o$) three phase field at 1150 °C (Figure 3b,c) and the second performed above the γ solvus temperature $T_{\gamma sol}$ in the ($\alpha + \beta$) two phase field at 1250 °C (Figure 3d,e). To get an impression of the microstructure immediately before deformation two samples were heated up, held at temperature, and then quenched (Figure 3b,d). At 1150 °C (Figure 3b), the microstructure still consists of lamellar colonies, however, the lamellae have coarsened and the amount of α has increased. At 1250 °C (Figure 3d), the microstructure consists of large globular α grains, *i.e.*, the former ($\alpha_2 + \gamma$) lamellar colonies, small globular β grains both within and between the α grains, and fine β precipitates that almost completely decorate the α grain boundaries.

At both temperatures the samples were deformed with a compression rate of $5 \times 10^{-3} \cdot \mathrm{s}^{-1}$ up to a total reduction of about 40% and subsequently quenched in order to retain the deformed microstructure (Figure 3c,e). After deformation at 1150 °C (Figure 3c) the lamellar microstructure has started to recrystallize. The lamellar colonies are elongated perpendicular to the compression direction and exhibit a high aspect ratio. They are surrounded by fine grained, dynamically recrystallized areas. It is clearly visible that DRX starts at kinks within lamellar colonies and at colony boundaries, both places with an increased dislocation density. After deformation at 1250 °C (Figure 3e) the microstructure is completely dynamically recrystallized and significantly refined. No aspect ratio is apparent, bended grain boundaries indicate a bulging mechanism.

3.3. Texture Formation

During the hot forming experiments the diffraction rings were continuously recorded. This has opened up the possibility to observe the formation and evolution of the crystallographic texture in the TiAl specimens both *in situ* and time resolved. Figure 4 shows unrolled Debye Scherrer rings at different stages during the deformation experiments at 1150 °C and 1250 °C. In order to ease analysis of the diffraction rings with respect to the different phases, the rings were unrolled to lines from 0° to 360° and the respective phase reflections identified, as indicated in Figure 4. The state immediately before deformation is represented by Figure 4a,d. The rings are spotty indicating the relatively coarse grained microstructure as shown in Figure 3b,d. The spots are equally distributed along the rings and no preferred orientation is visible. After 10% deformation (Figure 4b,e) the spots have azimuthally broadened (*i.e.*, along the ring) and thus the rings have become continuous. This can be attributed to the increase of crystal mosaicity and generation of small angle grain boundaries at the beginning of plastic deformation [19,20]. Additionally the grains also start to rotate due to dislocation motion. Depending on their specific slip systems the grains rotate into preferred orientations [28] which depend on the applied stress direction during deformation. This results in a shift of intensity along the rings. After 30% deformation, see Figure 4c,f, the intensity accumulation at preferred orientations is clearly visible which indicates the formation of the deformation texture.

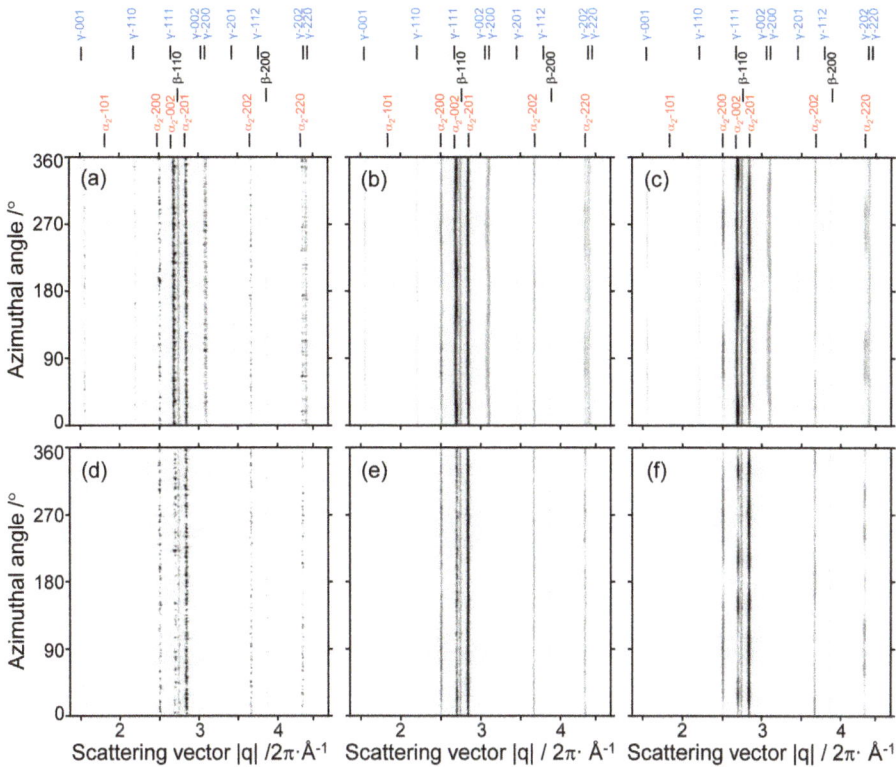

Figure 4. Unrolled diffraction rings taken during two hot forming experiments at (**a–c**) 1150 °C; (**d–f**) 1250 °C; (**a,d**) before deformation; (**b,e**) after 10%; and (**c,f**) after 30% deformation. The reflections of the various phases are indicated at the top of the figure.

A simple and effective way to illustrate the crystallographic evolution during deformation is through the use of azimuth angle *vs.* time plots (AT plots) as introduced by Liss *et al.* [20,29]. Such diagrams are constructed from a specific unrolled ring that is repeatedly plotted from the sequence of diffraction ring images that are collected over increasing time. The time resolved intensity changes along this ring represent the evolution of grain size and crystallographic texture during deformation. Figure 5 shows AT plots for the 002 reflection of the hexagonal α phase during hot forming experiments at 1150 °C, 1200 °C, and 1250 °C. The simultaneously measured parameters of force and deformation are displayed below the diagrams. In order to allow a quantitative comparison of different deformation conditions, the intensities have been normalized to multiples of a random distribution (mrd). As mentioned above, the sharp spots before deformation can be attributed to relatively coarse grains or lamellar colonies and indicate almost perfect crystals. They are equally azimuthally distributed and the intensity distribution does not significantly change during elastic deformation. As plastic flow starts, the spots become more and more diffuse indicating an increasing dislocation density, number of crystal defects, and tilting between crystallite blocks. Obviously, many individual spots move continuously to preferred orientations due to rotation of the crystal lattice of these grains. After about 15%–20% deformation, almost symmetrical intensity maxima are formed at certain angular distances from the load direction indicating the formation of the deformation texture.

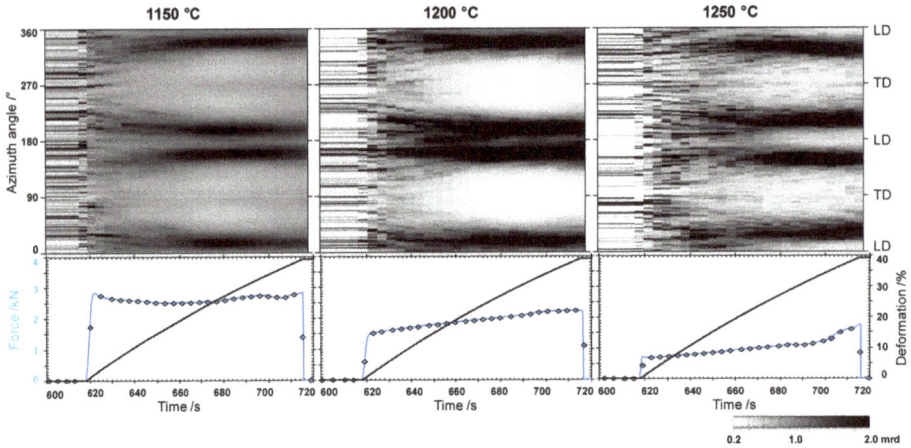

Figure 5. Azimuth angle *vs.* time diagrams for the α-002 reflection during hot compression at 1150 °C, 1200 °C, and 1250 °C using a compression rate of $5 \times 10^{-3} \cdot s^{-1}$. The azimuthal orientation distribution is coded using a greyscale. The process parameters of force and deformation are indicated below the diagrams. LD: Load direction. TD: Transverse direction.

Besides the relatively similar intensity evolution, some specific differences can be observed between the different deformation temperatures. In the 002-α AT plot for deformation at 1200 °C the intensity is concentrated at 20° from the load direction, and regions between these preferred orientations are almost intensity free. During deformation at 1250 °C, new spots continuously occur in between preferred orientations indicating the formation of dynamically recrystallized grains which then start to rotate again towards the preferred orientations. Interestingly, the dynamically recrystallized nuclei apparently have a different orientation to the highly deformed grains, while the texture of this specimen is determined by deformation. Additionally, the angle between the load direction and the preferred orientation increases to about 30°. This indicates a variation of the deformation mechanism most probably due to increased DRX and/or due to the change from a 3-phase to a 2-phase field. During deformation at 1150 °C a weak, additional preferred orientation arises in the 002-α AT plot at the transverse direction (TD). Unfortunately, the peaks of 002-α and 111-γ overlap. Thus it is not possible to separate the intensity contribution of each phase on the ring. Due to the increased amount of γ at lower temperatures, like 1150 °C, this AT plot shows a combination of α and γ during deformation texture evolution.

AT plots can reveal some unique, direct information which cannot be gathered by other methods, for example, different stages of texture formation during deformation or different predominant deformation mechanisms [29]. AT plots are suitable for single-phase materials, like Mg alloys [30], and multiphase alloys, as long as no overlapping reflections are used, and reflections with a low multiplicity, like 002-α. However, AT plots reach their limit when describing real textures. For real texture analysis one has to calculate the ODF and discuss the texture by means of inverse or recalculated pole figures. In our experiments this was possible because we performed uniaxial deformation and the load direction was parallel to the detector plane (Figure 1c). Under these circumstances the intensity distribution around the rings after texture formation is axially symmetric and thus the complete texture information is obtained within one single detector image, without any need for additional sample rotations.

Figure 6 shows the evolution of the α phase deformation texture during the above discussed experiments, using inverse pole figures in load direction. They show the frequency distribution of crystallographic directions in the load direction. The inverse pole figures after deformation of 15%, 20%, and 30% are presented for each forming temperature. Obviously, the intensity and sharpness of the deformation texture increases during each experiment. The inverse pole figures show that only directions within a certain distance to the c axis [0001] are aligned in load direction. This is the typical appearance of the tilted basal fiber texture as sketched on the right side in Figure 6. It is a typical deformation texture that is formed during compression of hexagonal phases like the α phase [31].

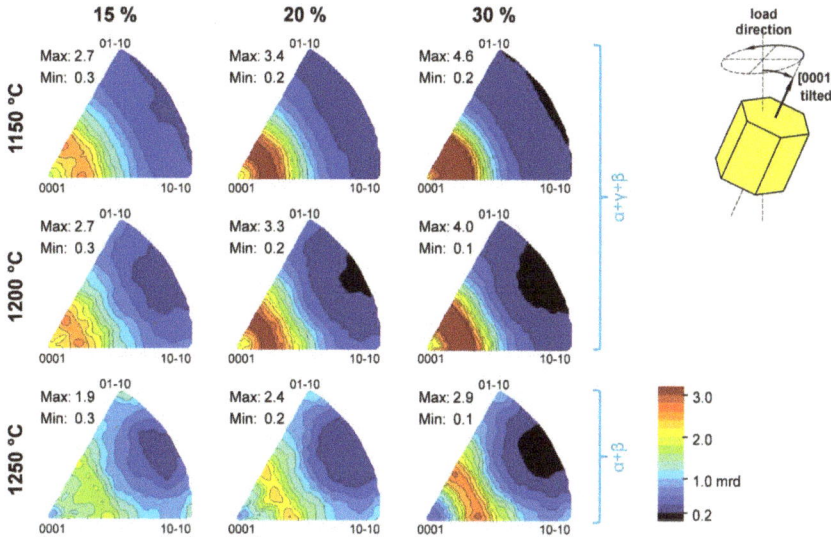

Figure 6. Evolution of the α phase deformation texture during hot compression at different forming temperatures represented as inverse pole figures in load direction after 15%, 20%, and 30% deformation. The sketch on the right illustrates the tilted basal fiber texture which is typically formed.

Figure 7 illustrates inverse pole figures, obtained after 30% deformation at different temperatures, as well as with different deformation rates. The upper part shows the α phase which is the dominant phase during deformation within the temperature range employed. The lower part of the figure shows inverse pole figures for the γ phase for tests performed at 1100 °C and 1150 °C and of the β phase at 1300 °C. The inverse pole figures for γ and β could be calculated due to their high volume fraction at these temperatures. The deformation texture of the tetragonal γ phase consists of weak <110> and <302> fibers. These are typical compression texture components of γ which have already been described in many previous articles [8,13]. The cubic β phase forms significant <001> and <111> fibers which are typical compression texture components of bcc metals. A significant difference in the α deformation texture can be observed between experiments performed in the 3-phase and in the 2-phase fields. With the higher temperatures in the 2-phase field, the texture becomes weaker and more diffuse and the tilt angle between the load direction and the intensity maximum increases. This can be attributed to the larger amount of ductile β phase which has a higher number of slip systems than the hexagonal α phase. Thus, the β phase takes over an increasing part of plastic deformation. This might also explain the strong β deformation texture formed during the experiments at 1300 °C.

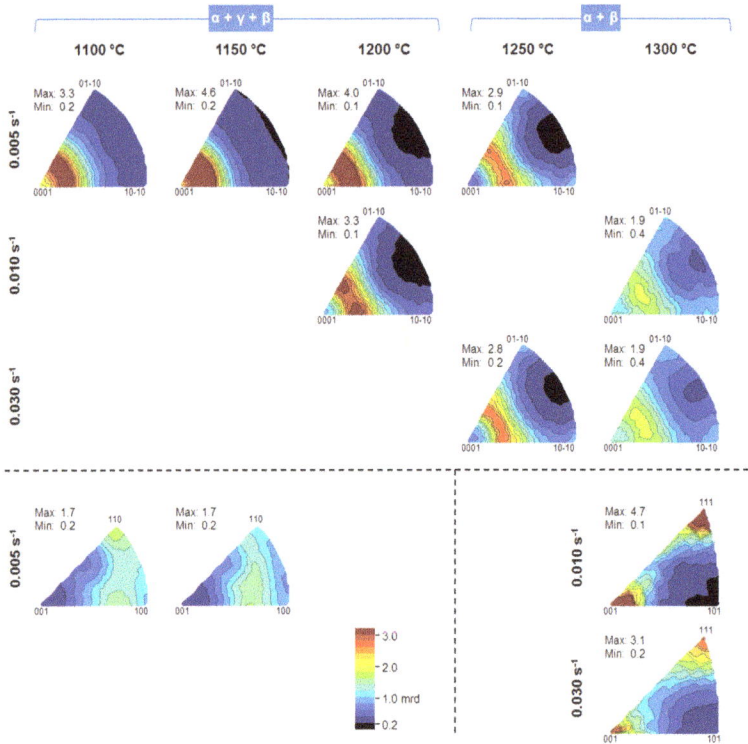

Figure 7. Inverse pole figures in load direction after 30% deformation showing the deformation textures formed during hot compression within different phase fields. Upper figures: the hexagonal α phase. Lower figures: the tetragonal γ phase and the cubic β phase, respectively.

From a technical point of view the texture of hot formed TiAl components should be as weak as possible. Thus, based on these results it is recommended that hot forming of novel low-Al, high-Nb γ-TiAl based alloys should be performed at temperatures just above the γ solvus temperature. In this temperature region the α deformation texture starts to weaken and the β phase fraction is small enough so that it does not contribute significantly to the deformation texture.

4. Summary

In situ high-energy XRD experiments have been performed on a Nb rich γ-TiAl based alloy with a nominal composition of Ti-42Al-8.5Nb during heating and hot compression. The experiments facilitate a direct observation of the high-temperature state that is not masked by post process alterations.

The phase constitution was directly recorded using HEXRD during heating, and different phase fields were identified for the hot forming experiments.

Formation of the deformation texture at elevated temperatures could be directly observed during the experiments. To our knowledge this study is one of very few *in situ* HEXRD texture investigations that have been published. The basic deformation texture is formed within the first 15%–20% of deformation. The deformation texture of the main component, the α phase, is significantly weaker in the high-temperature 2-phase field compared to the low temperature 3-phase field. This can be attributed to the higher amount of ductile bcc β phase as well as to increased DRX. It has been observed that dynamically recrystallized nuclei were formed with new orientations, which were rotated during further deformation towards orientations of the deformation texture.

This study demonstrates that *in situ* synchrotron radiation experiments can be a powerful tool in developing suitable process parameters especially for hot-forming of multiphase alloys, since it allows a simultaneous analysis of the constitution as well as microstructural and textural changes.

Acknowledgments: The authors thank Frank-Peter Schimansky and Dirk Matthiessen for producing the alloy powder, René Kirchhof for technical assistance at the beamline HEMS at DESY, and Jonathan Paul for fruitful discussions.

Author Contributions: Andreas Stark, Michael Oehring, Andreas Schreyer and Florian Pyczak designed the experiments. The *in situ* HEXRD experiments were performed by Andreas Stark, Michael Oehring, Marcus Rackel and Aristide Tchouaha Tankoua and they were supported by Lars Lottermoser, who programmed the special beamline makros, and Norbert Schell, who is responsible for the HEMS beamline. Aristide Tchouaha Tankoua prepared the samples and performed the SEM measurements together with Marcus Rackel. Marcus Rackel analyzed the XRD data and Andreas Stark performed the texture analysis. The manuscript was written by Andreas Stark and intensively discussed with Michael Oehring and Florian Pyczak.

Conflicts of Interest: The authors declare no conflict of interest.

References

1. Appel, F.; Paul, J.D.H.; Oehring, M. *Gamma Titanium Aluminide Alloys: Science and Technology*; Wiley-VCH: Weinheim, Germany, 2011.
2. Bewlay, B.P.; Weimer, M.; Kelly, T.; Suzuki, A.; Subramanian, P.R. The Science, Technology, and Implementation of TiAl Alloys in Commercial Aircraft Engines. *MRS Proc.* **2013**, *1516*, 49–58. [CrossRef]
3. Bolz, S.; Oehring, M.; Lindemann, J.; Pyczak, F.; Paul, J.; Stark, A.; Lippmann, T.; Schrüfer, S.; Roth-Fagaraseanu, D.; Schreyer, A.; *et al.* Microstructure and mechanical properties of a forged β-solidifying γ TiAl alloy in different heat treatment conditions. *Intermetallics* **2015**, *58*, 71–83. [CrossRef]
4. Schwaighofer, E.; Clemens, H.; Lindemann, J.; Stark, A.; Mayer, S. Hot-working behavior of an advanced intermetallic multi-phase γ-TiAl based alloy. *Adv. Mater. Sci. Eng. A* **2014**, *614*, 297–310. [CrossRef]
5. Cheng, L.; Xue, X.; Tang, B.; Kou, H.; Li, J. Flow characteristics and constitutive modeling for elevated temperature deformation of a high Nb containing TiAl alloy. *Intermetallics* **2014**, *49*, 23–28. [CrossRef]
6. Liang, X.-P.; Liu, Y.; Li, H.-Z.; Zhou, C.-X.; Xu, G.-F. Constitutive relationship for high temperature deformation of powder metallurgy Ti-47Al-2Cr-2Nb-0.2W alloy. *Mater. Des.* **2012**, *37*, 40–47. [CrossRef]
7. Clemens, H.; Mayer, S. Design, Processing, Microstructure, Properties, and Applications of Advanced Intermetallic TiAl Alloys. *Adv. Eng. Mater.* **2013**, *15*, 191–215. [CrossRef]
8. Stark, A.; Bartels, A.; Clemens, H.; Kremmer, S.; Schimansky, F.-P.; Gerling, R. Microstructure and Texture Formation During Near Conventional Forging of an Intermetallic Ti-45Al-5Nb Alloy. *Adv. Eng. Mater.* **2009**, *11*, 976–981.
9. Stark, A.; Schimansky, F.-P.; Clemens, H. Texture Formation during Hot-Deformation of High-Nb Containing γ-TiAl Based Alloys. *Solid State Phenom.* **2010**, *160*, 301–306. [CrossRef]
10. Fukutomi, H.; Hartig, C.; Mecking, H. Change of Microstructure in a TiAl Intermetallic Compound during High Temperature Deformation. *Z. Met.* **1990**, *81*, 272–277.
11. Hartig, C.; Fukutomi, H.; Mecking, H.; Aoki, K. Texture and Microstructure of Ti-49 at. % Al after Dynamic Recrystallization and Annealing. *ISIJ Int.* **1993**, *33*, 313–320. [CrossRef]
12. Mecking, H.; Hartig, C.; Kocks, U.F. Deformation modes in γ-TiAl as derived from the single crystal yield surface. *Acta Mater.* **1996**, *44*, 1309–1321. [CrossRef]
13. Bartels, A.; Schillinger, W.; Grassl, G.; Clemens, H. Texture formation in γ-TiAl sheets. In *Gamma Titanium Aluminides 2003*; Kim, Y.-W., Clemens, H., Rosenberger, A.H., Eds.; TMS: Warrendale, PA, USA, 2003; pp. 275–286.
14. Schillinger, W.; Bartels, A.; Gerling, R.; Schimansky, F.-P.; Clemens, H. Texture evolution of the γ- and the α/α₂-phase during hot rolling of γ-TiAl based alloys. *Intermetallics* **2006**, *14*, 336–347. [CrossRef]
15. Stark, A.; Bartels, A.; Gerling, R.; Schimansky, F.-P.; Clemens, H. Microstructure and Texture Formation during Hot Rolling of Niobium-Rich Gamma TiAl Alloys with Different Carbon Contents. *Adv. Eng. Mater.* **2006**, *8*, 1101–1108. [CrossRef]
16. Liss, K.-D.; Bartels, A.; Schreyer, A.; Clemens, H. High-Energy X-rays: A tool for Advanced Bulk Investigations in Materials Science and Physics. *Textures Microstruct.* **2003**, *35*, 219–252. [CrossRef]

17. Staron, P.; Fischer, T.; Lippmann, T.; Stark, A.; Daneshpour, S.; Schnubel, D.; Uhlmann, E.; Gerstenberger, R.; Camin, B.; Reimers, W.; *et al. In situ* experiments with synchrotron high-energy X-rays. *Adv. Eng. Mater.* **2011**, *13*, 658–663. [CrossRef]

18. Reimers, W.; Pycalla, A.R.; Schreyer, A.; Clemens, H. (Eds.) *Neutrons and Synchrotron Radiation in Engineering Materials Science*; Wiley-VCH: Weinheim, Germany, 2008.

19. Liss, K.-D.; Bartels, A.; Clemens, H.; Bystrzanowski, S.; Stark, A.; Buslaps, T.; Schimansky, F.-P.; Gerling, R.; Scheu, C.; Schreyer, A. Recrystallization and phase transitions in a γ-TiAl-based alloy as observed by *ex situ* and *in situ* high-energy X-ray diffraction. *Acta Mater.* **2006**, *54*, 3721–3735. [CrossRef]

20. Liss, K.-D.; Yan, K. Thermo-mechanical processing in a synchrotron beam. *Mater. Sci. Eng. A* **2010**, *528*, 11–27. [CrossRef]

21. Gerling, R.; Clemens, H.; Schimansky, F.-P. Powder metallurgical processing of intermetallic Gamma Titanium Aluminides. *Adv. Eng. Mater.* **2004**, *6*, 23–38. [CrossRef]

22. Schell, N.; King, A.; Beckmann, F.; Fischer, T.; Müller, M.; Schreyer, A. The High Energy Materials Science Beamline (HEMS) at PETRA III. *Mater. Sci. Forum* **2014**, *772*, 57–61. [CrossRef]

23. Hammersley, A.P.; Svensson, S.O.; Hanfland, M.; Fitch, A.N.; Häusermann, D. Two-dimensional detector software: From real detector to idealised image or two-theta scan. *High Press. Res.* **1996**, *14*, 235–248. [CrossRef]

24. Lutterotti, L.; Bortolotti, M.; Ischia, G.; Lonardelli, I.; Wenk, H.-R. Rietveld texture analysis from diffraction images. *Z. Krist.* **2007**, *26*, 125–130. [CrossRef]

25. Witusiewicz, V.T.; Bondar, A.A.; Hecht, U.; Velikanova, T.Y. The Al-B-Nb-Ti system: IV. Experimental study and thermodynamic re-evaluation of the binary Al-Nb and ternary Al-Nb-Ti systems. *J. Alloys Compd.* **2009**, *472*, 133–161. [CrossRef]

26. Stark, A.; Oehring, M.; Pyczak, F.; Schreyer, A. *In situ* observation of various phase transformation paths in Nb-rich TiAl alloys during quenching with different rates. *Adv. Eng. Mater.* **2011**, *13*, 700–704. [CrossRef]

27. Rackel, M.; Stark, A.; Gabrisch, H.; Schimansky, F.-P.; Schell, N.; Schreyer, A.; Pyczak, F. *In situ* synchrotron radiation measurements of orthorhombic phase formation in an advanced TiAl alloy with modulated microstructure. *MRS Proc.* **2015**. [CrossRef]

28. Yan, K.; Liss, K.-D.; Garbe, U.; Daniels, J.; Kirstein, O.; Li, H.; Dippenaar, R. From single grains to texture. *Adv. Eng. Mater.* **2009**, *11*, 771–773. [CrossRef]

29. Liss, K.-D.; Schmoelzer, T.; Yan, K.; Reid, M.; Peel, M.; Dippenaar, R.; Clemens, H. *In situ* study of dynamic recrystallization and hot deformation behavior of a multiphase titanium aluminide alloy. *J. Appl. Phys.* **2009**. [CrossRef]

30. Garces, G.; Morris, D.G.; Muñoz-Morris, M.A.; Perez, P.; Tolnai, D.; Mendis, C.L.; Stark, A.; Lim, H.K.; Kim, S.; Schell, N.; *et al.* Plasticity analysis by synchrotron radiation in a MgY_2Zn_1 alloy with bimodal grain structure and containing LPSO phase. *Acta Mater.* **2015**, *94*, 78–86. [CrossRef]

31. Wassermann, G.; Grewen, J. *Texturen Metallischer Werkstoffe*; Springer: Berlin, Germany, 1962.

metals

MDPI

Article

Hydrostatic Compression Behavior and High-Pressure Stabilized β-Phase in γ-Based Titanium Aluminide Intermetallics

Klaus-Dieter Liss [1,2,3,*], **Ken-Ichi Funakoshi** [4], **Rian Johannes Dippenaar** [3], **Yuji Higo** [5], **Ayumi Shiro** [6,†], **Mark Reid** [2,3], **Hiroshi Suzuki** [1,‡], **Takahisa Shobu** [6,‡] and **Koichi Akita** [1,‡]

1   Quantum Beam Science Center, Japan Atomic Energy Agency, Tokai, Ibaraki 319-1195, Japan; suzuki.hiroshi07@jaea.go.jp (H.S.); akita.koichi@jaea.go.jp (K.A.)
2   Australian Nuclear Science and Technology Organisation, Lucas Heights 2234, Australia; mark.reid@ansto.gov.au
3   School of Mechanical, Materials & Mechatronic Engineering, Faculty of Engineering and Information Sciences, University of Wollongong, Northfields Avenue, Wollongong 2522, Australia; rian@uow.edu.au
4   Neutron Science and Technology Center, Comprehensive Research Organization for Science and Society (CROSS-Tokai), Tokai, Ibaraki 319-1106, Japan; k_funakoshi@cross.or.jp
5   SPring-8, Japan Synchrotron Radiation Research Institute, Kouto, Sayo, Hyogo 679-5198, Japan; higo@spring8.or.jp
6   Quantum Beam Science Center, Japan Atomic Energy Agency, Kouto, Sayo, Hyogo 679-5148, Japan; shiro.ayumi@qst.go.jp (A.S.); shobu@sp8sun.spring8.or.jp (T.S.)
*   Correspondence: kdl@ansto.gov.au or liss@kdliss.de; Tel.: +61-2-9717-9479
†   Present Address: Quantum Beam Science Research Directorate, National Institute for Quantum and Radiological Science and Technology, Kouto, Sayo, Hyogo 679-5148, Japan.
‡   Present Address: Materials Sciences Research Center, Japan Atomic Energy Agency, Tokai, Ibaraki 319-1195, Japan and Kouto, Sayo, Hyogo 679-5148, Japan.

Academic Editor: Hugo F. Lopez
Received: 15 April 2016; Accepted: 13 June 2016; Published: 15 July 2016

Abstract: Titanium aluminides find application in modern light-weight, high-temperature turbines, such as aircraft engines, but suffer from poor plasticity during manufacturing and processing. Huge forging presses enable materials processing in the 10-GPa range, and hence, it is necessary to investigate the phase diagrams of candidate materials under these extreme conditions. Here, we report on an in situ synchrotron X-ray diffraction study in a large-volume press of a modern (α_2 + γ) two-phase material, Ti-45Al-7.5Nb-0.25C, under pressures up to 9.6 GPa and temperatures up to 1686 K. At room temperature, the volume response to pressure is accommodated by the transformation $\gamma \rightarrow \alpha_2$, rather than volumetric strain, expressed by the apparently high bulk moduli of both constituent phases. Crystallographic aspects, specifically lattice strain and atomic order, are discussed in detail. It is interesting to note that this transformation takes place despite an increase in atomic volume, which is due to the high ordering energy of γ. Upon heating under high pressure, both the eutectoid and γ-solvus transition temperatures are elevated, and a third, cubic β-phase is stabilized above 1350 K. Earlier research has shown that this β-phase is very ductile during plastic deformation, essential in near-conventional forging processes. Here, we were able to identify an ideal processing window for near-conventional forging, while the presence of the detrimental β-phase is not present under operating conditions. Novel processing routes can be defined from these findings.

Keywords: high pressure; high temperature; phase transformation; equation of states; plasticity; TiAl; intermetallics; synchrotron radiation; multi-anvil press; in situ diffraction

---

1. Introduction

Titanium aluminides exhibit significant potential as a low specific weight structural material for high-temperature automotive and aerospace propulsive applications [1–4]. Gamma-based TiAl intermetallics possess high strength and excellent oxidation resistance at half the specific weight of conventionally-employed nickel-based superalloys, up to a temperature range of 1000–1100 K [5]. The drawback, however, of the two-phase ($\gamma + \alpha/\alpha_2$) material is the poor ductility, reduced forgeability and a small deformation window for thermo-mechanical processing [3]. To overcome these limitations, material-design focuses on β-solidifying alloys, which are characterized by the presence of a ductile β/β_0-phase at the processing temperature. However, through appropriate heat treatments, this ductile phase can be transformed to mechanically-stronger structures at the expected operating temperatures [6]. In addition, heating into a multi-phase field results in sluggish grain growth, while the augmented fraction of β-phase allows for near-conventional forging [7]. This has been partly achieved by alloying γ-based TiAl alloys with β-stabilizing elements, such as Nb and Mo; however, some residual β/β_0-phase is still present at the operating temperature [4,8–10].

In the present study, we investigate the influence of pressure on the formation of the ductile β-phase in γ-based TiAl alloys, which is not only of fundamental interest, but is also most relevant to modelling high-pressure deformation techniques, such as high-pressure torsion, in order to achieve severe plastic deformation [11–15] and high-pressure near-net-shape forging [16–19]. These deformation processes operate at pressures up to 7 GPa and forces exceeding 1 GN, respectively. However, to date, no experimental studies have been reported of the attempts to assess the high-pressure performance of this two-phase ($\gamma + \alpha_2$) TiAl intermetallic, and very little has been done from a theoretical point of view [20]. Examinations of the pure-titanium temperature-pressure phase diagram [21,22] reveal that pressure-dependent phase transformations can be expected to occur, but the exact intermetallic chemistry and the multi-phase fields in the TiAl alloy system can significantly modify these trends.

In the binary, γ-based titanium aluminide alloy system, an ordered, close-packed and not necessarily stoichiometric *hcp*-based α_2 Ti$_3$Al co-exists with the tetragonal γ-TiAl-phase, which is an ordered, close-packed structure, derived from the *fcc* lattice. The α_2-phase disorders to *hcp* α at the eutectoid temperature T_{eu} and γ fully transforms into α at the γ-solvus temperature $T_{\gamma,solv}$, also referred to as α-transus temperature T_α. This high-temperature single-phase field is prone to excessive grain growth and, on deformation, inherits the anisotropic plastic behavior from the hexagonal lattice [8,23,24]. A variety of alloys based on TiAl with additions of niobium has been developed in the so-called TNB alloy series [5,25–27], where Nb substitutes Ti sites and eventually stabilizes the *bcc*-based β-phases. In the present study, we have investigated the TNB composition Ti-45Al-7.5Nb-0.25C, which has been produced through a powder-metallurgical route by Gerling et al. [28] using gas-atomization and subsequent hot-isostatic pressing for 2 h at 200 MPa and 1553 K. It is obtained from the same series as extensively characterized by Chladil et al. [29,30] and Yeoh et al. [31], showing globular γ-TiAl/α_2-Ti$_3$Al grains with an average size of about 15 μm (see Figure 1b in [29]). In situ investigations revealed conventional transformation behavior for this Ti-45Al-7.5Nb-0.25C alloy, with T_{eu} = 1453 K and T_α = 1565 K, with an additional peritectic ($\alpha + \beta$)-phase field at T_{per} ~1500 K, but no ($\alpha + \beta + \gamma$)-phase field has been observed.

The pure-titanium temperature-pressure phase diagram has been well investigated [21,32]. At atmospheric pressure and $T_\beta(0)$ = 1155 K, *hcp* α-Ti transforms into *bcc* β-Ti. Under hydrostatic pressure, the β-transus temperature T_β decreases linearly to meet the triple-point with the hexagonal ω-phase, at about 940 K and 9.0 GPa. Above this pressure, T_β increases again. The authors also reported that at and above room temperature, the ω-phase occurs only at higher pressure, appearing above 2 GPa. Aluminium, on the other hand, behaves as a simple solid, not transforming from its *fcc* structure under high pressure [33].

It is not possible to predict the behavior of intermetallic titanium aluminide alloys by using the pure-titanium temperature-pressure phase diagram. However, indications are that phase boundaries can shift with applied pressure and that the β-phase can be stabilized because the β-transus line has a negative slope. Moreover, there is a distinct likelihood that changes in lattice parameters and modified atomic packings, including atomic order and disorder, might influence the phase relationships.

In order to study the phase changes in TiAl intermetallic systems, in situ quantum beam diffraction [34], such as X-rays and neutrons, has been employed. The first in situ X-ray diffraction studies on these intermetallics have been undertaken by Shull and Cline [35], but the real breakthrough came when Liss et al. [36] used high-energy synchrotron radiation [37], to study these alloys. Combined time-resolved and multi-dimensional diffraction in a so-called materials oscilloscope [38] allows determination of the microstructural evolution of multi-phase titanium aluminides during thermo-mechanical processing [7,39]. In situ neutron diffraction is complementary to X-ray studies, due to the negative and positive scattering lengths of Ti and Al, respectively, and is therefore ultimately sensitive to the atomic order in a crystal structure [40,41].

High-pressure diffraction equipment has been well established for a number of decades at synchrotron radiation facilities, mainly driven by the geoscience community, but its application to structural materials science is still in its infancy [42,43]. The most common pressure device is a diamond-anvil cell, capable of reaching pressures up to 200 GPa and temperatures up to 5000 K [44]; however, with a maximum diameter of ~0.2 mm, the total volume of samples is limited. On the other hand, large-volume multi-anvil cells can be used for samples up to an 8-mm side length, with compromises on the maximum achievable pressure of up to approximately 30–50 GPa [45,46], but more typically in the 10-GPa range. Heating can be achieved by building furnaces into the hydrostatic volume, resulting in available sample volumes of 1–2 mm^3.

Because excessive grain growth in titanium alloys becomes a serious problem at very high temperature, a powder diffraction experiment might degenerate into a study of single crystals if the experimentally-accessible reciprocal space volume is collimated [47]. Moreover, studies of microstructural evolution, allowing for the segregation of the elements in a multi-phase field, such as the co-existing γ- and α$_2$-phases in titanium aluminides, require larger sample volumes. Furthermore, the chemistry of titanium alloys may change at the surface layer, such as the depletion of aluminium, as observed in a vacuum [40] or influenced by the pressure medium, as well as uni-axial stress components [48].

2. Experimental Section

In the current study, diamond anvil cells were deemed unsuitable, and large-volume multi-anvil cells have been utilized in order to ensure that a sufficiently large volume is available to allow for microstructural evolution and segregation. More specifically, the 15-MN force, Kawai-type multi-anvil machine, SPEED-Mk. II, at the SPring-8 bending magnet beamline BL04B1 was utilized. The beamline setup is depicted in Figure 1 [49], and a detailed description of the pressure apparatus was given by Katsura et al. [50]. A 10 mm-thick germanium solid-state detector of CANBERRA industries has been employed for energy-dispersive diffraction at a horizontal take-off angle of $2\theta = 5.9827°$. While the detector records energies from 3 keV onwards, the spectrum transmitted through the pressure cell ranges from ~34–140 keV, thereby covering a diffractogram from ~1.8–7.5 Å^{-1}, as shown in Figure 2. Energy calibration of the detector was done by recording the fluorescence lines of Mo, Pb, Au, Ag, Pt, Ta and Cu, before the cell was put into position. In addition to the main detector, a radiography setup was used, implemented by a YAG scintillator (Hamamatsu Photonics K.K., Hamamatsu, Japan), projected to a CCD camera. It images the transmitted beam through the pressure cell and allows for the control of the beam size via a slit system to focus on the sample or the pressure marker. Typical radiographs are shown in Figure 3. In order to reduce the effect of grain size on intensity readings, the whole SPEED-Mk.II apparatus (Sumitomo Heavy Industries, Ltd., Tokyo, Japan), containing the load frame and the pressure cell with the sample, has been oscillated during data acquisition by 8°, which is an important feature of the instrument used [50].

The heating pressure cell was designed and fabricated by the experimental team, and the details of the design are shown in Figure 4. It consists essentially of a sample space and a pressure marker of a 60MgO-40Au (mass %) mixture, located in a LaCrO$_3$ resistive furnace, contacted by platinum leads and controlled by an alternate-current power supply. A W-type thermo-couple of W-3Re and W-25Re (mass %) was used for temperature measurements. The remaining space is filled with insulating and pressure propagating ceramics, such as ZrO_2, BN and MgO. The furnace assembly sits in a 4.5-mm bore of the pressure octahedron, made of semi-sintered $MgO-5Cr_2O_3$ (mass %), which is then installed into the two-stage multi-anvil press (Figure 5). The gasket material is pyrophyllite.

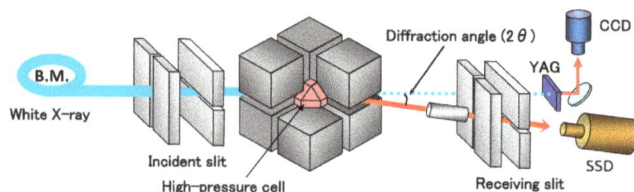

Figure 1. Schematic layout of the beamline. White X-rays emit from the bending magnet (B.M.), which are conditioned by incident slits to enter the high-pressure cell. Transmitted X-rays are used to image the sample cell, converted by a scintillator, YAG, and recorded by a camera, CCD. A fixed diffraction angle, 2θ, is selected by a collimator and receiving slits in front of the solid state X-ray detector, SSD. After [49].

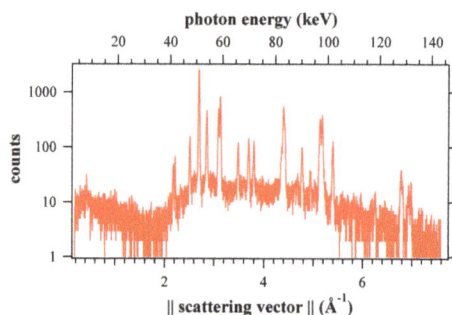

Figure 2. Full recorded detector spectrum (top axis), translating into a calibrated diffractogram (bottom axis). The bump in the background denotes the useful, transmitted energy spectrum between ~34 and 140 keV, i.e., ~1.8–7.5 $Å^{-1}$. The data shown are the signals off the specimen at 310 K and 0 GPa.

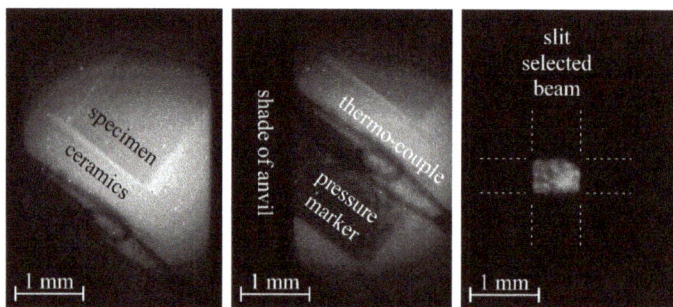

Figure 3. Typical radiographies of the pressure cell for sample alignment. The horizontal and vertical instrument axes show up and right in the figure. Translation of the entire apparatus with respect to the beam allows for different fields of view. The looped thermo-couple is a feature that can be well recognized in the left and central picture. To the right, beam size has been reduced to focus on a particular volume only, here the pressure marker.

14M TEL5

① Octahedron(14M TEL5)
② ZrO2 sleeve φ4.5xφ3.0x2.95t mm
③ Pt tube φ3.0xφ2.8x2.95t mm
④ MgO rod φ2.7x2.95t mm
⑤ BN sleeve φ4.5xφ3.5x5.5t mm
⑥ Graphite heater φ3.5xφ2.5x4.5t mm
⑦ Graphite cap φ3.5x0.5t mm
⑧ BN sleeve φ2.5xφ1.5x2.0t mm
⑨ BN cap φ2.5x0.5t mm
⑩ Sample φ1.5x2.0t mm
⑪ Pressure marker φ2.5x1.5t mm (MgO+Au (3:2wt%) powder)
⑫ W/Re thermocouple φ0.1 mm+φ0.25 mm

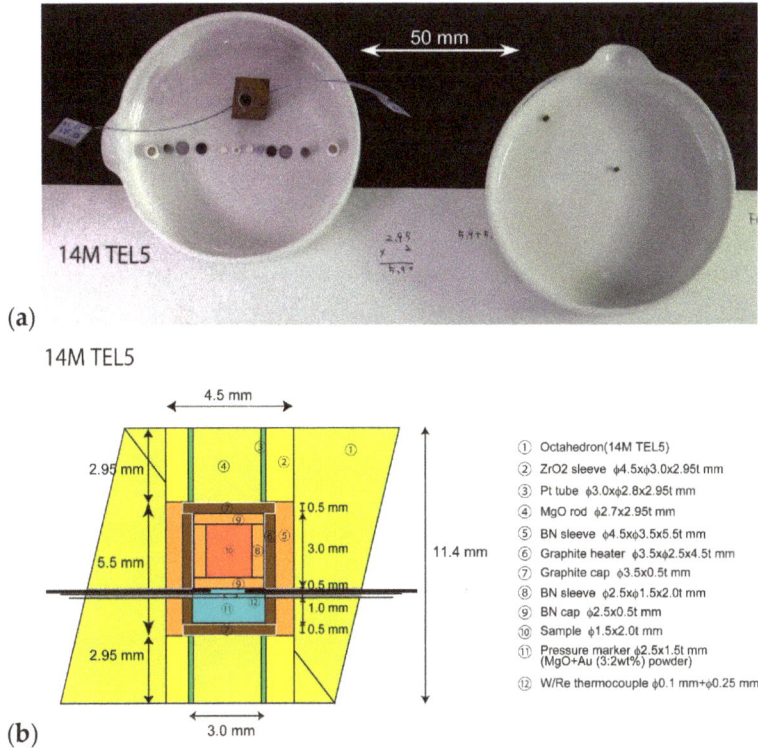

Figure 4. Design of pressure cell 14 M TEL5 (**b**) and photograph of the individual parts (**a**, left dish) and specimens (**a**, right dish).

Figure 5. Ken-ichi Funakoshi and Mark Reid loading the pressure cell and sample into the 15-MN press SPEED-Mk.II at SPring-8 beamline BL04B1.

The serial number of the experimental dataset is M1472. Radiographs were taken to locate the sample and the pressure marker positions and to achieve an optimal beam size of 0.5×0.7 mm^2 ($h \times v$); see Figure 3. Initially, the pressure was ramped stepwise to a target of 10 GPa, while the pressure was accurately determined by the use of the equation of state of the pressure markers [51,52], using the PDIndexer software (V4.32, Yusuke Seto, Kobe University, Kobe, Japan, 2016) [53]. Diffraction patterns were taken while holding at 0, 3.2 and 9.6 GPa, respectively, before the system was heated under maximum pressure.

The measured photon energy E is correlated with the wave number $k = 2\pi E/hc$; $k(\text{Å}^{-1}) = 0.506768\,E$ (keV), where h denotes the Planck constant and c the speed of light. With Bragg's relation $q = 2k\sin(\theta)$, this translates to the scale of momentum transfer or scattering vector in reciprocal space q. In the present case, the calibration is $q(\text{Å}^{-1}) = 5.2892 \cdot 10^{-2}\,E$ (keV).

Figure 6 displays the expanded diffractograms, taken at room temperature during pressurizing pressure steps, while Figure 7 shows all of the data in a zoomed range, both at room temperature and upon heating. A Rietveld refinement, using the MAUD program [54], was utilized for the analysis, allowing for the extraction of lattice parameters and phase fractions.

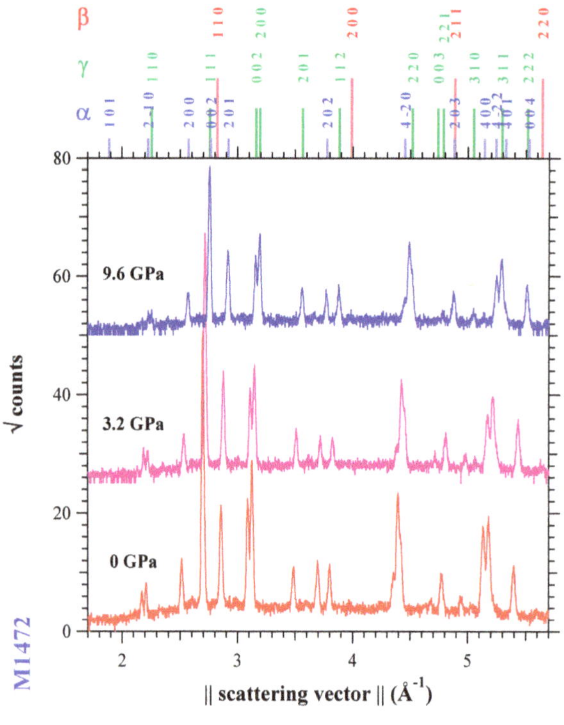

Figure 6. Diffraction patterns upon pressurizing. Line shifts to larger scattering vectors can be seen as the unit cell shrinks, accompanied by changes of peak intensity due to phase transformation.

Unless otherwise stated, we used the following choice of unit cell throughout the paper: both α- and α$_2$-phases (α/α$_2$) span a hexagonal lattice with the a-axis being twice and c equal to those of conventional *hcp*. α$_2$ and α differ by an ordered and disordered motif, respectively, the former making up the stoichiometry Ti$_3$Al. In order to compare the axis ratios, $2c/a$ is listed for the α-/α$_2$-phases, which directly compares to c/a of a simple *hcp* structure, in its ideal, close-packed case where $c/a_{hcp} = \sqrt{8/3} = 1.6329932$. The γ-lattice is constructed from the *fcc* structure by altering the Ti and Al layers along the *c*-direction. The unit cell is therefore tetragonal and slightly distorted, with a *c/a* ratio slightly larger than 1. The cubic β-phase is either disordered and *bcc* or ordered $β_o$ of Strukturbericht designation B2, where the two different kinds of atom sit on the corner and the body center of the cube. Accordingly, the unit cell volumes are $V_{Z,\alpha} = \sqrt{3}/2 \cdot a^2 c$, $V_{Z,\beta} = a^3$ and $V_{Z,\gamma} = a^2 c$. It is of interest to calculate the volume per atom V_A by dividing these volumes by the number of atoms in the unit cell, i.e., 8, 2 and 4, respectively, leading to $V_{A,\alpha} = \sqrt{3}/16 \cdot a^2 c$, $V_{A,\beta} = 1/2 \cdot a^3$ and $V_{A,\gamma} = 1/4 \cdot a^2 c$.

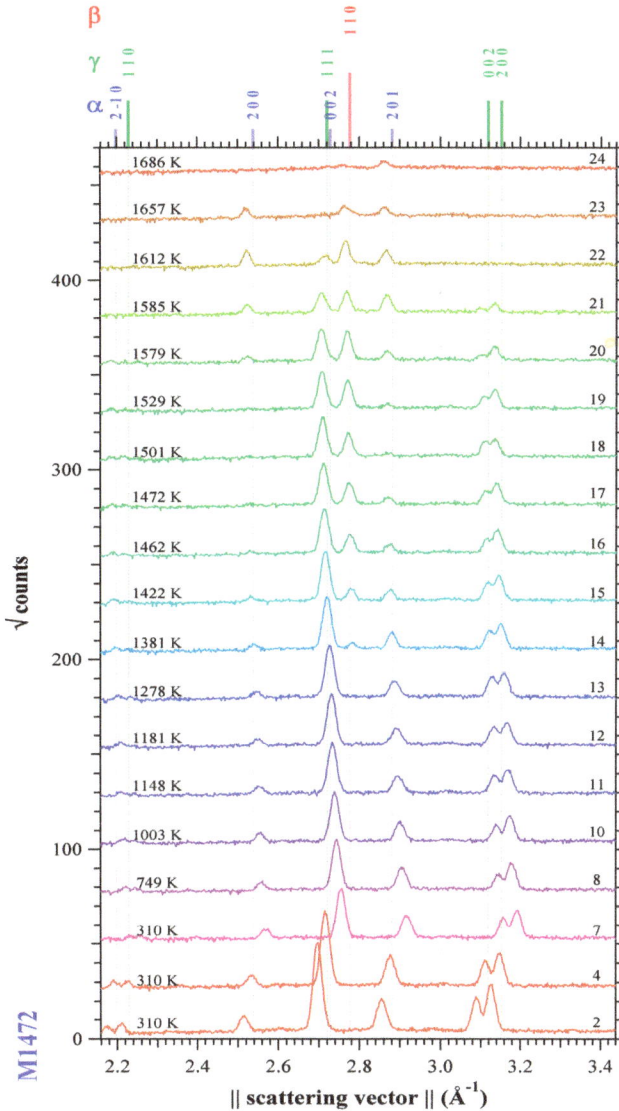

Figure 7. Zoomed compilation of all diffraction patterns upon pressurizing at room temperature (first 3) and on stepwise temperature ramping at the maximal achieved pressure. Temperature tags are given to the left, serial numbers to the right.

3. Results and Discussion

The large diffraction peak around 2.7 Å^{-1} in Figure 6 corresponds to the close-packed plane-stacking distance and is therefore common to both the *fcc*-based γ- and *hcp*-based α-/α$_2$-phases, indexing 111 and 002, respectively (there is a small irresolvable shift between the two, due to slight deviation from close-packing in each phase). Only α$_2$ and γ peaks can be identified at room temperature, while β peaks appear at 1381 K and increase in intensity above that temperature. The β-110 peak lies around 2.75 Å^{-1}, representing a near close-packed lattice spacing. The typical *fcc* peak positions of the γ-phase are split, due to the tetragonal deformation with an axes ratio $c/a > 1$, by only a few percent. Both the γ- and α$_2$-phases show superstructure reflections due to atomic order at room temperature, i.e., reflections, which would have vanishing structure factors in fully-disordered lattices, *fcc* and *hcp* in the present case.

3.1. Pressure Loading at Room Temperature

Pressure loading at room temperature has been evaluated at {0, 3.2, 9.6} GPa and leads to continuous peak shifts to larger scattering vectors due to the compressive strain on the unit cell. Furthermore, the splitting of the γ-002/200 and γ-112/211 peaks around ~3.1 Å^{-1} and ~5.2 Å^{-1}, respectively, diminishes slightly, resulting in a decreasing c/a ratio. Furthermore, relative changes in peak intensity are observed throughout the pressurization process. The volumetric strain allows the evaluation of the equation of state in this two-phase system.

3.1.1. Crystallographic Anisotropy, Disorder and Transformation Behavior

Linear lattice strain ε, for a given reflection at position $G = G_{hkl}$ and Miller indices h, k, l, is evaluated in reciprocal space by:

$$\varepsilon = \varepsilon_{hkl} = -\frac{\Delta G}{G_0} = \frac{G_0 - G}{G_0} \tag{1}$$

where G and G_0 are the strained and unstrained reciprocal lattice vectors, respectively. Here, we select the values of G_0 at ambient conditions (310 K, 0 GPa), the first pattern in Figures 6 and 7. The experimental results for lattice parameters and lattice strains along the crystal axes a and c are shown in Figure 8a–d, for both phases. Linear fittings to the function $y = a + bp$ with fit parameters a, b and pressure p are represented by the continuous lines and numerical results inserted into the figures. To a first approximation, the overall strain response to pressure is $\partial\varepsilon/\partial p \approx -2.2 \times 10^{-3}$ GPa^{-1} for both phases. However, details deviate from this average, resulting in anisotropy and inhomogeneity within and across the phases. This linear fitting leads to an approximation for the bulk modulus:

$$K_{lin} = -V\frac{\partial p}{\partial V} \approx -(3\partial\varepsilon/\partial p)^{-1} \approx 152 \text{ GPa} \tag{2}$$

The crystallographic strain anisotropy is demonstrated by varying c/a ratios, which are presented in Figure 8e,f for both the γ- and α$_2$-phases. For pure α-titanium, Zhang et al. [55] reported a constant, pressure-independent $2c/a$ ratio of 1.5868, which agrees well with the generally-accepted value of 1.5871 (from JCPDS Card 44-1294) [56]. Controversially, a large pressure dependence has been reported by Errandonea et al. [48], which Zhang et al. refute and interpret as a response to deviatoric stresses, occurring in their diamond-anvil cell. In alloys, especially titanium aluminides, lattice parameters and c/a ratios of both γ- and α$_2$-phases depend strongly on the composition of each phase, which varies as a function of temperature and segregation in multi-phase systems [31,57,58]. Increasing $2c/a$ ratios of {1.60333, 1.60432, 1.60476} with increasing Al content in single-phase α$_2$ Ti-{24, 28.4, 33.3}Al have been measured by Dubrovinskaia et al. [59]. In the present study, we observed a slight variation of the c/a ratios with increasing pressure, in both phases, as shown in Figure 8e,f. The c/a ratio in the γ-TiAl-phase decreases monotonically at an increasing rate with increasing pressure, towards a c/a ratio approaching unity, the value for a cubic *fcc* lattice. Such an *fcc*-Ti structure has been predicted by ab initio simulations [60,61], but it is energetically unstable and has not been experimentally observed. As a theoretically-postulated phase, it can be asymptotically used to interpret experimental data, such as for interpolating lattice parameters by Vegard's law [57].

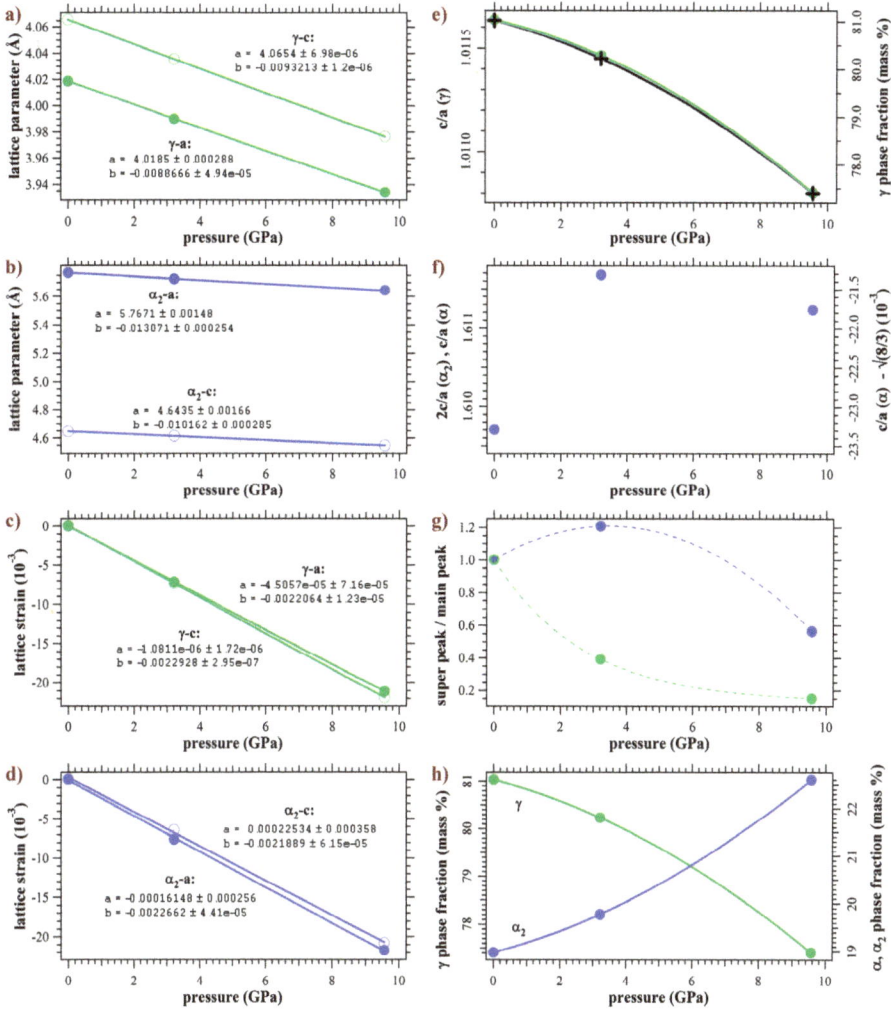

Figure 8. Evolution of structural parameters, phase fraction and diffraction intensity under pressure. Lattice parameters (**a,b**) and lattice strain (**c,d**) with linear fits $y = a + bp$; axis ratios (**e,f**); super-structure peak intensities, plotting ratios of super/main peak at pressure, normalized to the super/main peak at ambient pressure (**g**); and phase-fractions (**h**, superimposed to **e**).

In the case of γ-TiAl with a $L1_0$ structure, fully-disordered γ_d-TiAl would be *fcc*, as the break of symmetry is expressed by the order. However, it is lacking even in a massively-transformed γ-phase [36]. It has been reported that γ always tends to order, and when disordered, it transforms to the α-/α_2-phase [31]. Moreover, for alloys with no added nucleants, in situ temperature studies on cooling from the disordered α-phase consistently revealed significant undercooling below the α-transus, where the γ-phase should re-appear. Watson et al. [40] correlated the delayed re-appearance of the latter with the ordering of $\alpha \rightarrow \alpha_2$, where lattice planes first partially order before the stacking sequence changes. Asta et al. [60] argue that for equi-atomic compositions, the *hcp*-based random alloy is slightly more stable than the *fcc*-based random alloy, and therefore, the stability of the γ-phase can be attributed to a stronger tendency to ordering of the *fcc* lattice. In particular, the ordering energy is larger for the γ structure than it is for the α_2 structure.

In the present case, a continuous transformation from γ to α/α_2 takes place with pressure, as presented in Figure 8h, which interestingly scales directly with the c/a ratio of γ-TiAl, leading to the following scenario: with increasing pressure, atoms in the γ-phase are pressed towards the optimal close-packed *fcc* structure, reducing its c/a ratio. Inherent in this behavior is the introduction of disorder between the Ti and Al layers, facilitated by short range diffusion within the size of a unit cell. As disorder increases, the displacive character of the phase transformation $\gamma \rightarrow \alpha/\alpha_2$ becomes viable, in agreement with the arguments above and those advanced in the literature. Long-range diffusion and segregation do not occur at room temperature, while the pressure-driven disorder occurs throughout the bulk of the γ grains, leading to the conclusion that ultrafine lamellae form, with thicknesses at the nanometer scale, in a manner very similar to low-temperature annealing of α-quenched material [62–64]. The difference is that they may form on all four γ {111} habit planes and lead to a Widmanstätten-like or basket-weave arrangement, rather than a lamellar microstructure. Such formation and disorder on a nanometer scale is supported by the micrographs of pressure loaded and unloaded γ-TiAl-based material published by Srinivasarao et al. [65]. Phase stresses in these ultrafine structured grains build up a large amount of potential energy, preventing further transformation at a given pressure and retaining the given c/a ratio. The remaining disorder at the large density of interfaces leads to the observed decrease of superstructure reflections.

The anisotropic response of the α-/α_2-phase is qualitatively different. Its $2c/a$ ratio first increases when atoms are driven to a more ideally close-packed structure similar to the γ-phase, but then decreases slightly at higher pressure. In contrast to the γ-phase, the α-/α_2-phases are a sink for disorder, both stoichiometrically and crystallographically. According to the phase diagram [66], the α_2-Ti$_3$Al-phase is stable over a wide range of concentrations, between roughly Ti-20Al and Ti-35Al, even at room temperature. This indicates that up to 33% of either site can be accommodated by the wrong kind of atom, while the γ-TiAl formation begins at Ti-47.5Al on the Ti-rich side, allowing only 5% of Ti on Al sites. In the present two-phase system, the compositions of the α_2- and γ-phases are expected to be sitting on the phase boundary lines, i.e., Ti-35Al and Ti-47.5Al, respectively. Furthermore, α_2 fully disorders at high temperature (by a eutectoid reaction at $T_{eu}(0) = 1453$ K for our alloy), while γ never does.

By considering the experimentally-derived data, the events occurring during pressurization can be clarified: In the first step, while no or little γ-phase transforms into α/α_2, atoms of α_2 are pressed into their isotropic positions, i.e., towards ideally close-packed $2c/a$. A high density of the nano-Widmanstätten or basket-weave structure propagates upon further pressure loading throughout the bulk of the grains, to create a relatively isotropic stress state in each volume (as compared to a lamellar microstructure) and introduces disorder, as evidenced by the decrease of the superstructure reflections shown in Figure 8g. Concurrent with a smaller $2c/a$ in α-Ti and thermal disorder in α-TiAl [31], the $2c/a$ ratio tends to decrease again, an argument that is well supported by the experimental data.

In summary, crystallographic disorder is being introduced into the phases upon loading at room temperature, where no long-range diffusion and annealing can occur. As the γ-phase disorders, it transforms into the α-/α_2-phase, until transformation stresses built up to the extent that further transformation is impeded.

3.1.2. Volumetric Response and Equation of State

From a thermodynamic point of view, response to high pressure is given by the equation of state, relating the volume or density of the matter to pressure p, temperature T and potentially other variables. A well-known equation of state is the ideal gas law. In the case of solids, the interaction potentials of atoms have to be taken into account [67]. Put simply, at constant temperature, atoms sit in the equilibrium position of superimposed attractive and repulsive parts of the potential, defining the bond distance. As hydrostatic pressure is increased, the applied load acts against the repulsive potential, shortening the bond distances and decreasing the volume occupied per atom. An important parameter related to the equation of state is the bulk modulus:

$$K = -V \frac{\partial p}{\partial V} \tag{3}$$

which is evaluated at a given temperature. It relates to the compressibility, mechanical properties, sound velocity, etc. For isotropic, linear strain, this corresponds to Equation (2) and for orthorhombic systems:

$$\frac{\partial V}{V} = \sum_{i=a,b,c} \varepsilon_i \tag{4}$$

For solids, there exist a multitude of atomic potential models, corresponding to a multitude of equations of state [68], but all of them should lead to the same bulk modulus $K_0 = K(p = 0)$, when properly evaluated at $p \to 0$. Physically, $K(p)$ is not constant, as a solid is more difficult to compress the further it is compressed. The detail of the potential reflects in the shape of the compression curve $V(p)$ as it deviates from linearity. As the non-linearity is small, it suffices to consider:

$$K' = \frac{\partial K}{\partial p} \tag{5}$$

Sometimes, the details lie in the second derivative K'', for a second and third order fit, respectively.

The current case, with only three measured data points, allows the description of any second order curve in order of K', while fitting K'' would be an under-determined set of equations. Note, however, the change of derivative variable between Equations (3) and (5) and the functional behavior of a given equation of state, so that K'' may be implicitly determined in that equation. Therefore, the shape of the $V(p)$ curve depends on the equation used.

The most commonly-used equation is the Birch-Murnaghan equation of state [69,70], up to 200 GPa [71], although at ultra-high pressure [61], the Vinet equation [72] is probably more accurate. In the present instance, we have fitted both equations and found that there is no significant difference; hence, our subsequent interpretation is not affected, and we present the Birch-Murnaghan approach.

The Birch-Murnaghan equation of state derives from the Helmholtz free energy and is typically Taylor expanded to the n-th order. The third order expansion is widely employed for the current type of high-pressure studies [48,73,74] and provides three fitting parameters, V, K_0, K_0':

$$p = \frac{3}{2} K_0 \left\{ \left(\frac{V}{V_0} \right)^{-\frac{7}{3}} - \left(\frac{V}{V_0} \right)^{-\frac{5}{3}} \right\} \left\{ 1 + \frac{3}{4} (K_0' - 4) \left(\left(\frac{V}{V_0} \right)^{-\frac{2}{3}} - 1 \right) \right\} \tag{6}$$

Volume occupations per atom have been calculated from the experimental unit cell and are displayed in Figure 9 and listed in Table 1, together with comparable results from the literature. The markers represent the experimental data points for the two phases, as well as for the total volume weighted by the phase fraction, while the continuous lines are fits to Equation (6). For comparison, the $V(p)$ behaviors of pure α-Ti and ω-Ti are shown, as taken from Errandonea et al. [48], as well as for α_2 Ti-33.3Al from the study of Dubrovinskaia et al. [59]. The authors of the cited data also employed the Birch-Murnaghan equation, so that Equation (6) has been applied to the published parameters to compute the literature traces, transferred to volume per atom units.

Phases with a higher Al concentration consistently display a smaller volume per atom, due to the different atomic radii. To a first-order approximation, the compression curve of our α/α_2 data coincides very well with Dubrovinskaia's, which demonstrates that the measurements are accurate. In contrast to all of the examples given in the literature, i.e., α-Ti, ω-Ti and α_2-Ti-33.3Al, our curves start with a gentler slope of $V(p)$ and evolve more linearly, expressed by large K_0 and small K_0' values. The observed bulk moduli, lying around 146 GPa, are significantly larger than reported elsewhere for related systems. Pure metals have smaller bulk moduli than intermetallic compounds, which have stronger bonds. This is underlined by Asta's first-principles computation on *fcc*- and *hcp*-based Ti-Al compounds, showing that theoretically, γ-TiAl has one of the highest bulk moduli [60] (see Table 1), which still do not reach our experimental values. On the other hand, the presently-observed change of modulus K_0' with pressure is an order of magnitude smaller than otherwise earlier reported values.

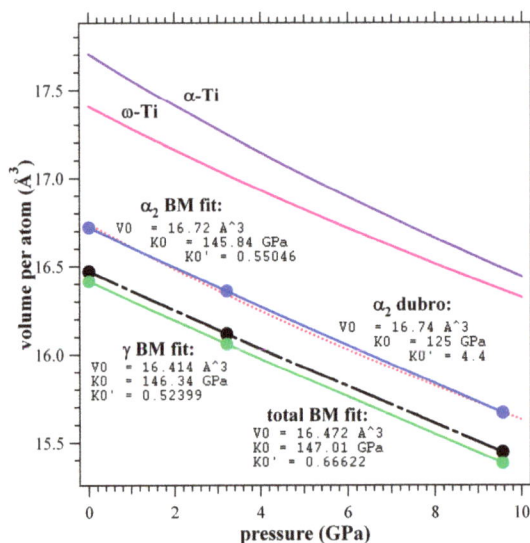

Figure 9. Atomic volumetric compression behavior of the investigated composition Ti-45Al-7.5Nb-0.25C with Birch–Murnaghan fits (experimental dots with continuous lines), as compared to α_2-single-phase compression, and α- and ω-titanium, reported by Dubrovinskaia [59] and Errandonea [48], respectively.

The salient difference between ours and earlier studies is that we are working in the $(\gamma + \alpha_2)$ two-phase field, where phases can transform continuously as a function of composition, temperature [66], disorder (see the discussion above) and pressure (see Figure 8h). A small change of these parameters will affect the relative phase fractions of γ and α/α_2. All of the other cited pressure studies have been conducted in compositional single-phase fields. Dubrovinskaia et al. reported the absence of any pressure-induced phase transformation in α_2 Ti-33.3Al, close to the border line with the $(\gamma + \alpha_2)$ two-phase-field. Therefore, we propose the following:

Upon loading, the atomic volume is reduced in each phase, and the phase transformation occurs, as discussed in Section 3.1.1, building up a nano-Widmanstätten (basket-weave) microstructure. The accumulating inter-phase stresses, highly interwoven and homogeneously distributed over the volumes of the initial grains, build up an additional potential acting against the applied pressure. As a consequence, the volume per atom, in each phase, is less compressed at any given pressure than it would be in its single-phase counterpart. As compression progresses, more and more material transforms, rather than pushing atoms against the repulsive potential, resulting in a seemingly increased bulk modulus. As such, a gradual phase transformation accommodates the response to hydrostatic pressure, rather than an increase of K, keeping K' small. It is interesting to note that the

γ-phase, with a smaller atomic volume, transforms to the α-/α_2-phase, with a larger volume per atom, driven by the strong ordering energy of γ [60], as discussed in Section 3.1.1. Nevertheless, the total volume of the sample decreases with pressure, as given by the average atomic volume:

$$V_A^{avg} = C_\alpha V_A^\alpha + C_\gamma V_A^\gamma \qquad (7)$$

C_γ and C_α are the phase fractions of γ and α/α_2, respectively, ($C_\gamma + C_\alpha = 1$), and are represented by the black markers and long-short-dashed line in Figure 9. The volume-increasing part $\gamma \rightarrow \alpha/\alpha_2$ of the transformation counteracts the volume compression in each phase, so that the total bulk modulus is even larger, and since the transformation rate itself is non-linear, it adds to the non-linearity of V, expressed as a slightly increased total K', compared to each individual phase; see Table 1.

3.2. Temperature Dependence at High Pressure

The heating experiment was designed to determine if phase field shifts occur under pressure. In the current study, the sample was heated to the point of partial melting and data recorded during heating at various temperature holding points, while the maximum pressure of 9.6 GPa was applied. Figure 7 shows the diffractograms as a function of room temperature loading and subsequent heating under constant load. The phase fractions of the alloy under investigation, Ti-45Al-7.5Nb-0.25C, obtained by Rietveld analysis, are depicted in Figure 10. Also included in the figure is the data of Yeoh et al. for a Ti-45Al-7.5Nb-0.5C alloy. Both alloys have been manufactured at the GKSS research center, Geesthacht, Germany, in a similar way and in consecutive batches [28] in order to study compositional effects only. The thermo-calorimetric and in situ diffraction studies undertaken by Chladil et al. [29] and Yeoh et al. [31] have been combined [30]. For Ti-45Al-7.5Nb-xC alloys, increasing carbon additions of $x \in \{0, 0.25, 0.5\}\%$ increase the eutectoid temperature T_{eu} to $\{1432, 1453, 1473\}$ K, respectively, but otherwise, the transformation behavior is the same, specifically the γ-solvus or α-transus remains stable at $T_\alpha = 1565$ K. These expected transition temperatures T_{eu} and T_α are marked in Figure 10 for the Ti-45Al-7.5Nb-0.25C used in this study. Although niobium is a β-phase stabilizer, β-containing phase fields were not detected in concentrations up to 7.5 at. % [31], while higher concentrations above 8% Nb rendered a detectable amount of this phase [30,75].

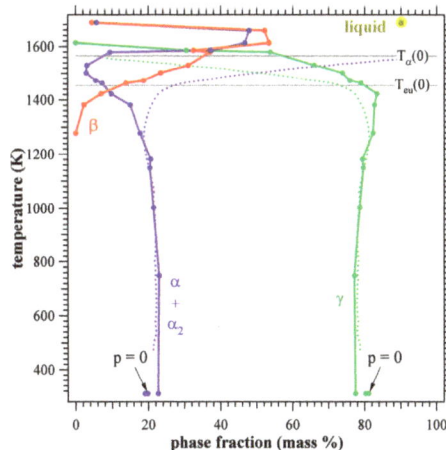

Figure 10. Phase evolution of Ti-45Al-7.5Nb-0.25C under high pressure at 9.6 GPa (continuous lines) in comparison to ambient pressure ($p = 0$ and dotted lines for Ti-45Al-7.5Nb-0.5C from Yeoh et al. [31]). The ambient pressure transition temperatures, reported by Chladil et al. [29], are marked for the investigated alloy.

The salient difference between the high and the ambient pressure experiments lies in the appearance of a large fraction of β-phase above ~1350 K in the high-pressure experiments, while no trace of this phase was experimentally observed in Yeoh's study [31]. Below 1350 K, only α-/α$_2$- and γ-phases are present.

At ambient pressure, the γ mass fraction is 81%; it decreases gradually to 78.0% at 600 K before increasing above 1000 K to recover the maximum of 81% at 1250 K. From this temperature, it decreases gradually up to the eutectoid temperature, T_{eu}, then diminishes at a high rate, up to T_α, where it becomes fully α-phase [31]. The cause of the minimization of the γ-phase between 600 K and 1000 K is attributed to local stacking rearrangements, where disorder, transitional structures [62,63] or even intermediate phase transitions [76,77] may occur, depending on the composition and prior heat treatment. The induced disorder favors the γ → α/α$_2$ transition, due to the high ordering potential in the γ-phase [60]. Order is expected to be highest at temperatures between 1200 and 1300 K, resulting in the maximum amount of γ-phase [31].

Under a pressure of 9.6 GPa, the γ-fraction was reduced to 77.4% at room temperature; see Section 3.1.1. During heating, it reduces further to 77.1% at 750 K, lying underneath the values of the ambient pressure heating, and then recovers while the difference vanishes at 810 K. As lattice disorder and planar distortions were already induced at room temperature upon pressurizing, the γ-phase had transformed to α/α$_2$, which then reverts between 750 K and 810 K, a temperature region known to show kinetics and planar re-arrangements [62,63]. Subsequently, mass fraction curves at both ambient and high-pressure overlap. While at ambient pressure, the γ-fraction begins to diminish at 1250 K, it increases to a maximum of 83.4% at 1420 K. Subsequently, with increasing temperature, γ monotonically decreases until it vanishes at $T_{γ,solv}$ = 1590 K.

Table 1. Compilation of experimental lattice parameters a_0 and c_0 under ambient conditions, as well as the derived quantities; their axis ratios and volume per atom V_A, compression parameters K_0, K_0' (first 3 rows) and data from the literature. The α-phase lattice is given in α$_2$ cell notation, and therefore, $2c/a$ is noted. The first V_A column is computed from a_0 and c_0, while the second results from the fit of pressure data to Equation (6). The original data of Yeoh's publication [31] has been re-visited to extract the listed values at 300 K. Literature values are reported from their experimental findings, in addition to Ghosh's first-principles study [61]. Further listed references are Dubrovinskaia [59], Errandonea [48], Asta [60], Zhang [55], JCPDS [56] and Menon [78]. a, c in (Å); V in (Å3); K_0 in (GPa).

Phase	a_0	c_0	Axis Ratio	V_A	V_A	K_0	K_0'	Reference
γ	4.01867	4.06542	1.0116332	16.4138371	16.414	146.34	0.52399	this work
α/α$_2$	5.76803	4.64241	1.60970383	16.7201111	16.72	145.84	0.55046	this work
total				16.4720291	16.472	147.01	0.66622	this work
α$_2$-Ti-33.3Al	5.7763	4.6348	1.6047643	16.7406041	16.74	125	4.4	Dubrovinskaia
α$_2$-Ti-28.4Al	5.7829	4.6388	1.60431617	16.7933623	16.79375	131	3.6	Dubrovinskaia
α$_2$-Ti-24.0Al	5.8083	4.6563	1.60332627	17.0051191	17.005	133	2.6	Dubrovinskaia
α-Ti			1.583		17.7013462	117	3.9	Errandonea
ω-Ti			0.609		17.4024491	138	3.8	Errandonea
γ			1.012			128		Asta
α$_2$			1.698			126		Asta
γ	3.9814	4.0803	1.02484051	16.1697657	16.181	112.1	3.91	Ghosh
α$_2$	5.7372	4.6825	1.63232936	16.6847003	16.584	111.9	3.83	Ghosh
α-Ti			1.5868			114	4	Zhang
α-Ti	5.901	4.6826	1.58705304	17.651391				JCPDS
γ-Ti-50Al	3.9973	4.0809	1.02091412	16.3015706				Menon
γ-Ti-45Al-7.5Nb-0.5C	4.02421	4.07335	1.01221109	16.4912285				Yeoh
α$_2$-Ti-45Al-7.5Nb-0.5C	5.77568	4.65646	1.61243698	16.8152283				Yeoh

At a pressure of 9.6 GPa, a coexisting, third phase, β, with a = 3.20 Å, appears at 1350 K, in stark contrast to heating at ambient pressure. It is apparent that in large measure, α$_2$ transforms to β (α$_2$ → β), evidenced by the increasing fraction of γ, below 1420 K. As the fraction of γ decreases on further heating, the fraction of β increases due to the intermediate formation of α/α$_2$. A minimum fraction of α/α_2, 2.9% at 1510 K, suggests that a disordering transition $\alpha_2 \rightarrow \alpha$ occurs at T_{eu}, above which the α-phase fraction increases again, due to the transformation of γ. The suggested transformation sequence $\gamma \rightarrow \alpha \rightarrow \alpha + \beta$ is evidenced as supported by the rates of transformation: Initially, α transforms to β, then as γ transforms more rapidly, the α-phase fraction catches up, while the β formation decreases. When γ disappears abruptly at $T_{\gamma,solv} = 1590$ K, the α-phase fraction increases sharply, and the β-phase decreases temporarily from 36.9% to 32.4%, which immediately recovers to 53% β + 47% α in the two-phase-field. Above 1660 K, the alloy starts to melt, while the ~50/50 solid-phase ratio is maintained.

The appearance of the β-phase at high pressure has important implications for the deformation behavior of this material, in particular during high-pressure near-conventional forging and shaping. Liss et al. [7] have demonstrated by in situ experiments that the plastic deformation behavior of a β-stabilized Ti-43.5Al-4Nb-1Mo-0.1B alloy in the (α + β) two-phase field at 1573 K occurs by fast dynamic recovery of the β-phase, while the co-existing α-grains deform primarily by slip, with slow dynamic recovery and results in even lesser dynamic recrystallization. In this scenario, the dynamically-recovering β-phase allows for deformation to accommodate the harder α-phase grains embedded therein. Deformation of the same material at 1493 K in the (α + β + γ)-phase field leads, qualitatively, to the conclusion that β is the actively deforming phase, while α and γ grains display greater stiffness [79]. This high kinetic activity of the β-phase is due to phonon softening, relating to extremely high self-diffusion, in the relatively open *bcc* lattice [47,80]. This results in rapid dynamic recrystallization, recovery and grain growth, which in essence, eradicates almost any deformation texture of this phase [7]. Microscopic and electron-backscatter-diffraction studies have subsequently confirmed this behavior [79,81].

These compositionally β-stabilized alloys are being developed for near-conventional, near-net-shape forging production since conventional (α + γ) TiAl alloys are prone to brittleness and highly anisotropic plastic behavior, which is related to the strongly ordered γ- and the hexagonal, *hcp* α-phases. Clemens and Mayer [10] proposed the development of an alloy that contains a relatively large fraction of β-phase at processing temperatures, while the fraction of β-phase is minimized at the expected operation temperatures in, for example, a Ti-43.5Al-4Nb-1Mo-0.1B alloy. However, the residual β-fraction would reduce the alloy's resistance to creep, and it is therefore essential to minimize the amount of β below 1100 K.

Although the β-phase is usually prone to rapid grain growth, Liss [7,39] and Kabra [82] found, by conducting in situ diffraction studies, that grain growth is inhibited in a multi-phase alloy, even in a β-phase containing titanium or zirconium alloy. This is in large part due to segregation, as different phases have distinct compositions and require long-range diffusion for grain growth to occur. Therefore, in two- or multiple-phase materials, the grain size is relatively stable, even at elevated forging temperatures.

Under the conditions pertaining to the present study, all of these requirements are optimally fulfilled, as no β-phase is present in the (α/α_2 + γ) two-phase field, up to ~1350 K, well above the expected operating temperatures of this material. Conversely, the ductile β-phase is abundant under high pressure in both three- and two-phase regions, potentially giving rise to a large thermo-mechanical processing window. Even a small β-phase fraction, between 5% and 15%, leads to good plasticity in both (α + β + γ)- and (α + β)-phase fields [79]. The present study highlights the potential to develop new processing routes by optimizing the fraction of the ductile β-phase by a judicious selection of a temperatures and pressure, in all three (α_2 + β + γ)-, (α + β + γ)- and (α + β)-phase fields. For example, grains of the ordered structures α_2 and γ may be largely conserved while being rearranged through plastic deformation of the β-phase. Alternatively, a minimum α fraction can be chosen for larger and superplastic deformations, minimizing anisotropy, i.e., texture, or even deformation without the γ-phase for extreme and rapid forging.

4. Conclusions

The response of a γ-titanium aluminide-based, two-phase alloy of composition Ti-45Al-7.5Nb-0.25C to high pressure and temperature has been investigated, leading to unprecedented results, regarding the evolution of lattice parameters, atomic volume and phase transformations.

The pressure-induced appearance of the *bcc* β-phase above 1350 K and at 9.6 GPa is the salient result for potential manufacturing applications, such as high-pressure forging, while at conventional pressures, this phase is not abundant under operational conditions, such as in the turbines of a jet-engine. It renders the possibility of good forgeability at achievable processing temperatures coupled with phase stability and creep resistance at the expected operating conditions of the alloy.

The large amount of β-phase found at high pressure and temperature, supported by the pressure-temperature phase diagram of pure titanium, suggests that this phase is stabilized at much lower concentrations of niobium, and this can ultimately lead to the development of niobium-free, binary titanium aluminides.

Under room temperature pressure loading, γ-phase gradually transforms to α/α_2, driven by local disordering of γ, while atoms are pressed towards their close-packed *fcc* positions. Since the γ $\rightarrow \alpha/\alpha_2$ transition is largely displacive, planar faults build up on all of the equivalent γ {111} planes, producing ultrafine lathes in a nano-Widmanstätten or basket-weave arrangement, introducing more disorder into both co-existing phases.

No phase transformations have been reported in the literature on pressurizing single-phase γ or α_2 titanium aluminides. The fact that we are working in a two-phase field allows gradual transformation between the two during loading.

Due to the ordering energy in the γ-phase being the highest, the system is forced to transform, γ $\rightarrow \alpha/\alpha_2$, even though the newly-formed product phase occupies a larger volume per atom than the mother phase. However, the total volume of the specimen, when weighted by the phase fraction and the individual bulk moduli, is still reduced with increasing pressure.

A Murnaghan-Birch equation of state has been fitted to the two individual, co-existing phases γ and α/α_2, as well as to the total weighted volume reduction. With values around 146 GPa, the bulk moduli *K*, fitted to the individual phases, are significantly higher than the values reported in the literature (see Table 1). Conversely, its pressure derivative, *K'*, is an order of magnitude smaller than the values reported in the literature, since the volume response to pressure is accommodated by the γ $\rightarrow \alpha/\alpha_2$ phase transformation, underlined by the even higher total bulk modulus, fitted to the weighted atomic volume.

Accordingly, a Murnaghan-Birch type equation, based on solely attractive and repulsive atomic potentials, is insufficient to describe the present case: a term describing the phase transformations together with the above-mentioned ordering energies needs to be introduced.

The room-temperature transformation builds up a large amount of mechanically-stored energy, which eventually balances the disorder energy in the γ-phase. Upon heating under pressure, the system largely recovers between 750 K and 810 K, following the ambient-pressure phase fractions up to 1250 K.

Under high pressure, the solid-solid transition temperatures are shifted to higher values. The fraction of γ-phase increases to a maximum at 1420 K, while at ambient pressure, the maximum is reached at 1250 K. The sharp γ-phase dissolution temperature $T_{\gamma,\text{solv}}$ is increased, as well as the eutectoid temperature T_{eu} for $\alpha_2 \rightarrow \alpha$ is expected at the minimum of the α fraction (see Table 2).

Table 2. Transition and other distinct temperatures. Ambient pressure values after [29].

Ambient Pressure [29]			High Pressure: 9.6 GPa				
Max γ	Eutectoid	γ-Solvus	Max γ	Min α	γ-Solvus	β Start	Solidus
$T_{\gamma,\text{max}}$ (0)	T_{eu} (0)	T_α $T_{\gamma,\text{solv}}$ (0)	$T_{\gamma,\text{max}}$	$T_{\alpha,\text{min}}$ (T_{eu})	$T_{\gamma,\text{solv}}$	$T_{\beta,\text{start}}$	T_{m}
1250 K	1453 K	1565 K	1420 K	1510 K	1590 K	1350 K	1660 K

Starting at ~1350 K, approximately half of the α-/α_2-phase transforms to the high-pressure β-phase, which plays an important role in the transformation kinetics. At T_{eu}, the α-/α_2-phase is at a minimum, and above T_γ, it increases again, before the solid starts to melt above >1660 K.

The ductile β-phase is present in three regions, namely $(\alpha_2 + \beta + \gamma)$, $(\alpha + \beta + \gamma)$ and $(\alpha + \beta)$, while the highly anisotropic phases α/α_2 show a minimum of 2.9%, which potentially allows for the adjustment of processing parameters to property requirements and optimization.

The present study is the first of its kind in a two-phase titanium aluminide intermetallic, a composition of considerable technological importance for high-temperature applications and which has been the focus of fundamental research for over 25 years. Recently, titanium aluminides have been successfully implemented as high-temperature components in airplane jet engines. Regarding the development of giga-newton super presses, near-conventional forging of such material under high pressure becomes feasible. While the first fundamental results under high pressure and high temperature have been presented here, there remains a number of questions, such as more detailed studies of the pressurization process and an equation of state incorporating pressure, temperature, composition and atomic order. The latter should lead to neutron scattering experiments, feasible at the PLANET beamline of the J-PARC facility [83], not only to investigate ordered α_2, but also ordered β_o. Based on the pressure-temperature phase diagram of pure titanium, we predict that the high-temperature, high-pressure-induced β-phase is stabilized over a wide range of TiAl-based alloys, including binary $(\alpha_2 + \gamma)$ titanium aluminides, which opens an extraordinary variety of alloy designs, processing and applications.

Note that essential to this study is the employment of a large volume pressure apparatus, which allows the evolution of a realistic microstructure in a realistic sample volume, rather than a micro- to nanometer-sized sample in a diamond anvil cell. Modern multi-anvil apparatuses, such as PLANET, also allow the addition of uni-axial stress components [84], enabling the study of in situ plastic deformation under high pressure. Similar complex high-pressure machines evolve at synchrotron sources [42,43], and the usage of two-dimensional detectors [85] has also been realized, which then will allow the investigation of various deformation mechanisms and kinetics, by evaluation, such as in a materials oscilloscope [38,39].

Acknowledgments: The authors are grateful to Dr. habil. Arno Bartels, Technische Universität Hamburg-Harburg, Germany, for providing the sample material. M.R. greatly acknowledges travel funding from the School of Mechanical, Mechatronics & Materials Engineering, University of Wollongong, Australia. The synchrotron radiation experiments were performed at the beamline BL04B1 of SPring-8 with the approval of the Japan Synchrotron Radiation Research Institute (JASRI) (Proposal No. 2013B1157).

Author Contributions: The research was conceived of by K.-D.L., K.A., R.J.D. and H.S., who applied for the beam time. K.-I.F. and Y.H. designed the experimental setup and the pressure cell, which was manufactured by A.S., K.-I.F. and T.S., H.S. manufactured the samples. H.S., K.A., M.R. and K.-I.F. assembled the sample cells, while K.-I.F., M.R., Y.H. and K.-D.L. prepared the instrument. The diffraction experiments were run by all, but R.J.D., K.-D.L. undertook all post-experimental data analysis and wrote the first version of the manuscript. All authors checked and improved the manuscript. Figure 11 shows the investigators and support staff (Y.H. and R.J.D. are missing).

Figure 11. The experimentalists: support staff, Shuoyuan Zhang, and some of the authors, Koichi Akita, Mark Reid, Ayumi Shiro, Ken-ichi Funakoshi, Takahisa Shobu, Hiroshi Suzuki and Klaus-Dieter Liss.

Conflicts of Interest: The authors declare no conflict of interest.

Abbreviations

fcc	face-centered cubic
bcc	body-centered cubic
hcp	hexagonal close-packed
YAG	yttrium aluminium garnet, $Y_3Al_5O_{12}$
CCD	charged coupled device
SSD	solid-state detector
TNB	TiAl-Nb alloy

References

1. Lipsitt, H.A. Titanium Aluminides—An Overview. *MRS Proc.* **1984**, *39*, 351–364. [CrossRef]
2. Kothari, K.; Radhakrishnan, R.; Wereley, N.M. Advances in gamma titanium aluminides and their manufacturing techniques. *Prog. Aerosp. Sci.* **2012**, *55*, 1–16. [CrossRef]
3. Gupta, R.K.; Ramkumar, P. Titanium Aluminides for Metallic Thermal Protection System of Reusable Space Transportation Vehicle: A Review. *Front. Aerosp. Eng.* **2015**, *4*, 14–19. [CrossRef]
4. Clemens, H.; Mayer, S. Intermetallic titanium aluminides in aerospace applications—Processing, microstructure and properties. *Mater. High Temp.* **2016**. [CrossRef]
5. Yang, R. Advances and challenges of TiAl base alloys. *Acta Metall. Sin.* **2015**, *51*, 129–147.
6. Clemens, H.; Schloffer, M.; Schwaighofer, E.; Werner, R.; Gaitzenauer, A.; Rashkova, B.; Schmoelzer, T.; Pippan, R.; Mayer, S. Advanced β-Solidifying Titanium Aluminides—Development Status and Perspectives. In *Symposium JJ—Intermetallic-Based Alloys—Science, Technology and Applications*; Materials Research Society: Warrendale, PA, USA, 2013; Volume 1516, pp. 3–16.
7. Liss, K.-D.; Schmoelzer, T.; Yan, K.; Reid, M.; Peel, M.; Dippenaar, R.; Clemens, H. In situ study of dynamic recrystallization and hot deformation behavior of a multiphase titanium aluminide alloy. *J. Appl. Phys.* **2009**, *106*, 113526:1–113526:6. [CrossRef]
8. Appel, F.; Oehring, M.; Paul, J.D.H.; Klinkenberg, C.; Carneiro, T. Physical aspects of hot-working gamma-based titanium aluminides. *Intermetallics* **2004**, *12*, 791–802. [CrossRef]
9. Erdely, P.; Schmoelzer, T.; Schwaighofer, E.; Clemens, H.; Staron, P.; Stark, A.; Liss, K.-D.; Mayer, S. In Situ Characterization Techniques Based on Synchrotron Radiation and Neutrons Applied for the Development of an Engineering Intermetallic Titanium Aluminide Alloy. *Metals* **2016**, *6*. [CrossRef]
10. Clemens, H.; Mayer, S. Design, Processing, Microstructure, Properties, and Applications of Advanced Intermetallic TiAl Alloys. *Adv. Eng. Mater.* **2013**, *15*, 191–215. [CrossRef]
11. Zhilyaev, A.P.; Langdon, T.G. Using high-pressure torsion for metal processing: Fundamentals and applications. *Prog. Mater. Sci.* **2008**, *53*, 893–979. [CrossRef]
12. Yan, K.; Bhattacharyya, D.; Lian, Q.; Kabra, S.; Kawasaki, M.; Carr, D.G.; Callaghan, M.D.; Avdeev, M.; Li, H.; Wang, Y.; et al. Martensitic Phase Transformation and Deformation Behavior of Fe-Mn-C-Al Twinning-Induced Plasticity Steel during High-Pressure Torsion. *Adv. Eng. Mater.* **2014**, *16*, 927–932. [CrossRef]
13. Edalati, K.; Toh, S.; Iwaoka, H.; Watanabe, M.; Horita, Z.; Kashioka, D.; Kishida, K.; Inui, H. Ultrahigh strength and high plasticity in TiAl intermetallics with bimodal grain structure and nanotwins. *Scr. Mater.* **2012**, *67*, 814–817. [CrossRef]
14. Alhamidi, A.; Edalati, K.; Horita, Z. Production of nanograined intermetallics using high-pressure torsion. *Mater. Res.* **2013**, *16*, 672–678. [CrossRef]
15. Krämer, L.; Kormout, K.S.; Setman, D.; Champion, Y.; Pippan, R. Production of Bulk Metallic Glasses by Severe Plastic Deformation. *Metals* **2015**, *5*, 720–729. [CrossRef]
16. France-Métallurgie. Blog Archive China's 80,000-Ton Press Forge Almost Ready for Use (US). Available online: http://www.france-metallurgie.com/index.php/2013/03/18/chinas-80000-ton-press-forge-almost-ready-for-use-us/ (accessed on 10 June 2016).
17. Altan, T.; Semiatin, S.L. *Feasibility of Using a Large Press (80,000–200,000 Ton) for Manufacturing Future Components on Army Systems*; U.S. Army Tank-Automotive Command Research and Development Center: Warren, MI, USA, 1983.

18. The Machines that Made the Jet Age /Boing Boing. Available online: http://boingboing.net/2012/02/13/machines.html (accessed on 10 June 2016).

19. Huber, D.; Werner, R.; Clemens, H.; Stockinger, M. Influence of process parameter variation during thermo-mechanical processing of an intermetallic β-stabilized γ-TiAl based alloy. *Mater. Charact.* **2015**, *109*, 116–121. [CrossRef]

20. Li, X.-S.; Wang, H.-Y.; Li, C.-Y.; Mi, G.-F.; Hu, Q.-K. Structural and Thermodynamic Properties of TiAl intermetallics under High Pressure. *Commun. Theor. Phys.* **2012**, *57*, 141–144. [CrossRef]

21. Young, D.A. *Phase Diagrams of the Elements*; University of California Lawrence Livermore Laboratory: Livermore, CA, USA, 1975.

22. Jayaraman, A.; Klement, W.; Kennedy, G.C. Solid-Solid Transitions in Titanium and Zirconium at High Pressures. *Phys. Rev.* **1963**, *131*, 644–649. [CrossRef]

23. Bystrzanowski, S.; Bartels, A.; Stark, A.; Gerling, R.; Schimansky, F.-P.; Clemens, H. Evolution of microstructure and texture in Ti-46Al-9Nb sheet material during tensile flow at elevated temperatures. *Intermetallics* **2010**, *18*, 1046–1055. [CrossRef]

24. Appel, F.; Brossmann, U.; Christoph, U.; Eggert, S.; Janschek, P.; Lorenz, U.; Müllauer, J.; Oehring, M.; Paul, J.D.H. Recent Progress in the Development of Gamma Titanium Aluminide Alloys. *Adv. Eng. Mater.* **2000**, *2*, 699–720. [CrossRef]

25. Appel, F.; Oehring, M.; Wagner, R. Novel design concepts for gamma-base titanium aluminide alloys. *Intermetallics* **2000**, *8*, 1283–1312. [CrossRef]

26. Appel, F. Recent Developments in the Design and Processing of Gamma-Based Titanium Aluminide Alloys. *Mater. Sci. Forum* **2003**, *426–432*, 91–98. [CrossRef]

27. Bystrzanowski, S. *Creep Behavior and Microstructure Stability of the Ti-46Al-9Nb Sheet Material*; Shaker Verlag GmbH: Aachen, Germany, 2005.

28. Gerling, R.; Clemens, H.; Schimansky, F.P. Powder Metallurgical Processing of Intermetallic Gamma Titanium Aluminides. *Adv. Eng. Mater.* **2004**, *6*, 23–38. [CrossRef]

29. Chladil, H.F.; Clemens, H.; Leitner, H.; Bartels, A.; Gerling, R.; Schimansky, F.-P.; Kremmer, S. Phase Transformations in High Niobium and Carbon Containing γ-TiAl Based Alloys. *Intermetallics* **2006**, *14*, 1194–1198. [CrossRef]

30. Chladil, H.F.; Clemens, H.; Zickler, G.A.; Takeyama, M.; Kozeschnik, E.; Bartels, A.; Buslaps, T.; Gerling, R.; Kremmer, S.; Yeoh, L.; et al. Experimental studies and thermodynamic simulation of phase transformations in high Nb containing gamma-TiAl based alloys. *Int. J. Mater. Res.* **2007**, *98*, 1131–1137. [CrossRef]

31. Yeoh, L.A.; Liss, K.-D.; Bartels, A.; Chladil, H.; Avdeev, M.; Clemens, H.; Gerling, R.; Buslaps, T. In situ high-energy X-ray diffraction study and quantitative phase analysis in the alpha plus gamma phase field of titanium aluminides. *Scr. Mater.* **2007**, *57*, 1145–1148. [CrossRef]

32. Pecker, S.; Eliezer, S.; Fisher, D.; Henis, Z.; Zinamon, Z. A multiphase equation of state of three solid phases, liquid, and gas for titanium. *J. Appl. Phys.* **2005**, *98*, 043516:1–043516:12. [CrossRef]

33. Greene, R.G.; Luo, H.; Ruoff, A.L. Al as a Simple Solid: High Pressure Study to 220 GPa (2.2 Mbar). *Phys. Rev. Lett.* **1994**, *73*, 2075–2078. [CrossRef] [PubMed]

34. *Quantum Beam Science—An Open Access Journal from MDPI*; Rittman, M., Liss, K.-D., Eds.; MDPI: Basel, Switzerland, 2016.

35. Shull, R.D.; Cline, J.P. High Temperature X-ray Diffractometry of Ti-Al Alloys. In *Materials Chemistry at High Temperatures*; Hastie, J.W., Ed.; Humana Press: Totowa, NJ, USA, 1990; pp. 95–117.

36. Liss, K.-D.; Bartels, A.; Clemens, H.; Bystrzanowski, S.; Stark, A.; Buslaps, T.; Schimansky, F.-P.; Gerling, R.; Scheu, C.; Schreyer, A. Recrystallization and phase transitions in a gamma-TiAl-based alloy as observed by ex situ and in situ high-energy X-ray diffraction. *Acta Mater.* **2006**, *54*, 3721–3735. [CrossRef]

37. Liss, K.-D.; Bartels, A.; Schreyer, A.; Clemens, H. High-energy X-rays: A tool for advanced bulk investigations in materials science and physics. *Textures Microstruct.* **2003**, *35*, 219–252. [CrossRef]

38. Liss, K.-D. In situ diffraction studies related to thermo-mechanical processes. In Proceedings of the ICCE-23: Annual International Conference on Composites and Nano-Engineering, Chengdu, China, 12–18 July 2015.

39. Liss, K.-D.; Yan, K. Thermo-mechanical processing in a synchrotron beam. *Mater. Sci. Eng. A* **2010**, *528*, 11–27. [CrossRef]

40. Watson, I.J.; Liss, K.-D.; Clemens, H.; Wallgram, W.; Schmoelzer, T.; Hansen, T.C.; Reid, M. In Situ Characterization of a Nb and Mo Containing gamma-TiAl Based Alloy Using Neutron Diffraction and High-Temperature Microscopy. *Adv. Eng. Mater.* **2009**, *11*, 932–937. [CrossRef]
41. Kabra, S.; Yan, K.; Mayer, S.; Schmoelzer, T.; Reid, M.; Dippenaar, R.; Clemens, H.; Liss, K.-D. Phase transition and ordering behavior of ternary Ti-Al-Mo alloys using in situ neutron diffraction. *Int. J. Mater. Res.* **2011**, *102*, 697–702. [CrossRef]
42. Chen, B.; Lin, J.-F.; Chen, J.; Zhang, H.; Zeng, Q. Synchrotron-based High Pressure Research in Materials Science. *MRS Bull.* **2016**, *41*, 473–478. [CrossRef]
43. Liss, K.-D.; Chen, K. Frontiers of synchrotron research in materials science. *MRS Bull.* **2016**, *41*, 435–441. [CrossRef]
44. Errandonea, D. Transition metals: Can metals be a liquid glass? *Nat. Mater.* **2009**, *8*, 170–171. [CrossRef] [PubMed]
45. Utsumi, W.; Funakoshi, K.; Urakawa, S.; Yamakata, M.; Tsuji, K.; Konishi, H.; Shimomura, O. SPring-8 Beamlines for High Pressure Science with Multi-Anvil Apparatus. *Rev. High Press. Sci. Technol.* **1998**, *7*, 1484–1486. [CrossRef]
46. Utsumi, W.; Funakoshi, K.; Katayama, Y.; Yamakata, M.; Okada, T.; Shimomura, O. High-pressure science with a multi-anvil apparatus at SPring-8. *J. Phys. Condens. Matter* **2002**, *14*. [CrossRef]
47. Hattori, T.; Saitoh, H.; Kaneko, H.; Okajima, Y.; Aoki, K.; Utsumi, W. Does Bulk Metallic Glass of Elemental Zr and Ti Exist? *Phys. Rev. Lett.* **2006**, *96*. [CrossRef] [PubMed]
48. Errandonea, D.; Meng, Y.; Somayazulu, M.; Häusermann, D. Pressure-induced transition in titanium metal: A systematic study of the effects of uniaxial stress. *Phys. B Condens. Matter* **2005**, *355*, 116–125. [CrossRef]
49. Tange, Y.; Higo, Y. *BL04B1: In Situ Observation of High Pressure Phase Change of Simple Material*; SPring-8, Asia Oceania Forum for Synchrotron Radiation Research: Kouto, Japan, 2007.
50. Katsura, T.; Funakoshi, K.; Kubo, A.; Nishiyama, N.; Tange, Y.; Sueda, Y.; Kubo, T.; Utsumi, W. A large-volume high-pressure and high-temperature apparatus for in situ X-ray observation, 'SPEED-Mk.II'. *Phys. Earth Planet. Int.* **2004**, *143–144*, 497–506. [CrossRef]
51. Speziale, S.; Zha, C.-S.; Duffy, T.S.; Hemley, R.J.; Mao, H. Quasi-hydrostatic compression of magnesium oxide to 52 GPa: Implications for the pressure-volume-temperature equation of state. *J. Geophys. Res. Solid Earth* **2001**, *106*, 515–528. [CrossRef]
52. Shim, S.-H.; Duffy, T.S.; Takemura, K. Equation of state of gold and its application to the phase boundaries near 660 km depth in Earth's mantle. *Earth Planet. Sci. Lett.* **2002**, *203*, 729–739. [CrossRef]
53. Seto, Y.; Nishio-Hamane, D.; Nagai, T.; N. Sata, N. Development of a software suite on X-ray diffraction experiments. *Rev. High Pressure Sci. Technol.* **2010**, *20*, 269–276. [CrossRef]
54. Lutterotti, L. Total pattern fitting for the combined size-strain-stress-texture determination in thin film diffraction. *Nucl. Instrum. Methods Phys. Res. Sect. B* **2010**, *268*, 334–340. [CrossRef]
55. Zhang, J.; Zhao, Y.; Hixson, R.S.; Gray, G.T.; Wang, L.; Utsumi, W.; Saito, H.; Hattori, T. Thermal equations of state for titanium obtained by high pressure/temperature diffraction studies. *Phys. Rev. B* **2008**, *78*, 054119:1–054119:7. [CrossRef]
56. *Titanium-Card No 44-1294*; Joint Committee on Powder Diffraction Standards (JCPDS)—International Center for Diffraction Data (ICDD): Washington, DC, USA, 1999.
57. Liss, K.-D.; Whitfield, R.E.; Xu, W.; Buslaps, T.; Yeoh, L.A.; Wu, X.; Zhang, D.; Xia, K. In situ synchrotron high-energy X-ray diffraction analysis on phase transformations in Ti-Al alloys processed by equal-channel angular pressing. *J. Synchrotron Radiat.* **2009**, *16*, 825–834. [CrossRef] [PubMed]
58. Yan, K.; Carr, D.G.; Kabra, S.; Reid, M.; Studer, A.; Harrison, R.P.; Dippenaar, R.; Liss, K.-D. In Situ Characterization of Lattice Structure Evolution during Phase Transformation of Zr-2.5Nb. *Adv. Eng. Mater.* **2011**, *13*, 882–886. [CrossRef]
59. Dubrovinskaia, N.A.; Vennström, M.; Abrikosov, I.A.; Ahuja, R.; Ravindran, P.; Andersson, Y.; Eriksson, O.; Dmitriev, V.; Dubrovinsky, L.S. Absence of a pressure-induced structural phase transition in Ti_3Al up to 25 GPa. *Phys. Rev. B* **2000**, *63*, 024106:1–024106:5. [CrossRef]
60. Asta, M.; de Fontaine, D.; van Schilfgaarde, M. First-principles study of phase stability of Ti-Al intermetallic compounds. *J. Mater. Res.* **1993**, *8*, 2554–2568. [CrossRef]
61. Ghosh, G.; Asta, M. First-principles calculation of structural energetics of Al-TM (TM = Ti, Zr, Hf) intermetallics. *Acta Mater.* **2005**, *53*, 3225–3252. [CrossRef]

62. Liss, K.-D.; Stark, A.; Bartels, A.; Clemens, H.; Buslaps, T.; Phelan, D.; Yeoh, L.A. Directional atomic rearrangements during transformations between the alpha- and gamma-phases in titanium aluminides. *Adv. Eng. Mater.* **2008**, *10*, 389–392. [CrossRef]

63. Liss, K.-D.; Bartels, A.; Clemens, H.; Stark, A.; Buslaps, T.; Phelan, D.; Yeoh, L.A. In situ characterization of phase transformations and microstructure evolution in a gamma-TiAl based alloy. In *Structural Aluminides for Elevated Temperature Applications*; Kim, Y.-W., Morris, D., Yang, R., Leyens, C., Eds.; TMS (The Minerals, Metals & Materials Society): Warrendale, PA, USA, 2008; pp. 137–144.

64. Cha, L.; Scheu, C.; Clemens, H.; Chladil, H.F.; Dehm, G.; Gerling, R.; Bartels, A. Nanometer-scaled lamellar microstructures in Ti-45Al-7.5Nb-(0; 0.5)C alloys and their influence on hardness. *Intermetallics* **2008**, *16*, 868–875. [CrossRef]

65. Srinivasarao, B.; Zhilyaev, A.P.; Muñoz-Moreno, R.; Pérez-Prado, M.T. Effect of high pressure torsion on the microstructure evolution of a gamma Ti-45Al-2Nb-2Mn-0.8 vol. % TiB2 alloy. *J. Mater. Sci.*, **2013**, *48*, 4599–4605. [CrossRef]

66. Schuster, J.C.; Palm, M. Reassessment of the binary Aluminum-Titanium phase diagram. *J. Phase Equilib. Diffus.* **2006**, *27*, 255–277. [CrossRef]

67. Anderson, D.L. Thermodynamics and Equations of State. In *Theory of the Earth*; Blackwell Scientific Publications: Boston, MA, USA, 1989; pp. 79–102.

68. Singh, R.S. High pressure properties of metal using various equations of state. *J. Integr. Sci. Technol.* **2013**, *1*, 48–53.

69. Birch, F. Elasticity and constitution of the Earth's interior. *J. Geophys. Res.* **1952**, *57*, 227–286. [CrossRef]

70. Murnaghan, F.D. The Compressibility of Media under Extreme Pressures. *Proc. Natl. Acad. Sci. USA* **1944**, *30*, 244–247. [CrossRef] [PubMed]

71. Dubrovinsky, L.; Dubrovinskaia, N.; Bykova, E.; Bykov, M.; Prakapenka, V.; Prescher, C.; Glazyrin, K.; Liermann, H.; Hanfland, M.; Ekholm, M.; et al. The most incompressible metal osmium at static pressures above 750 gigapascals. *Nature* **2015**, *525*, 226–229. [CrossRef] [PubMed]

72. Vinet, P.; Rose, J.H.; Ferrante, J.; Smith, J.R. Universal features of the equation of state of solids. *J. Phys. Condens. Matter* **1989**, *1*, 1941–1963. [CrossRef]

73. Nishihara, Y.; Nakajima, Y.; Akashi, A.; Tsujino, N.; Takahashi, E.; Funakoshi, K.; Higo, Y. Isothermal compression of face-centered cubic iron. *Am. Mineral.* **2012**, *97*, 1417–1420. [CrossRef]

74. Seto, Y. Seto's Home Page. Available online: http://pmsl.planet.sci.kobe-u.ac.jp/~seto (accessed on 15 April 2016).

75. Stark, A.; Rackel, M.; Tankoua, A.T.; Oehring, M.; Schell, N.; Lottermoser, L.; Schreyer, A.; Pyczak, F. In Situ High-Energy X-ray Diffraction during hot-forming of a multiphase TiAl Alloy. *Metals* **2015**, *5*, 2252–2265. [CrossRef]

76. Stark, A.; Bartels, A.; Clemens, H.; Schimansky, F. On the Formation of Ordered ω-phase in High Nb Containing γ-TiAl Based Alloys. *Adv. Eng. Mater.* **2008**, *10*, 929–934. [CrossRef]

77. Rackel, M.; Stark, A.; Gabrisch, H.; Schimansky, F.; Schell, N.; Schreyer, A.; Pyczak, F. In situ synchrotron radiation measurements of orthorhombic phase formation in an advanced TiAl alloy with modulated microstructure. *MRS Proc.* **2015**, *1760*. [CrossRef]

78. Menon, E.S.K.; Fox, A.G.; Mahapatra, R. Accurate determination of the lattice parameters of γ-TiAl alloys. *J. Mater. Sci. Lett.* **1996**, *15*, 1231–1233. [CrossRef]

79. Schmoelzer, T.; Liss, K.; Kirchlechner, C.; Mayer, S.; Stark, A.; Peel, M.; Clemens, H. An in situ high-energy X-ray diffraction study on the hot-deformation behavior of a β-phase containing TiAl alloy. *Intermetallics* **2013**, *39*, 25–33. [CrossRef]

80. Petry, W.; Heiming, A.; Trampenau, J.; Alba, M.; Herzig, C.; Schober, H.R.; Vogl, G. Phonon dispersion of the bcc phase of group-IV metals. I. *bcc* titanium. *Phys. Rev. B* **1991**, *43*, 10933–10947. [CrossRef]

81. Schmoelzer, T.; Liss, K.; Rester, M.; Yan, K.; Stark, A.; Reid, M.; Peel, M.; Clemens, H. Dynamic Recovery and Recrystallization during Hot-Working in an Advanced TiAl Alloy. *Prakt. Metallogr. Pract. Metallogr.* **2011**, *48*, 632–642. [CrossRef]

82. Kabra, S.; Yan, K.; Carr, D.G.; Harrison, R.P.; Dippenaar, R.J.; Reid, M.; Liss, K. Defect dynamics in polycrystalline zirconium alloy probed in situ by primary extinction of neutron diffraction. *J. Appl. Phys.* **2013**, *113*, 063513:1–063513:8. [CrossRef]

83. Hattori, T.; Sano-Furukawa, A.; Arima, H.; Komatsu, K.; Yamada, A.; Inamura, Y.; Nakatani, T.; Seto, Y.; Nagai, T.; Utsumi, W.; et al. Design and Performance of High-Pressure PLANET Beamline at Pulsed Neutron Source at J-PARC. *Nucl. Instrum. Methods Phys. Res. Sect. Accel. Spectrom. Detect. Assoc. Equip.* **2015**, *780*, 55–67. [CrossRef]

84. Sano-Furukawa, A.; Hattori, T.; Arima, H.; Yamada, A.; Tabata, S.; Kondo, M.; Nakamura, A.; Kagi, H.; Yagi, T. Six-axis multi-anvil press for high-pressure, high-temperature neutron diffraction experiments. *Rev. Sci. Instrum.* **2014**, *85*, 113905:1–113905:8. [CrossRef] [PubMed]

85. Funakoshi, K.; Higo, Y.; Nishihara, Y. High-pressure two-dimensional angle-dispersive X-ray diffraction measurement system using a Kawai-type multianvil press at SPring-8. *J. Phys. Conf. Ser.* **2010**, *215*. [CrossRef]

metals

MDPI

Article

Forge-Hardened TiZr Null-Matrix Alloy for Neutron Scattering under Extreme Conditions

Takuo Okuchi [1],*, Akinori Hoshikawa [2] and Toru Ishigaki [2]

[1] Institute for Study of the Earth's Interior, Okayama University, 827 Yamada, Misasa, Tottori 682-0193, Japan
[2] Frontier Research Center for Applied Atomic Sciences, Ibaraki University, 162-1 Shirakata, Tokai, Naka, Ibaraki 319-1106, Japan; akinori.hoshikawa.eml@vc.ibaraki.ac.jp (A.H.); toru.ishigaki.01@vc.ibaraki.ac.jp (T.I.)
* Author to whom correspondence should be addressed; okuchi@misasa.okayama-u.ac.jp; Tel.: +81-858-43-1215; Fax: +81-858-43-2184.

Academic Editor: Klaus-Dieter Liss
Received: 8 October 2015; Accepted: 7 December 2015; Published: 9 December 2015

Abstract: For neutron scattering research that is performed under extreme conditions, such as high static pressures, high-strength metals that are transparent to the neutron beam are required. The diffraction of the neutron beam by the metal, which follows Bragg's law, can be completely removed by alloying two metallic elements that have coherent scattering lengths with opposite signs. An alloy of Ti and Zr, which is known as a TiZr null-matrix alloy, is an ideal combination for such purposes. In this study, we increased the hardness of a TiZr null-matrix alloy via extensive mechanical deformation at high temperatures. We successfully used the resulting product in a high-pressure cell designed for high-static-pressure neutron scattering. This hardened TiZr null-matrix alloy may play a complementary role to normal TiZr alloy in future neutron scattering research under extreme conditions.

Keywords: neutron scattering; null-matrix alloy; high pressure; TiZr

1. Introduction

To conduct neutron scattering research under extreme conditions, such as neutron diffraction at high static pressures, metals that are transparent to neutrons and have a large mechanical strength are required for creating suitable sample environments [1]. The ideal metals for such applications should have small coherent and incoherent scattering lengths to reduce both the Bragg diffraction and background signals. The coherent scattering length of a metal can be minimized by preparing an alloy of two elements that have coherent scattering lengths with opposite signs. Such alloys are collectively called "null-matrix alloys", and the representatives are Mn (68 atom %) and Cu (32 atom %), which are ductile and soft [2], and Ti (68 atom %) and Zr (32 atom %), which are brittle and hard [3]. The latter $Ti_{68}Zr_{32}$ alloy (hereinafter referred to as TiZr) has been extensively used to fabricate metal gaskets for compressing samples in high-pressure environments, which are essential components of typical high-pressure cells designed for crystal structure analysis [4–6]. TiZr has the highest mechanical strength as a null-matrix alloy, which is roughly comparable to that of stainless steels [1]. TiZr also exhibits a moderate elongation of 6%–8% to be used in the aforementioned application [7].

In this study, by considering the unique properties of TiZr, we attempted to further enhance its mechanical properties while maintaining its null-scattering behavior via extensive mechanical deformation at high temperatures at ambient pressure. After preparing and analyzing the resulting product, we tested the applicability of the treated alloy to high-static-pressure experiments by using it in a high-pressure cell designed for neutron scattering.

2. Experimental Methods

While a structural phase transition of TiZr null-matrix alloy was observed to occur as the effect of high static pressure [8], the effects of extensive deformation on its physical properties has not yet been studied. The metallographic textures and mechanical properties of Ti–Zr two-component α-alloy system can be modified and improved by subjecting them to plastic deformations (via mechanical forging) [9]. Therefore, here we prepared a cast billet of TiZr and applied extensive plastic deformation for it at ambient pressure [10,11]. A total of 85 kg of the TiZr alloy was cast and forged as a custom-made commercial order to Daido Bunseki Research, Inc. (Nagoya, Japan). In detail, the weighted Ti- and Zr-containing precursors were first melted together in a plasma progressive-casting furnace, where an electrode for the subsequent vacuum arc remelting (VAR) procedure was prepared. The electrode was then subjected to the VAR procedure, which produced a homogenized ingot after the remelting and recooling. The appearance of the ingot is shown in Figure 1, while the composition is shown in Table 1.

Figure 1. Photograph of the VAR-fabricated TiZr ingot, which was 240 mm in diameter, 384 mm in length, and 85.4 kg in weight.

Table 1. Chemical composition of the vacuum arc remelting (VAR)-fabricated TiZr ingot.

Composition wt. %							
Zr	C	Al	B	O	N	Fe	Ti *
47.1	0.008	0.01	<0.01	0.053	0.004	0.03	52.8

* Ti as the balance.

The VAR-fabricated TiZr ingot was then polished to remove any macroscopic cracks in the surface. An evacuated steel sheath was used to subject the billet to vacuum conditions during the forging process, which is an effective method for preventing the oxidation of the ingot. The ingot-containing steel sheath was heated to approximately 1100 °C and then forged immediately. The diameter of the ingot was then reduced to approximately 100 mm as the temperature decreased to around 700 °C.

The thinned ingot (Figure 2) was then cut, re-sealed in another steel sheath, and forged again into a number of cylinders with diameters of 20–30 mm. This process was conducted at temperatures within the β-phase field, and the final products were air-cooled into the α-phase [12]. The final products had a forging ratio (F_0/F) of more than 60, or $e = \ln(F_0/F) = 4$–5, where F_0 and F are the initial and final cross-sectional areas, respectively. These values were sufficient to produce the desired metallographic texture. Hereinafter, we will refer to the final products as "f-TiZr", where the "f" indicates that the resulting products were both "forged" and "finely textured".

Figure 2. Photograph of the forged ingot that is approximately 100 mm in diameter, which was mechanically thinned from the VAR-fabricated TiZr ingot that was 240 mm in diameter.

The metallographic textures and elemental concentration maps of the f-TiZr alloy were obtained with a field-emission scanning electron microscopy (FE-SEM; JEOL JSM-7001F, Akishima, Tokyo, Japan) setup that was capable of performing energy dispersive X-ray spectroscopy (EDS). The micro-Vickers hardness of the alloy was measured using a micro hardness tester with 300 g to 500 g of the applied force (Mitsutoyo HM-221, Kawasaki, Japan). Powder X-ray diffraction (XRD; Rigaku Smartlab, Akishima, Tokyo, Japan) was used to analyze the crystal structure, while neutron powder diffraction (NPD) was used to evaluate the coherent neutron scattering. The XRD patterns were obtained by using the para-focusing method with Cu Kα$_1$ radiation operated at 40 kV and 30 mA. The NPD patterns were obtained with an iMATERIA pulsed-neutron powder diffractometer at the Materials and Life Science Experimental Facility of the Japan Proton Accelerator Research Complex (J-PARC; Tokai, Japan). The neutron diffractometer covered a wide range of d-values (0.26–40 Å) in the double-frame operating mode [13].

We expected that the toughness and elongation of the f-TiZr would not be sufficient for certain applications that involve extensive deformations of the metal. Therefore, some of the samples were cut from the forged TiZr ingot with 100 mm in diameter (Figure 2) and were annealed and furnace-cooled at 660 °C for 6 h in an Ar atmosphere using a gas convertible vacuum furnace (FUA112DB, Advantec Toyo Kaisya, Ltd., Tokyo, Japan), which is an effective method for increasing the toughness and elongation of the alloy while reducing its hardness [9].

3. Results and Discussion

Figure 3 shows typical cross-sectional SEM images of the VAR-fabricated TiZr ingot before it was forged. The back-scattered electron images show a contrast on the averaged atomic number, where the compositional heterogeneity is heavily emphasized. Upon cooling the ingot after its solidification, the temperature passes through the phase boundary between the body-centered cubic *Im3m* (β) phase and the hexagonal close-packed $P6_3/mmc$ (α) phase, causing the recrystallization of α-platelets. This in turn creates the Widmanstätten patterns, which are similar to that observed in some furnace-cooled Ti–Zr alloys with different compositions [9]. The typical lamellar width in the VAR-fabricated TiZr ingot was 5–50 µm. TiZr alloys with such textures have been used as null-matrix alloys for neutron scattering.

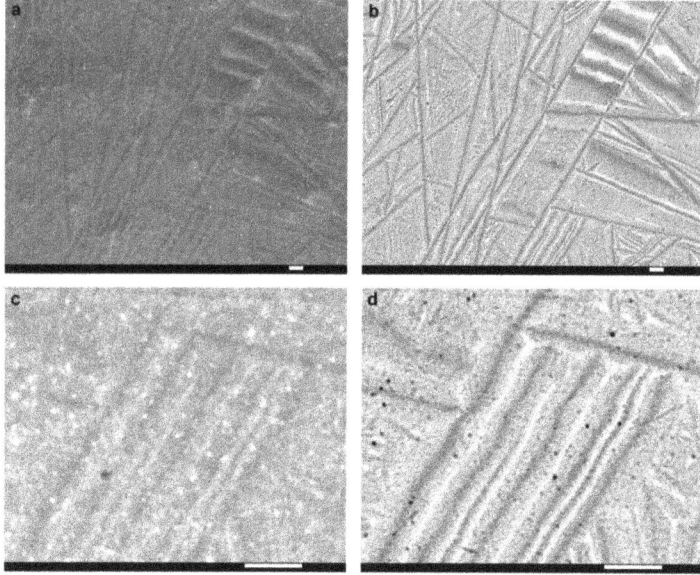

Figure 3. Cross-sectional (**a**) secondary-electron and (**b**) back-scattered-electron SEM images of the VAR-fabricated TiZr ingot before it was subjected to the mechanical forging. (**c,d**) Magnified images of areas in (**a**) and (**b**), respectively (magnification = 4×). All scale bars indicate a length of 10 µm.

Figure 4 shows SEM images and elemental concentration maps of a polished cross section of the f-TiZr alloy, which was cut from a forged cylinder that was 20 mm in diameter. The very fine metallographic textures shown in these images should be effective to increase the hardness and strength of the metal [14], which has been repeatedly confirmed by previous studies, including those of Ti-based alloy systems [15–18]. This is also shown in our measurements of micro-Vickers hardness of the f-TiZr alloy at H_v = 339, which is significantly higher than that of non-forged $Ti_{60}Zr_{40}$ or $Ti_{70}Zr_{30}$ alloy with H_v = 230~280 [19]. The typical width of the α-platelets in the f-TiZr alloy was 0.1–0.5 µm, which is approximately two orders of magnitude finer than that observed in the VAR-fabricated TiZr ingot (Figure 3).

Figure 4. Cross-sectional (**a**) secondary-electron and (**b**) back-scattered-electron SEM images of the f-TiZr alloy along with the scale bars with 10 μm in length. (**c**) Ti and (**d**) Zr concentration maps of the f-TiZr alloy obtained by EDS along with intensity scale bars (below each image where the left-end color shows no intensity) and averaged intensity (black line in each bar). All these images have identical spacial scale to that of Figure 3c,d.

Figure 5 shows a typical XRD pattern of the f-TiZr alloy. All of the reflections were successfully indexed to those of the α-phase, as indicated by the indices of each peak. A significant broadening of these reflections is consistent with a large internal strain within the α-platelets; this strain is essential for extensively hardening α-Ti alloys [9,19,20]. Figure 6 shows the NPD patterns of the f-TiZr alloy. The patterns show that there are no diffraction peaks from the f-TiZr for the range of *d*-values measured. Therefore, we have confirmed that the forging process performed on the TiZr alloy has not degraded its null-scattering properties.

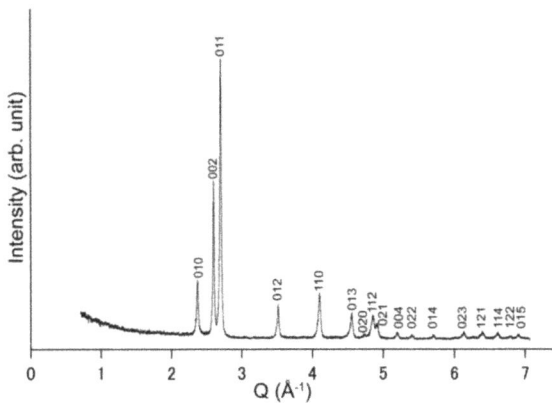

Figure 5. X-ray diffraction (XRD) pattern of the f-TiZr alloy. The Miller indices of the α-phase are shown next to the corresponding reflections.

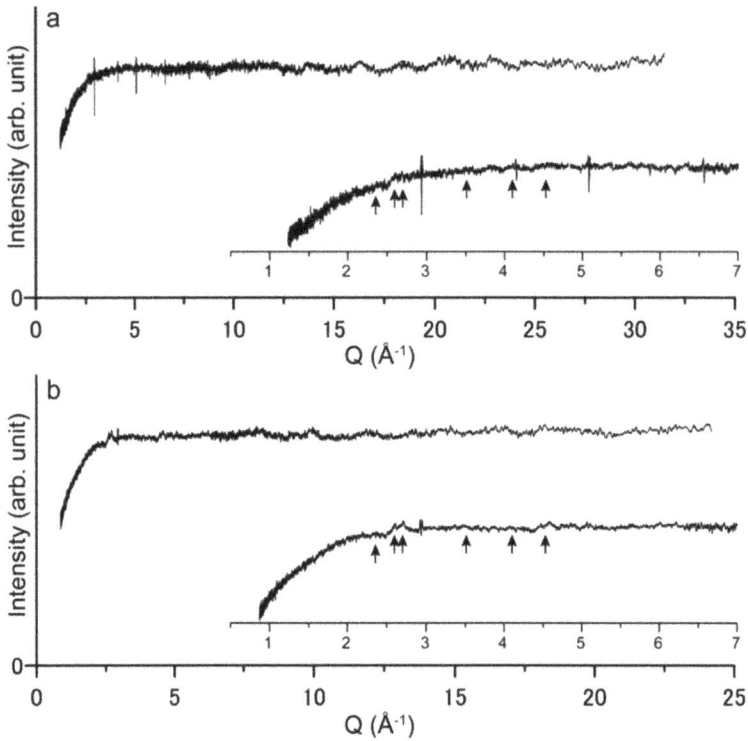

Figure 6. Neutron powder diffraction (NPD) patterns of the f-TiZr alloy: (**a**) NPD pattern from the high-resolution bank ($\Delta d/d \approx 0.16\%$) and (**b**) that from the special environment bank ($\Delta d/d \approx 0.5\%$). The intensities of both profiles initially increase with increasing Q, which is due to inelastic scattering. The inset profiles show the lowest Q regime which have the same vertical scale with the main figures. Arrows under both of the inset profiles indicate the positions of representative Bragg reflections that are observable when TiZr has a non-ideal composition; their indices include 010, 002, 011, 012, 110, and 013 for the α phase (see Figure 5). These Bragg reflections could be more evident for the TiZr used in the previous research (*i.e.*, Figure 2.5 in Reference [1]). A few spurious, narrow peaks involving both negative and positive anomalies present in the profiles (at 2.93, 4.16 and 5.11 Å^{-1}), which were not completely removed by the background-subtraction procedures. These are caused by the diffraction of the sample holder, which was composed of V.

Figure 7 shows SEM images and elemental concentration maps of the annealed f-TiZr alloy. The width of the lamellae is smaller than 10 μm, and thus, the lamellae have not returned to the original dimensions of the VAR-fabricated TiZr ingot. On the other hand, its well-ordered metallographic texture is similar to that of the VAR-fabricated TiZr ingot or a furnace-cooled alloy [9]. The measured micro-Vickers hardness of the annealed f-TiZr was $H_v = 279$, which was reduced from the original hardness of the f-TiZr ($H_v = 339$).

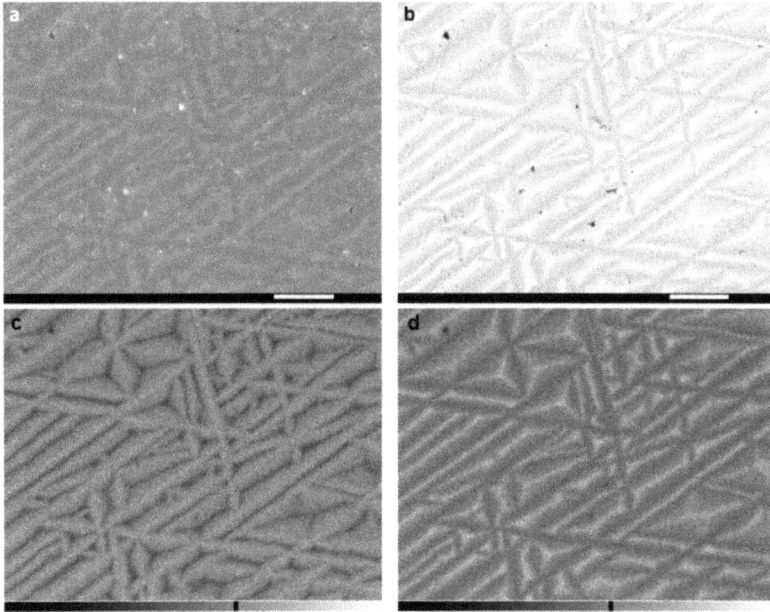

Figure 7. Cross-sectional (**a**) secondary-electron and (**b**) back-scattered-electron SEM images of the annealed f-TiZr alloy along with the scale bars with 10 μm in length. (**c**) Ti and (**d**) Zr concentration maps of the annealed f-TiZr alloy obtained by EDS along with intensity scale bars (below each image where the left-end color shows no intensity) and averaged intensity (black line in each bar). All these images have identical spacial scale to that of Figure 3c,d.

4. Application of the f-TiZr Alloy

A series of experiments were conducted by fabricating deformable metallic gaskets from the f-TiZr alloy, and static pressures greater than 15 GPa were successfully generated without the gaskets fracturing. We used a newly designed high-pressure cell that was optimized for neutron-scattering measurements with strong pulsed-neutron sources [21,22]. In brief, two anvils composed of single-crystal moissanite and sintered polycrystalline diamond with 5 mm culets were used to compress a powdered $Mg(OD)_2$ sample, which was confined within the f-TiZr gasket (initial thickness = 1 mm) and then compressed between the two anvils to 9 GPa. The total sample volume compressed to this high pressure was approximately 3 mm^3, which was sufficient for performing a precise structure refinement via NPD at J-PARC. The gasket was recovered after this compression experiment, and it is shown in Figure 8. Similar to the other successful cases, the gasket has not cracked, even after the extensive deformation.

The annealed f-TiZr alloy was also applied to high-pressure NPD measurements with various types of anvil geometry, the results of which were partially published elsewhere [23].

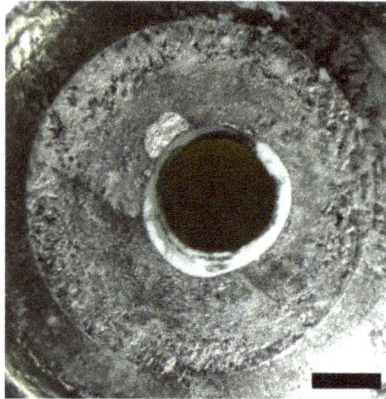

Figure 8. The recovered f-TiZr gasket, which experienced extensive deformation in the high-pressure cell. It was used to compress a powdered $Mg(OD)_2$ sample to 9 GPa. The scale bar indicates a length of 1 mm.

5. Conclusions

By applying extensive mechanical deformation at high temperatures to a TiZr alloy, the metallographic textures of the alloy became much finer, with the lamellae width decreasing by two orders of magnitude. The mechanical hardness of the TiZr was increased because of this change in the metallographic texture, which suggests that its strength was also increased, as previously observed in the general Ti–Zr two-component alloy system [9]. This f-TiZr alloy was successfully applied to experiments involving the generation of high static pressures, which requires the extensive deformation of the hard metal. In addition, the f-TiZr alloy was also used to fabricate a thin-walled micro-sample container, which enabled us to obtain a background-free NPD pattern with a pulsed-neutron source [24,25]. The successful application of the f-TiZr alloy has demonstrated that forged TiZr products may play a complementary role in future neutron scattering research.

Acknowledgments: The authors wish to thank T. Iikubo at Daido Bunseki Research Co, Ltd. and T. Matsumoto at the University of Tokyo for providing technical assistance. The machine time for the pulsed-neutron diffraction experiments was provided through the J-PARC MLF user program (No. 2013B0279). N. Tomioka at Okayama University is acknowledged for helping these experiments. Four anonymous referees are acknowledged for their comments to improve the manuscript. This work was supported in part by JSPS Grants-in-Aid for Scientific Research (No. 19GS0205, 26287135 and 15K13593).

Author Contributions: T.O. wrote this manuscript and performed the experiments to obtain all of the results except for the neutron diffraction patterns. T.O., A.H., and T.I. performed the experiments to obtain the neutron diffraction patterns. All authors contributed to the interpretation of the experimental data.

Conflicts of Interest: The authors declare no conflicts of interest.

References

1.   Klotz, S. *Techniques in High Pressure Neutron Scattering*; CRC Press: Boca Raton, FL, USA, 2012.

2.   Smith, J.H.; Vance, E.R.; Wheeler, D.A. A null-matrix alloy for neutron diffraction. *J. Phys. E Sci. Instrum.* **1968**, *1*, 945–947. [CrossRef]

3.   Sidhu, S.S.; Heaton, L.; Zauberis, D.D.; Campos, F.P. Neutron diffraction study of titanium-zirconium system. *J. Appl. Phys.* **1956**, *27*, 1040–1042. [CrossRef]

4.   Besson, J.M.; Nelmes, R.J.; Hamel, G.; Loveday, J.S.; Weill, G.; Hull, S. Neutron powder diffraction above 10 GPa. *Phys. B Condens. Matter* **1992**, *180–181*, 907–910. [CrossRef]

5.   Klotz, S.; Besson, J.M.; Hamel, G.; Nelmes, R.J.; Loveday, J.S.; Marshall, W.G.; Wilson, R.M. Neutron powder diffraction at pressures beyond 25 GPa. *Appl. Phys. Lett.* **1995**, *66*, 1735–1737. [CrossRef]

6. Xu, J.; Ding, Y.; Jacobsen, S.D.; Mao, H.K.; Hemley, R.J.; Zhang, J.; Qian, J.; Pantea, C.; Vogel, S.C.; Williams, D.J.; *et al.* Powder neutron diffraction of wustite ($Fe_{0.93}O$) to 12 GPa using large moissanite anvils. *High Press. Res.* **2004**, *24*, 247–253. [CrossRef]
7. Williams, D.N.; Wood, R.A.; Jaffee, R.I.; Ogden, H.R. The effects of zirconium in titanium-base alloys. *J. Less Common Met.* **1964**, *6*, 219–225. [CrossRef]
8. Zeidler, A.; Guthrie, M.; Salmon, P.S. Pressure-dependent structure of the null-scattering alloy $Ti_{0.676}Zr_{0.324}$. *High Press. Res.* **2005**, *35*, 239–246. [CrossRef]
9. Imgram, A.G.; Williams, D.N.; Ogden, H.R. Tensile properties of binary titanium-zirconium and titanium-hafnium alloys. *J. Less Common Met.* **1962**, *4*, 217–225. [CrossRef]
10. Meyers, M.A.; Mishra, A.; Benson, D.J. Mechanical properties of nanocrystalline materials. *Prog. Mater. Sci.* **2005**, *51*, 427–556. [CrossRef]
11. Sakai, T.; Belyakov, A.; Kaibyshev, R.; Miura, H.; Jonas, J.J. Dynamic and post-dynamic recrystallization under hot, cold and severe plastic deformation conditions. *Prog. Mater. Sci.* **2014**, *60*, 130–207. [CrossRef]
12. Blacktop, J.; Crangle, J.; Argent, B.B. The $\alpha \rightarrow \beta$ transformation in the Ti–Zr system and the influence of additions of up to 50 at. % Hf. *J. Less Common Met.* **1985**, *109*, 375–380. [CrossRef]
13. Ishigaki, T.; Hoshikawa, A.; Yonemura, M.; Morishima, T.; Kamiyama, T.; Oishi, R.; Aizawa, K.; Sakuma, T.; Tomota, Y.; Arai, M.; *et al.* IBARAKI materials design diffractometer (iMATERIA)—Versatile neutron diffractometer at J-PARC. *Nucl. Instrum. Meth. A* **2009**, *600*, 189–191. [CrossRef]
14. Hall, E.O. *Yield Point Phenomena in Metals and Alloys*; Plenum Press: New York, NY, USA, 1970; p. 316.
15. Herebtsov, S.; Lojkowski, W.; Mazurb, A.; Salishcheva, G. Structure and properties of hydrostatically extruded commercially pure titanium. *Mater. Sci. Eng. A* **2010**, *527*, 5596–5603. [CrossRef]
16. Salishchev, G.A.; Galeev, R.M.; Malysheva, S.P.; Zherebtsov, S.V.; Mironov, S.Y.; Valiakhmetov, O.R.; Ivanisenko, E.I. Formation of submicrocrystalline structure in titanium and titanium alloys and their mechanical properties. *Met. Sci. Heat Treat.* **2006**, *48*, 63–69. [CrossRef]
17. Valiev, R.Z.; Alexandrov, I.V. Paradox of strength and ductility in metals processed by severe plastic deformation. *J. Mater. Res.* **2002**, *17*, 5–8. [CrossRef]
18. Kobayashi, Y.; Tanaka, Y.; Matsuoka, K.; Kinoshita, K.; Miyamoto, Y.; Murata, H. Effect of forging ratio and grain size on tensile and fatigue strength of pure titanium forgings. *J. Soc. Mater. Sci. Jpn.* **2005**, *54*, 66–72. [CrossRef]
19. Kobayashi, E.; Matsumoto, S.; Doi, H.; Yoneyama, T.; Hamanaka, H. Mechanical properties of the binary titanium-zirconium alloys and their potential for biomedical materials. *J. Biomed. Mater. Res.* **1995**, *29*, 943–950. [CrossRef] [PubMed]
20. Dobromyslov, A.V.; Taluts, N.I. Structure investigation of quenched and tempered alloys of the Zr–Ti system. *Phys. Met. Metallogr.* **1987**, *63*, 114–120.
21. Okuchi, T.; Sasaki, S.; Ohno, Y.; Abe, J.; Arima, H.; Osakabe, T.; Hattori, T.; Sano-Furukawa, A.; Komatsu, K.; Kagi, H.; *et al.* Neutron powder diffraction of small-volume samples at high pressure using compact opposed-anvil cells and focused beam. *J. Phys. Conf. Ser.* **2012**. [CrossRef]
22. Okuchi, T.; Yoshida, M.; Ohno, Y.; Tomioka, N.; Purevjav, N.; Osakabe, T.; Harjo, S.; Abe, J.; Aizawa, K.; Sasaki, S. Pulsed neutron powder diffraction at high pressure by a capacity-increased sapphire anvil cell. *High Press. Res.* **2013**, *33*, 777–786. [CrossRef]
23. Okuchi, T.; Tomioka, N.; Purevjav, N.; Abe, J.; Harujo, S.; Gong, W. Structure refinement of sub-cubic-mm volume sample at high pressures by pulsed neutron powder diffraction: Application to brucite in an opposed anvil cell. *High Press. Res.* **2014**, *34*, 273–280. [CrossRef]
24. Purevjav, N.; Okuchi, T.; Tomioka, N.; Abe, J.; Harjo, S. Hydrogen site analysis of hydrous ringwoodite in mantle transition zone by pulsed neutron diffraction. *Geophys. Res. Lett.* **2014**, *41*, 6718–6724. [CrossRef]
25. Tomioka, N.; Okuchi, T.; Purevjav, N.; Abe, J.; Harjo, S. Hydrogen sites in the dense hydrous magnesian silicate phase E: A pulsed-neutron powder diffraction study. *Phys. Chem. Miner.*. in press.

![metals logo] *metals*

MDPI

Article

Monte Carlo Modelling of Single-Crystal Diffuse Scattering from Intermetallics

Darren J. Goossens

School of Physical, Environmental and Mathematical Sciences, University of New South Wales,
Canberra ACT 2600, Australia; d.goossens@adfa.edu.au; Tel.: +61-2-6268-8422; Fax: +61-2-6268-8786

Academic Editor: Klaus-Dieter Liss
Received: 2 December 2015; Accepted: 27 January 2016; Published: 4 February 2016

Abstract: Single-crystal diffuse scattering (SCDS) reveals detailed structural insights into materials. In particular, it is sensitive to two-body correlations, whereas traditional Bragg peak-based methods are sensitive to single-body correlations. This means that diffuse scattering is sensitive to ordering that persists for just a few unit cells: nanoscale order, sometimes referred to as "local structure", which is often crucial for understanding a material and its function. Metals and alloys were early candidates for SCDS studies because of the availability of large single crystals. While great progress has been made in areas like *ab initio* modelling and molecular dynamics, a place remains for Monte Carlo modelling of model crystals because of its ability to model very large systems; important when correlations are relatively long (though still finite) in range. This paper briefly outlines, and gives examples of, some Monte Carlo methods appropriate for the modelling of SCDS from metallic compounds, and considers data collection as well as analysis. Even if the interest in the material is driven primarily by magnetism or transport behaviour, an understanding of the local structure can underpin such studies and give an indication of nanoscale inhomogeneity.

Keywords: diffuse scattering; single crystal; short-range order; CePdSb; Kondo

1. Introduction

Short-range order (SRO) is present in almost all families of crystalline compounds, from metals to proteins [1–8]. SRO can influence electrical, magnetic and most other physical properties, including ferroelectricity, superconductivity and multiferroic behaviour.

SRO manifests in the diffuse scattering, the coherent scattered intensity which is not localised on the reciprocal lattice; in other words, it is found throughout reciprocal space, not just on the Bragg reflections at integer *hkl*. Thus, to best investigate the diffuse scattering it is necessary to survey a large region (area or volume) of reciprocal space with low noise and high dynamic range. This is not a trivial exercise, and much effort has gone into data collection and reduction [6,9–11].

Data are typically presented as reciprocal space cuts or sections, which essentially plot diffracted intensity as a function of position in reciprocal space.

Metals were an early test-bed for ways of modelling SRO, in particular chemical SRO as modelled by, for example, Cowley SRO parameters [12–15]. Cowley realised that Fourier transforming the diffuse intensity could give atomic pair correlations when the scattering admitted a direct interpretation, for example when looking at a diffuse peak that would sharpen to a Bragg spot on going through a phase transition. Warren and co-workers showed how the atomic size effect (the dependence of interatomic spacing on species, most simply conceptualised as thinking about atoms as being of different radii) caused asymmetries in the scattering [16]. When the system is relatively simple, sometimes an analytical form can be found to yield the distribution of scattering.

If the underlying crystallography is simple, it may be possible to use an essentially analytic analysis, as for example can be obtained by expanding the diffraction equations [17] and using conditional probabilities to express the various terms. These probabilities can then be adjusted and the expected scattering calculated.

However, in more complex cases, in particular systems containing many atomic species and/or in which the atoms form into clusters with their own structure factors that then conflate with the scattering from the defects and the local ordering, it is often difficult or impossible to interpret the scattering directly or to meaningfully invert it to get the real space structures. These, and cases where we must allow for displacive relaxation around defects, require a more model-based approach. When contrast between scatterers is weak (atoms nearby on the periodic table will have very similar X-ray scattering factors), it may be necessary to use neutron and/or X-ray single-crystal diffuse scattering (SCDS) data. Neutron diffraction requires larger crystals, which may be difficult to obtain, so it may be that X-ray SCDS is coupled with neutron pair distribution function analysis (PDF; [18–20]), obtained from polycrystalline specimens.

A wide range of local structures have been observed in metallic compounds, from classic examples like chemical substitution and resulting clustering or anti-clustering in alloys, through to subtle phenomena related to the atomic size effect and even the rotation of large motifs, such as the cages of atoms seen in complex intermetallics [21]. For relatively simple systems, recent advances allow almost direct interpretation of the diffuse scattering, while developments in detailed calculation methods, like density functional theory and molecular dynamics, allow direct calculation of low energy short-ranged order configurations when not too many atoms are required [22–25].

However, when many atoms are involved and the correlation lengths encompass many unit cells, the number of atoms involved is beyond the scope of such methods. Then, the ability to model a crystal of $>10^5$ atoms becomes useful. Methods like 3D-ΔPDF [26] offer what are almost "direct methods" for such systems and are currently a fascinating field of development. The reverse Monte Carlo (RMC) approach [19,27,28] offers a means to directly fit the diffuse scattering data, but can be limited in the size of simulation that can be implemented because of the way in which a single atomic move must have a significant effect on the goodness of fit of the model.

Thus, at this time, the most flexible approach remains the forward Monte Carlo (MC), though it has its own weaknesses, in particular one must posit the nature of the disorder and then find a means of introducing that disordered structure into the model, before calculating the Fourier transform of the model and testing the theory. The process can be slow; models are difficult to optimise; and knowing what to include in the model (what forms of disorder and how to induce them) requires considerable insight. Further, since disorder can take on so many forms, it is often necessary to write bespoke computer code to tackle a given problem, something which is time consuming and not conducive to broad acceptance of the technique.

This paper aims to very briefly look at Monte Carlo analysis of diffuse scattering, particularly as it pertains to metallic materials, alloys and the like. The fascinating field of quasicrystals, many of which are metallic, will not be covered. This field has been surveyed in a range of detailed and high quality presentations, which need not be repeated here [29–31].

2. Data Collection

The experiments considered here use large slices of reciprocal space, rather than collecting intensity at a few key scattering vectors. This allows elucidation of SRO that is anisotropic or only affects small regions of reciprocal space. Similarly, the use of pair distribution function and powder diffraction is not discussed, though both are very important techniques [3,18,19,32].

The quality and quantity of data required depends, of course, on the experiment being undertaken. Ideally, the different scatterers will have well-differentiated cross-sections for the radiation being used. If the disorder is anisotropic, then data that extend in three dimensions are desirable. If local ordering is only significant in, say, the *ab* plane, then collection of the *hk*0 section of reciprocal space may be sufficient. If quantitative comparison of the calculated SCDS with the observed is desired [33], the observed data must show low noise, few artefacts, and a background that can be removed either by subtraction of "blank" runs or some other method, like fitting a function to it. For qualitative comparison with calculations, showing whether features are present or not, for example, noisier data may be acceptable, and the less quantitative results of electron diffraction are also useful. Analysis of SCDS is often limited by the data that can be obtained, but as long as features in the scattering can be identified as "real", then some insight can be gained.

2.1. X-ray

Assuming that the X-ray source is a constant wavelength, monochromated source, volumes of diffuse scattering are collected by rotating a sample in front of an area detector. Earlier work often made use of a line counter [34], but the modern prevalence of area detectors has rendered this approach largely redundant.

The main variation is in the choice of detectors. In particular, while much important data collection has made use of image plates [31,35–42], the use of electronic counters that can provide a high dynamic range has become possible [43–45]. These have a much improved duty-cycle. Experiments with image plates at synchrotrons, where beams are very intense, can follow an exposure of a few seconds, rarely more than 30 s, with a readout time of a minute or more, which is not good use of the intense and expensive beam.

Figure 1 presents a generic schematic diagram of a constant wavelength experiment. The main parameters include the sample to detector distance, the wavelength and whether the beam path is enclosed in a vacuum or He-filled vessel, which reduces noise, or is through air, which tends to result in intense forward scattering that requires careful correction and collections of "blank" runs, which can then be subtracted from the data. Other corrections may be required depending on the nature of the detector and the stability of the beam and the nature of the beam. If a laboratory source is being used, the compromise between intensity and quality of monochromation can result in the beam possessing a white component, which is much weaker than the characteristic radiation, but nevertheless results in a radial streak through the Bragg peaks, because of the long exposures required to reveal the diffuse scattering. Other artefacts that would not be apparent in an experiment using shorter exposure times may also be revealed. These include X-rays that pass through the image plate and scatter off components of the detector and re-enter the image plate from behind (this was discovered when the shadows of the image plate mounting screws were projected onto the detector(!)), as well as resolution streaks, discussed in Figure 15 of [46].

The high intensities at a synchrotron can cause problems when the area detector intercepts a Bragg reflection; depending on the design of the detector, a wire or a pixel can become saturated. In CCD devices, charge can spill over and contaminate surrounding pixels (deep depletion devices overcome this somewhat); in a wire detector, a bright spot anywhere on the wire may force the removal from the data set of all "pixels" measured by that wire [11].

Other issues include ghosting, when a pixel value on a measurement is partly influenced by the previous measurement. This can happen in image plates, where a very highly exposed pixel may not be fully "reset" by the readout, and thus, its value on the next exposure is not correct.

Traditionally, flat reciprocal space cuts have been reconstructed from the curved sections collected in an experiment such as that in Figure 1. Flat sections generally admit to easier visual interpretation, as the normal is everywhere the same and corresponds to a particular reciprocal space direction. However, from a computational point of view there is little difference between calculating the scattering in a flat or curved section. Further, at high X-ray energies the radius of the Ewald sphere is so large that each exposure is almost a flat section in reciprocal space anyway. In such cases, it is sensible to align the crystal carefully, such that useful data can be obtained with relatively few exposures. This leads to the ability to do parametric studies of diffuse scattering, which is an area under-exploited at this time.

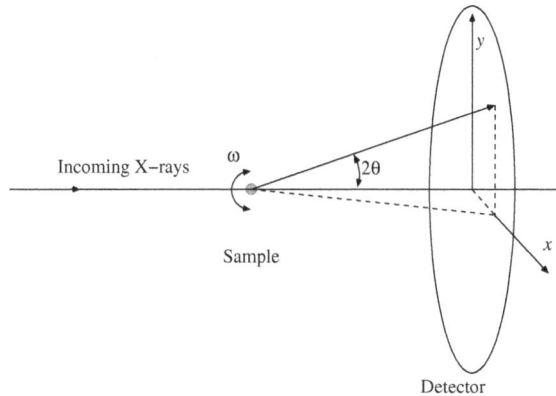

Figure 1. A schematic diagram of a diffuse scattering collection using a 2D detector. The sample angle is ω; incoming X-rays are of known wavelength, λ; and the scattering angle is as usual 2θ; but because we wish to transform the detector coordinates into *hkl*'s, we work with x and y coordinates on the detector. During a single exposure, the sample is typically rocked through an angle $d\omega \sim 0.25°$, then ω is incremented by $d\omega$ and the measurement repeated. After $180/d\omega$ such exposures, enough data points have been collected to reconstruct most of reciprocal space [10] out of the the maximum value, which is given by the radius of the detector, the sample-detector distance, and λ.

It may be noted that static and dynamic displacements cannot be distinguished with an X-ray experiment because, compared to the high energies of the X-rays, all atomic motions are of very low energy (seem very "slow") and are seen as "static"; this is an area where neutrons may be preferable.

2.2. Neutron

Neutron diffraction comes in essentially two varieties: constant wavelength and time of flight. The latter is most commonly found at a spallation neutron source, while the former is found at reactor sources or a steady-state spallation source, like SINQ, the Swiss Spallation Neutron Source.

A constant wavelength experiment essentially uses the same configuration as for the X-ray case (Figure 1). A typical example is the Wombat instrument at the Bragg Institute at the Australian Nuclear Science and Technology Organisation (ANSTO) [47,48]. This instrument uses a two-dimensional detector to collect a sort of "cake slice" of diffraction space, such that data collected at multiple sample angles can be combined to give a volume from which sections can be extracted. Such an instrument does not select for neutron energy, so scattering from dynamic effects like phonons overlaps with that from static structures like chemical short-range order. This is much as for X-rays, except that the neutron energy is much lower, and inelastic effects may change the neutron wavelength substantially, which has the effect of "moving" the scattered beam around on the detector and, thus, shifting the inelastic scattered intensity to different positions in the reciprocal space map. Such effects can in some cases be interpreted usefully [49]. They do lead to a reduction of the symmetry of the pattern and may limit the ability to quantitatively model the scattering. If diffuse scattering is measured using an instrument that can select for neutron energy, for example a chopper spectrometer, then static can be separated from dynamic, although that depends on the energy resolution of the instrument; quasi-elastic scattering may be binned in with the "strictly elastic" scattering.

At a spallation neutron source, the time structure of the pulse collapses an entire diffraction pattern into a single pixel on a detector, meaning that such instruments, for example SXD (single crystal diffractometer) at ISIS [50,51] and TOPAZ at the Spallation Neutron Source [52], collect very large volumes of reciprocal space with a single sample setting. Rotating the sample leads to rapidly scanning a large volume, generally much larger than that accessible at a constant wavelength source. On the other hand, instrument resolution can vary dramatically from forward- to back-scattering detectors, and since the experiment is essentially imaging reciprocal space, this can affect the interpretability of some patterns. Further, such instruments are often "open" in geometry, without collimation between sample and detector. Thus, they effectively image the sample onto the detector, meaning that anisotropic sample shape can lead to odd-shaped features. This is not an issue when the feature is to be integrated up to get an intensity for conventional Bragg analysis, but when reciprocal space maps are being looked at, it can have an effect.

It is possible to use energy discrimination on spallation instruments [49,53], and again, this yields the possibility of separating dynamic from static effects.

Whether constant wavelength or spallation, polarisation analysis can be used to separate magnetic from structural diffuse scattering [54–57].

3. Basic Principles of Monte Carlo Modelling of SRO

This topic is dealt with in great detail elsewhere [17,58–60], so a simple outline will suffice; Figure 2 summarises the process.

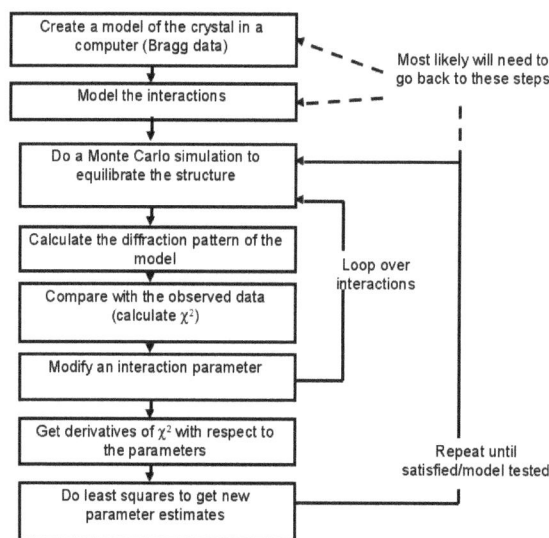

Figure 2. The overall MC modelling procedure. The flow chart illustrated in Figure 3 is an expansion of the box labelled "Do a Monte Carlo simulation to equilibrate the structure". This diagram assumes a least squares procedure based on calculating a χ^2 statistic for the model (or perhaps a kind of R-factor [61,62]); but often, the comparison will be done heuristically by the investigator, and the results will be more qualitative. The initial model is based on the average structure from Bragg data.

At its simplest, the type of MC modelling considered here has just a few steps.

- Decide on a starting configuration for the model. This usually means creating (in a computer) a $M \times N \times P$ array of unit cells, typically 32 on a side, and populating it with atoms based on the average structure determined by conventional studies.
- Choose some interactions between atoms. To set up chemical SRO when there are two species, a typical interaction is a Ising-like potential for the energy associated with the occupancy of site i, E^i_{occ}:

$$E^i_{occ} = -J_{NN} \sum_{NN} S_i S_j \tag{1}$$

where j indexes nearest neighbours and the sign of J determines whether a positive or negative nearest neighbour occupancy correlation, C_{NN}, is energetically favourable. Further, such terms may be present for more distant neighbours. $S_j = \pm 1$.

If it is displacements that are of interest, the simplest choice is to connect atoms with Hooke's law springs The program ZMC [63] is designed to induce correlations amongst atomic and/or molecular displacements by causing the atoms to interact with surrounding atoms via Hooke's law springs of the form:

$$E_{\text{inter}} = \sum_{\text{cv}} F_i (d_i - d_{0i}(1 + \epsilon_i))^2 \tag{2}$$

where d_i is the length of vector i connecting atoms, d_{0i} is its equilibrium length and F_i is its force constant. The sum is over all contact vectors (cv). ϵ_i is the "size-effect" term, which allows that the equilibrium length required for the calculation may not be the average length as determined from Bragg scattering; this is particularly likely to be the case in occupationally-disordered materials, where the Bragg-refined intermolecular distance is in fact an average over several different distances resulting from differing atomic or molecular species (or vacancies).

- The actual MC part happens as follows (summarised in Figure 3). An atom is chosen at random, and its energy is calculated. Its configuration is changed, and the energy calculation repeated. The new configuration is kept or rejected based on a simple criterion: if new energy is lower, it is kept, and it may be kept if new energy is higher, with some probability based on simulation "temperature".
- Note that the configuration may be changed by adding small random variations to an atom's variables (e.g., moving it slightly) or by swapping the variables of one site with those of another. Swapping is particularly useful as a means of maintaining an initial population of displacements or chemical species, while inducing correlations within that population.
- Once every site has been visited, on average, some large number of times, which could be ten, hundreds or thousands, depending on the needs of the simulation, the simulation is complete, and the atomic coordinates are read out.
- A Fourier transform program DIFFUSE [64] then calculates the diffuse scattering for comparison with the experiment.
- It is possible to embed this process within a procedure that automatically modifies the interaction parameters to try to improve the fit between calculated and observed diffuse scattering, although often useful results can be obtained by qualitative comparison, which can be used to reveal key aspects of the local order without comprehensive fitting.

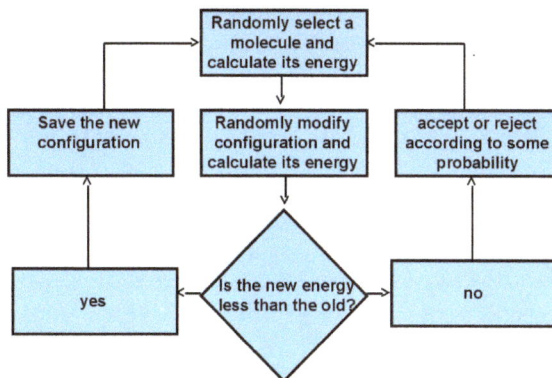

Figure 3. A simple representation of a single forward MC step; a molecule may be a single atom or a more complex motif.

The advantage of this approach is that the "energy" can be anything as long as it is quick to calculate. It may be relatively realistic or quite abstract, whatever suits the problem. However, knowing what disorder is present and then how to parameterise the interactions to induce it is not simple.

4. A Model System

In this section a model system, CePdSb, is considered from the point of view of inducing a range of local orderings and their resulting diffraction effects. No comparison with the observations is made, as we are looking simply to show how the disorder is modelled and some of the forms it can take.

CePdSb and related compounds form a family demonstrating a wide range of unusual magnetic phenomena, including the Kondo effect, heavy fermion behaviour and half-metal behaviour [65–71].

CePdSb itself shows a crystal structure in which the Pd and Sb lie on ordered sub-lattices at coordinates $(1/3, 2/3, 0.4684)$ and $(2/3, 1/3, 0.516)$ [66], with space-group $P6_3mc$ and lattice parameters approximately $a = 4.935$Å and $c = 7.890$Å. This is different from an earlier structure in which the z coordinates of both Pd and Sb were taken as 0.5 [67] and the Pd and Sb were considered to be randomly mixed across the Pd/Sb sites.

The Ce atoms lie at $(0, 0, 1/3)$ 2a positions, forming chains along the c axis. The structure is represented in Figure 4. For the purposes of demonstrating various diffraction effects, we will explore what happens when the Ce2 site ($z = 2/3$) is occupied by approximately 67% Ce atoms and 33% vacancies.

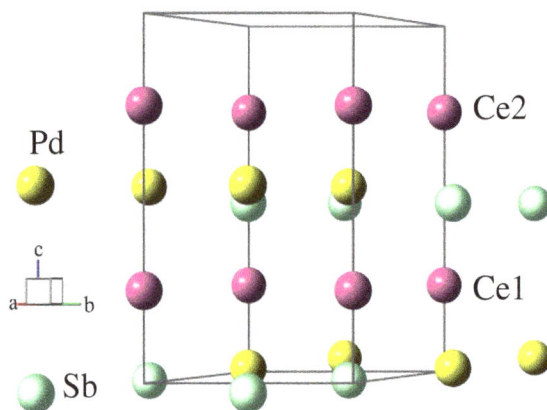

Figure 4. A schematic diagram of the structure of CePdSb, showing the Ce layers and the Pb/Sb layers, the latter of which are not flat, but "puckered" [72].

If we take the average structure of CePdSb [66] and calculate the diffuse scattering, we of course see nothing of interest, as there are no short-range correlations. However, we may, for example, connect atoms with Hooke's law springs (Equation (2)) and run a simulation. Figure 5 shows three sections through the diffuse scattering from CePdSb. The first row of images comes from a model in which there are no Ce vacancies and the atoms are connected by Hooke's law springs. The interactions induce streaks, most apparent in the $hk5.5$ layer. The second row shows the same cuts, but for a model in which there are 33% vacancies on the Ce2 layers, and they are forced to cluster. In the third row, the vacancies anti-cluster, and we can see in Figure 5i that this induces sharp spots in the half-layer, $hk5.5$, where previously, there were only streaks (the streaks are in fact sections through planes of scattering that can also be seen in the $hk5$ layer, though being less obvious due to the bright spots). We can also see that the clustering has little effect on the $hk5$ layer, while in $hk0$ it causes the spots that are present in hexagonal motifs around each Bragg peak (one hexagon is noted by white lines in Figure 5a) to extend closer to the origin. These spots actually come from the fact that the Pd and Sb atoms are not on idealised positions, such as $(1/3, 2/3, 1/2)$. When the vacancies cluster, we have large regions of the crystal where the scattering from the Ce2 layer is absent (effectively, these are like crystallites of composition $Ce_{0.5}PdSb$), giving different cancellation and allowing the spots to persist. When the vacancies anti-cluster (in the third row), the average scattering from Ce2 is preserved on the local scale, as well, and the cancellation is more like that seen in Figure 5a, though not identical.

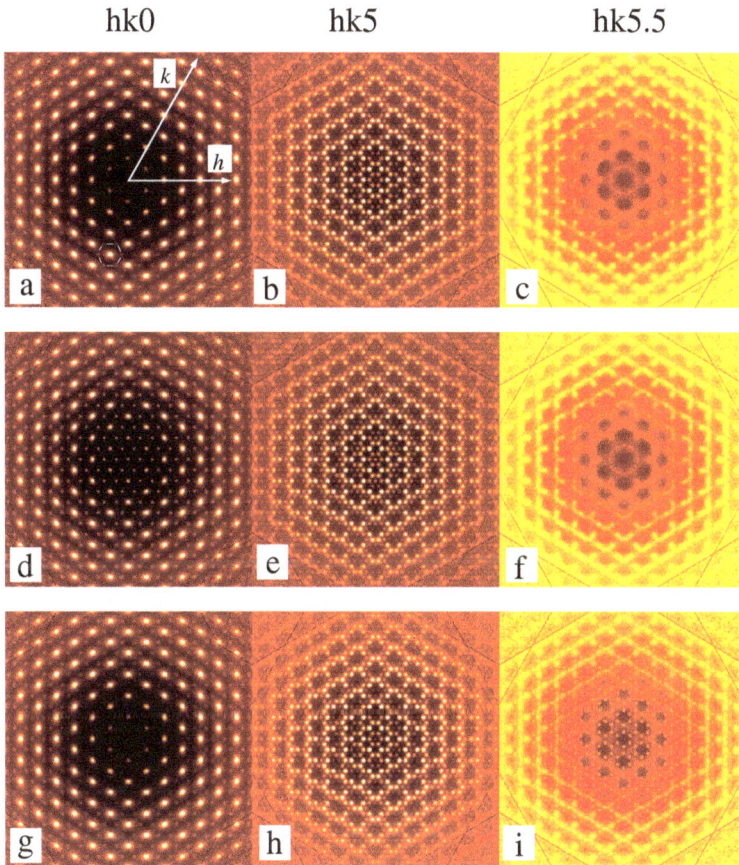

Figure 5. Slices of calculated diffuse scattering from different models of CePdSb. Row 1: no vacancies. Row 2: 33% vacancies on Ce2 site, clustering. Row 3: 33% vacancies on Ce2 site, anti-clustering. *hk*5.5 layers are normalised more brightly to bring out the details. For details, see the text. *h* and *k* axes noted on (**a**) to indicate directions.

In Figure 6, in rows 1 and 2, the displacive and occupancy effects are combined: the average distance atom-vacancy has been made 20% bigger than the average, while atom-atom is 10% smaller and vacancy-vacancy is 40% bigger. This is to mimic the effect sometimes seen where atoms move away from vacancies due to the lack of a bond, rather than moving into the gap. Row 1 is the model where the vacancies cluster; row 2 is where they avoid each other.

However, the third row of images in Figure 6 is the same as the second, but the size-effect signs have been reversed. Examining the two rectangles in Figure 6d,g shows how the brightness of consecutive spots is reversed by the change in size effect (white rectangles). Note however that other spots are relatively independent of this effect; this is one way in which this kind of modelling is useful, as it allows for the combined effects of the different structure factors (in a sense, each correlation has its own "structure factor") and form factors and how they interact.

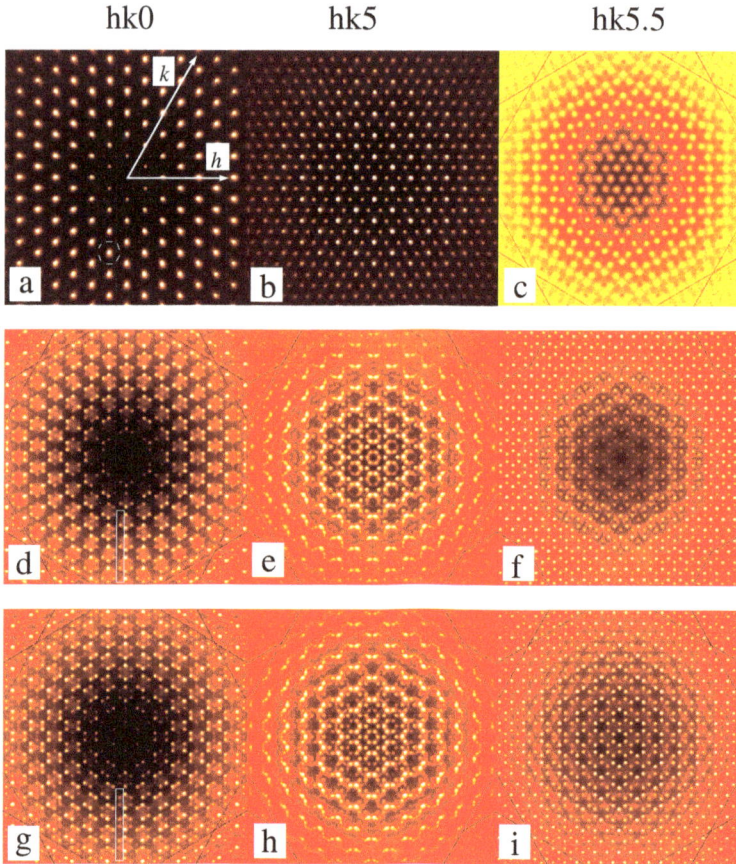

Figure 6. Slices of calculated diffuse scattering from different models of CePdSb, this time incorporating the atomic size effect. Row 1: Same as row 2 of Figure 5, but atoms move away from vacancies and vacancies away from each other. Row 2: Same as row 3 of Figure 5, but atoms move away from vacancies and vacancies away from each other. Row 3: Same as row 3 of Figure 5, but atoms move toward vacancies and vacancies toward each other. For details, see the text. h and k axes noted on one figure to indicate directions.

Note how the size effect is very different when applied to the clustering model (row 1) and the anti-clustering models (rows 2 and 3). Rows 2 and 3 of Figure 5 are different, but relatively subtly. Compare then rows 1 and 2 of Figure 6, which are the same two rows, now with the same kinds of size effects applied. Because the fraction of atom-vacancy bonds and atom-atom bonds is very different in the two models, the scattering is very different. This shows how strongly these effects can interact, something that can be difficult to disentangle without this kind of modelling to lean on.

Hence, even these relatively simple effects can have interesting and complex influences on the diffraction patterns of metallic systems. The MC model allows insight to be gained when the system is too complex to use direct inversion of the diffuse scattering to determine the correlations. In particular, exploring a range of representative models that look at various possible forms of SRO and their resulting diffraction is a useful guide to finding out what kinds of SRO are present in the real system.

5. Conclusions

Complex metallic systems, such as intermetallics, alloys, quasicrystals and Hume-Rothery phases, can all show detailed local ordering, which gives rise to highly structured and often very anisotropic single-crystal diffuse scattering. This paper reviews some of the issues associated with collecting and analysing such scattering and uses hypothetical calculation on the intermetallic CePdSb to illustrate some of the effects that may be observed in real systems.

Local order is important in determining many materials' properties and should not be ignored when trying to relate structure to function, especially when phenomena on the nanoscale are to be considered.

By qualitatively inducing various orderings in an MC model, the signatures of these orderings can be determined and compared to the observed data, providing guidance as to what structures are present in the real material.

Acknowledgments: Many thanks to T.R. Welberry and A.P. Heerdegen of the Australian National University for many useful discussions; they bear no responsibility for the opinions expressed herein. Thanks to Klaus-Dieter Liss for the invitation to write this article.

Conflicts of Interest: The author declares no conflict of interest.

References

1. Hukins, D.W.L. *X-ray Diffraction by Ordered and Disorderd Systems*; Pergamon Press: New York, NY, USA, 1981.
2. Krivoglaz, M.A. *Diffuse Scattering of X-rays and Neutrons by Fluctuations*; Springer-Verlag: Berlin, Germany, 1996.
3. Billinge, S.J.L.; Thorpe, M.F. *Local Structure from Diffraction*; Plenum: New York, NY, USA, 1998.
4. Welberry, T.R. Diffuse X-ray scattering and models of disorder. *Rep. Prog. Phys.* **1985**, *48*, 1543–1593.
5. Schweika, W. Disordered Alloys—Diffuse Scattering and Monte Carlo Simulation. In *Springer Tracts in Modern Physics*; Springer: Heidelberg, Germany, 1997; Volume 141.
6. Wall, M.; Adams, P.; Fraser, J.; Sauter, N. Diffuse X-ray Scattering to Model Protein Motions. *Structure* **2014**, *22*, 182–184.
7. Barabash, R.I.; Ice, G.E.; Turchi, P.E.A. *Diffuse Scattering and the Fundamental Properties of Materials*, 1st ed.; Momentum Press: New York, NY, USA, 2009.
8. Welberry, T.; Weber, T. One hundred years of diffuse scattering. *Crystallogr. Rev.* **2016**, *22*, 2–78.
9. Bürgi, H.B.; Weber, T. The structural complexity of a polar, molecular material brought to light by synchrotron radiation. *Mol. Cryst. Liq. Cryst.* **2003**, *390*, 1–4.
10. Estermann, M.A.; Steurer, W. Diffuse scattering data acquisition techniques. *Phase Transit.* **1998**, *67*, 165–195.
11. Welberry, T.R.; Goossens, D.J.; Heerdegen, A.P.; Lee, P.L. Problems in Measuring Diffuse X-ray Scattering. *Z. Krist.* **2005**, *220*, 1052–1058.
12. Cowley, J.M.; Gonnes, J. Diffuse scattering in electron diffraction. In *International Tables for Crystallography Volume B*; Springer: Dordrecht, The Netherlands, 1993; pp. 434–440.
13. Cowley, J.M. Kinematical Diffraction from Solid Solutions with Short Range Order and Size Effect. *Acta Crystallogr.* **1968**, *24*, 557–563.
14. Cowley, J.M. Short-Range Order and Long-Range Order Parameters. *Phys. Rev.* **1965**, doi:10.1103/PhysRev.138.A1384.
15. Cowley, J.M. Short- and Long-Range Order Parameters in Disordered Solid Solution. *Phys. Rev.* **1960**, *120*, 1648–1657.
16. Warren, B.E.; Averbach, B.L.; Roberts, B.W. Atomic Size Effect in the X-ray Scattering by Alloys. *J. Appl. Phys.* **1951**, *22*, 1493–1496.
17. Welberry, T.R. *Diffuse X-ray Scattering and Models of Disorder*; Oxford University Press: Oxford, UK, 2004.
18. Whitfield, R.E.; Goossens, D.J.; Welberry, T.R. Total scattering and pair distribution function analysis in modelling disorder in PZN ($PbZn_{1/3}Nb_{2/3}O_3$). *IUCrJ* **2016**, *3*, 20–31.

19. Neder, R.B.; Proffen, T. *Diffuse Scattering and Defect Structure Simulations: A Cook Book Using the Program DISCUS*; OUP: Oxford, UK, 2008.

20. Proffen, T.; Billinge, S.J.L. PDFFIT, a program for full profile structural refinement of the atomic pair distribution function. *J. Appl. Crystallogr.* **1999**, *32*, 572–575.

21. Henderson, R. A Cavalcade of Clusters: The Interplay Between Atomic and Electronic Structure in Complex Intermetallics. Ph.D. Thesis, Cornell University, Ithaca, NY, USA, January 2013.

22. Bosak, A.; Chernyshov, D. On model-free reconstruction of lattice dynamics from thermal diffuse scattering. *Acta Crystallogr. Sect. A* **2008**, *64*, 598–600.

23. Bosak, A.; Chernyshov, D.; Vakhrushev, S.; Krisch, M. Diffuse scattering in relaxor ferroelectrics: True three-dimensional mapping, experimental artefacts and modelling. *Acta Crystallogr. Sect. A* **2012**, *68*, 117–123.

24. Paściak, M.; Welberry, T.R. Diffuse scattering and local structure modeling in ferroelectrics. *Z. Krist.* **2011**, *226*, 113–125.

25. Maisel, S.B.; Schindzielorz, N.; Müller, S.; Reichert, H.; Bosak, A. An accidental visualization of the Brillouin zone in an Ni–W alloy via diffuse scattering. *J. Appl. Crystallogr.* **2013**, *46*, 1211–1215.

26. Simonov, A.; Weber, T.; Steurer, W. Yell: A computer program for diffuse scattering analysis *via* three-dimensional delta pair distribution function refinement. *J. Appl. Crystallogr.* **2014**, *47*, 1146–1152.

27. Nield, V.M.; Keen, D.A.; McGreevy, R.L. The interpretation of single-crystal diffuse scattering using reverse Monte Carlo modelling. *Acta Crystallogr. Sect. A* **1995**, *51*, 763–771.

28. Tucker, M.G.; Keen, D.A.; Dove, M.T.; Goodwin, A.L.; Hui, Q. RMCProfile: Reverse Monte Carlo for polycrystalline materials. *J. Phys. Condens. Matter* **2007**, *19*, 335218.

29. Steurer, W. Twenty years of structure research on quasicrystals. Part 1. Pentagonal, octagonal, decagonal and dodecacagonal quasicrystals. *Z. Krist.* **2004**, *219*, 391–446.

30. Estermann, M.; Lemster, K.; Haibach, T.; Steurer, W. Towards the real structure of quasicrystals and approximants by analysing diffuse scattering and deconvolving the patterson. *Z. Krist.* **2000**, *215*, 584–596.

31. Estermann, M.; Steurer, W. Surveying the Entire Reciprocal Space of Quasicrystals with Imaging Plate Technology. In *Quasicrystals*; Janot, C., Mosseri, R., Eds.; World Scientific: Singapore, 1995.

32. Egami, T.; Billinge, S.J.L. *Underneath the Bragg Peaks, Structural Analysis of Complex Materials*; Pergamon: Oxford, UK, 2003.

33. Welberry, T.R. Diffuse X-ray Scattering and Disorder in *p*-methyl-*N*-(*p*-chlorobenzylidene)aniline $C_{14}H_{12}ClN$ (ClMe): Analysis *via* Automatic Refinement of a Monte Carlo Model. *Acta Crystallogr.* **2000**, *56*, 348–358.

34. Osborn, J.C.; Welberry, T.R. A Position-Sensitive Detector System for the Measurement of Diffuse X-ray Scattering. *J. Appl. Crystallogr.* **1990**, *23*, 476–484.

35. Templer, R.H.; Warrender, N.A.; Seddon, J.M.; Davis, J.M. The Intrinsic Resolution of X-ray Imaging Plates. *Nucl. Instrum. Methods* **1991**, *310*, 232–235.

36. Miyahara, J.; Takahashi, K.; Amemiya, Y.; Kamiya, N.; Satow, Y. A New Type of X-ray Area Detector Utilizing Laser Stimulated Luminescence. *Nucl. Instrum. Methods* **1986**, *246*, 572–578.

37. Gibaud, A.; Harlow, D.; Hastings, J.B.; Hill, J.P.; Chapman, D. A High-Energy Monochromatic Laue (MonoLaue) X-ray Diffuse Scattering Study of $KMnF_3$ Using an Image Plate. *J. Appl. Crystallogr.* **1997**, *30*, 16–20.

38. Amemiya, Y.; Matsushita, T.; Nakagawa, A.; Satow, Y.; Miyahara, J.; Chikawa, J. Design and Performance of an Imaging Plate System for X-ray Diffraction Study. *Nucl. Instrum. Methods* **1988**, *266*, 645–653.

39. Bourgeois, D.; Moy, J.P.; Svensson, S.O.; Kvick, A. The Point-Spread Function of X-ray Image-Intensifiers/CCD-Camera and Imaging-Plate Systems in Crystallography: Assessment and Consequences for the Dynamic Range. *J. Appl. Crystallogr.* **1994**, *27*, 868–877.

40. Iwasaki, H.; Matsuo, Y.; Ohshima, K.I.; Hashimoto, S. Time-Resolved Two-Dimensional Observation of the Change in X-ray Diffuse Scattering from an Alloy Single Crystal Using an Imaging Plate on a Synchrotron-Radiation Source. *J. Appl. Crystallogr.* **1990**, *23*, 509–514.

41. Thomas, L.H.; Welberry, T.R.; Goossens, D.J.; Heerdegen, A.P.; Gutmann, M.J.; Teat, S.J.; Wilson, C.C.; Lee, P.L.; Cole, J.M. Disorder in pentachloronitrobenzene, $C_6Cl_5NO_2$: A diffuse scattering study. *Acta Crystallogr. B* **2007**, *63*, 663–673.

42. Welberry, T.R.; Goossens, D.J.; Haeffner, D.R.; Lee, P.L.; Almer, J. High-energy diffuse scattering on the 1-ID beamline at the Advanced Photon Source. *J. Synchrotron Radiat.* **2003**, *10*, 284–286.

43. Arndt, U.W. X-ray Position-Sensitive Detectors. *J. Appl. Crystallogr.* **1986**, *19*, 145–163.
44. Henrich, B.; Bergamaschi, A.; Broennimann, C.; Dinapoli, R.; Eikenberry, E.; Johnson, I.; Kobas, M.; Kraft, P.; Mozzanica, A.; Schmitt, B. PILATUS: A single photon counting pixel detector for X-ray applications. *Nucl. Instrum. Methods Phys. Res. Sect. A* **2009**, *607*, 247–249.
45. Seeck, O.H.; Murphy, B. *X-ray Diffraction: Modern Experimental Techniques*, 1st ed.; CRC Press: Singapore, 2015.
46. Liss, K.D.; Bartels, A.; Schreyer, A.; Clemens, H. High-Energy X-rays: A tool for Advanced Bulk Investigations in Materials Science and Physics. *Textures Microstruct.* **2003**, *35*, 219–252.
47. Studer, A.J.; Hagen, M.E.; Noakes, T.J. Wombat: The high-intensity powder diffractometer at the OPAL reactor. *Phys. B Condens. Matter* **2006**, *385–386*, 1013–1015.
48. Whitfield, R.E.; Goossens, D.J.; Studer, A.J.; Forrester, J.S. Measuring Single-Crystal Diffuse Neutron Scattering on the Wombat High-Intensity Powder Diffractometer. *Metall. Mater. Trans. A* **2012**, *43A*, 1423–1428.
49. Welberry, T.R.; Goossens, D.J.; David, W.I.F.; Gutmann, M.J.; Bull, M.J.; Heerdegen, A.P. Diffuse neutron scattering in benzil, $C_{14}D_{10}O_2$, using the time-of-flight Laue technique. *J. Appl. Cryst.* **2003**, *36*, 1440–1447.
50. Keen, D.A.; Gutmann, M.J.; Wilson, C.C. SXD—The single-crystal diffractometer at the ISIS spallation neutron source. *J. Appl. Crystallogr.* **2006**, *39*, 714–722.
51. Welberry, T.R.; Gutmann, M.J.; Woo, H.; Goossens, D.J.; Xu, G.; Stock, C.; Chen, W.; Ye, Z.G. Single-crystal neutron diffuse scattering and Monte Carlo study of the relaxor ferroelectric $PbZn_{1/3}Nb_{2/3}O_3$ (PZN). *J. Appl. Crystallogr.* **2005**, *38*, 639–647.
52. Koetzle, T.F.; Bau, R.; Hoffmann, C.; Piccoli, P.M.B.; Schultz, A.J. Topaz: A single-crystal diffractometer for the spallation neutron source. *Acta Crystallogr. Sect. A* **2006**, *62*, s116.
53. Rosenkranz, S.; Osborn, R. Corelli: Efficient single crystal diffraction with elastic discrimination. *Pramana J. Phys.* **2008**, *71*, 705–711.
54. Schweika, W.; Böni, P. The instrument DNS: Polarization analysis for diffuse neutron scattering. *Physica B* **2001**, *297*, 155–159.
55. Ersez, T.; Kennedy, S.; Hicks, T.; Fei, Y.; Krist, T.; Miles, P. New features of the long-wavelength polarisation analysis spectrometer LONGPOL. *Phys. B Condens. Matter* **2003**, *335*, 183–187.
56. Stewart, J.R.; Deen, P.P.; Andersen, K.H.; Schober, H.; Barthélémy, J.F.; Hillier, J.M.; Murani, A.P.; Hayes, T.; Lindenau, B. Disordered materials studied using neutron polarization analysis on the multi-detector spectrometer, D7. *J. Appl. Crystallogr.* **2009**, *42*, 69–84.
57. Klose, F.; Constantine, P.; Kennedy, S.J.; Robinson, R.A. The Neutron Beam Expansion Program at the Bragg Institute. *J. Phys. Conf. Ser.* **2014**, *528*, 012026.
58. Welberry, T.R.; Goossens, D.J. The interpretation and analysis of diffuse scattering using Monte Carlo simulation methods. *Acta Crystallogr. Sect. A* **2008**, *64*, 23–32.
59. Schweika, W. *Disordered Alloys: Diffuse Scattering and Monte Carlo Simulations*; Springer: Berlin, Germany, 1998.
60. Binder, K. *Monte Carlo Methods in Statistical Physics*; Springer: Berlin, Germnay, 1979.
61. Chan, E.J.; Goossens, D.J. Study of the single-crystal X-ray diffuse scattering in paracetamol polymorphs. *Acta Cryst. B* **2012**, *B68*, 80–88.
62. Welberry, T.R.; Goossens, D.J.; Edwards, A.J.; David, W.I.F. Diffuse X-ray scattering from benzil, $C_{14}D_{10}O_2$: Analysis via automatic refinement of a Monte Carlo model. *Acta Cryst.* **2001**, *A57*, 101–109.
63. Goossens, D.J.; Heerdegen, A.P.; Chan, E.J.; Welberry, T.R. Monte Carlo Modelling of Diffuse Scattering from Single Crystals: The Program ZMC. *Metall. Mater. Trans. A* **2010**, *42A*, 23–31, doi:10.1007/s11661-010-0199-1.
64. Butler, B.D.; Welberry, T.R. Calculation of Diffuse Scattering from Simulated Crystals: A Comparison with Optical Transforms. *J. Appl. Crystallogr.* **1992**, *25*, 391–399.
65. Ślebarski, A. Half-metallic ferromagnetic ground state in CePdSb. *J. Alloy. Compd.* **2006**, *423*, 15–20.
66. Riedi, P.; Armitage, J.; Lord, J.; Adroja, D.; Rainford, B.; Fort, D. A ferromagnetic Kondo compound: CePdSb. *Phys. B Condens. Matter* **1994**, *199–200*, 558–560.
67. Malik, S.; Adroja, D. Magnetic behaviour of RPdSb (R = rare earth) compounds. *J. Magn. Magn. Mater.* **1991**, *102*, 42–46.
68. Katoh, K.; Ochiai, A.; Suzuki, T. Magnetic and transport properties of CePdAs and CePdSb. *Phys. B Condens. Matter* **1996**, *223–224*, 340–343.

69. Malik, S.K.; Adroja, D.T. CePdSb: A possible ferromagnetic Kondo-lattice system. *Phys. Rev. B* **1991**, *43*, 6295–6298.

70. Lord, J.S.; Tomka, G.J.; Riedi, P.C.; Thornton, M.J.; Rainford, B.D.; Adroja, D.T.; Fort, D. A nuclear magnetic resonance investigation of the ferromagnetic phase of CePdSb as a function of temperature and pressure. *J. Phys. Condens. Matter* **1996**, *8*, 5475.

71. Neville, A.; Rainford, B.; Adroja, D.; Schober, H. Anomalous spin dynamics of CePdSb. *Phys. B Condens. Matter* **1996**, *223–224*, 271–274.

72. Ozawa, T.C.; Kang, S.J. Balls & Sticks: Easy-to-use structure visualization and animation program. *J. Appl. Crystallogr.* **2004**, *37*, 679, doi:10.1107/S0021889804015456.

metals

MDPI

Article

Investigation of Elastic Deformation Mechanism in As-Cast and Annealed Eutectic and Hypoeutectic Zr–Cu–Al Metallic Glasses by Multiscale Strain Analysis

Hiroshi Suzuki [1,*], Rui Yamada [2], Shinki Tsubaki [3], Muneyuki Imafuku [3], Shigeo Sato [4], Tetsu Watanuki [5], Akihiko Machida [5] and Junji Saida [2]

[1] Quantum Beam Science Center, Japan Atomic Energy Agency, Tokai, Naka, Ibaraki 319-1195, Japan
[2] Frontier Research Institute for Interdisciplinary Sciences, Tohoku University, Sendai, Miyagi 980-8578, Japan; rui-yamada@fris.tohoku.ac.jp (R.Y.); jsaida@fris.tohoku.ac.jp (J.S.)
[3] Faculty of Engineering, Tokyo City University, Setagaya, Tokyo 158-8857, Japan; shinki_tsubaki@yahoo.co.jp (S.T.); imafukum@tcu.ac.jp (M.I.)
[4] Graduate School of Science and Engineering, Ibaraki University, Hitachi, Ibaraki 316-8511, Japan; shigeo.sato.ar@vc.ibaraki.ac.jp
[5] Quantum Beam Science Center, Japan Atomic Energy Agency, Sayo, Hyogo 679-5148, Japan; wata@spring8.or.jp (T.W.); machida@spring8.or.jp (A.M.)
* Correspondence: suzuki.hiroshi07@jaea.go.jp; Tel.: +81-29-282-5478

Academic Editor: Klaus-Dieter Liss
Received: 24 November 2015; Accepted: 30 December 2015; Published: 5 January 2016

Abstract: Elastic deformation behaviors of as-cast and annealed eutectic and hypoeutectic Zr–Cu–Al bulk metallic glasses (BMG) were investigated on a basis of different strain-scales, determined by X-ray scattering and the strain gauge. The microscopic strains determined by Direct-space method and Reciprocal-space method were compared with the macroscopic strain measured by the strain gauge, and the difference in the deformation mechanism between eutectic and hypoeutectic Zr–Cu–Al BMGs was investigated by their correlation. The eutectic $Zr_{50}Cu_{40}Al_{10}$ BMG obtains more homogeneous microstructure by free-volume annihilation after annealing, improving a resistance to deformation but degrading ductility because of a decrease in the volume fraction of weakly-bonded regions with relatively high mobility. On the other hand, the as-cast hypoeutectic $Zr_{60}Cu_{30}Al_{10}$ BMG originally has homogeneous microstructure but loses its structural and elastic homogeneities because of nanocluster formation after annealing. Such structural changes by annealing might develop unique mechanical properties showing no degradations of ductility and toughness for the structural-relaxed hypoeutectic $Zr_{60}Cu_{30}Al_{10}$ BMGs.

Keywords: hypoeutectic Zr–Cu–Al bulk metallic glass; structural relaxation; X-ray scattering; pair distribution function; elastic modulus

1. Introduction

Bulk metallic glasses (BMG) exhibit interesting mechanical features such as high strength with high ductility (low Young's modulus), which is a different trend from typical metallic materials. On the other hand, structural relaxation has been known as a thermal behavior of metallic glasses that changes various mechanical properties with a few percent volume shrinkage by a heat treatment below the glass transition temperature T_g. Especially, structural relaxation-induced embrittlement would be a factor to degrade unique mechanical properties of the metallic glass. Meanwhile, Yokoyama *et al.* recently found that a hypoeutectic Zr–Cu–Al BMG with a Zr composition of 10% more than the eutectic composition shows no degradations of ductility and toughness after complete structural relaxation [1,2]. Furthermore, the fatigue property on the hypoeutectic BMG is independent of the annealing temperature, while that on a eutectic BMG changes after annealing. In addition, crystal-like ordering and icosahedral-like contrast are partially recognized in the amorphous glassy matrix after annealing in the hypoeutectic BMG, while the annealed eutectic BMG has homogeneous amorphous, glassy structure. Consequently, the microstructural changes after structural relaxation might be a crucial factor to affect the mechanical and physical properties of BMGs after annealing.

The atomic pair distribution function (PDF) obtained by X-ray scattering, which can evaluate a neighbor atomic distance, can quantitatively estimate the local strain of nanostructures with no or less crystal periodicity. The PDF technique has been utilized so far for the deformation analysis of metallic glasses [3–6]. For instance, it was clarified that the local atomic strain of the metallic glass obtained by the PDF technique is smaller than the macroscopic bulk strain, and various deformation models have been suggested, based on their observations. Therefore, the difference in the deformability between eutectic and hypoeutectic BMGs can be accessed by the PDF technique on a basis of the microscopic deformation behavior in an atomic level. In this study, the microscopic deformation behaviors of the eutectic and hypoeutectic Zr–Cu–Al BMGs are evaluated by the PDF technique with synchrotron high energy X-ray scattering, and a change in the mechanical properties, induced by structural relaxation for their BMGs, was discussed on a basis of the correlation between microscopic and macroscopic deformations.

2. Experimental Procedure

Specimens used in this study were as-cast and annealed eutectic $Zr_{50}Cu_{40}Al_{10}$ and hypoeutectic $Zr_{60}Cu_{30}Al_{10}$ BMGs. Conditions of annealing were 697 K for 2 min for the $Zr_{50}Cu_{40}Al_{10}$ BMG (T_g = 706K) and 661 K for 2 min for the $Zr_{60}Cu_{30}Al_{10}$ BMG (T_g = 671K). Hereafter, the $Zr_{50}Cu_{40}Al_{10}$ BMG and the $Zr_{60}Cu_{30}Al_{10}$ BMG call Z50 and Z60, respectively.

The X-ray scattering experiments were performed using high energy X-rays of 69.8 keV at BL22XU in SPring-8, Hyogo, Japan [7]. Figure 1a shows the schematic layout of the optical system used in this study. The dog-bone shaped specimen with 1.2 mm in thickness (see Figure 1b) was mounted on a load frame and was irradiated by an incident beam with a size of 0.3 mm × 0.3 mm. Diffraction from the specimen was measured by an Imaging Plate (IP) with 400 mm × 400 mm in size. An aluminum plate with 4 mm in thickness was set in front of the IP to reduce the background by fluorescent X-rays. Tensile loadings were applied to the specimen by using the load frame until 500 to 600 MPa with a crosshead speed of 0.1 mm/min, and the diffraction patterns were measured while holding each applied stress at seven different steps. The distances, L from the IP to the specimen were set to be 300 mm and 700 mm, and exposure times were 300 s and 120 s, respectively. Diffraction patterns in the loading and transverse directions were extruded by circumferentially integrating a range of ±5° in the corresponding direction of the two-dimensional scattering image using the WinPIP software [8].

Two novel techniques, suggested by Poulsen *et al.* [3], were utilized for the strain analysis of amorphous metallic glasses, *i.e.*, Reciprocal-space (Q-space) method (QSM), which can measure the local strain from the peak shift of the first peak of the intensity function $I(Q)$ or the structure function $S(Q)$, and Direct-space method (DSM), which can measure the local strain directly from a change in the atomic distance obtained from the pair distribution function $G(r)$. The diffraction patterns measured at L = 700 mm and 300 mm were provided for QSM and DSM, respectively. The PDF was produced by the PDFgetX3 program [9] with a Q-range of Fourier transform from 1.4 to 17 Å^{-1}.

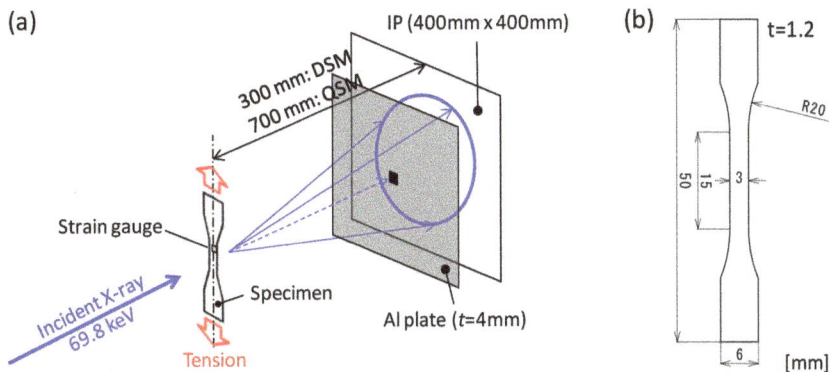

Figure 1. (**a**) Schematics of the optical layout and (**b**) the specimen used in this study.

3. Results

3.1. PDF and Microstructure

Figure 2 shows $I(Q)$, $S(Q)$ and $G(r)$ for Z50 and Z60. The radius, r in $G(r)$ indicates the distance from an average atom located at the origin. All functions for Z50 in Figure 2a,c,e seems to be unchanged by annealing. It is known that the free-volume of Z50 decreases due to structural relaxation by annealing [10]; however, any changes cannot be observed in their patterns. On the other hand, the first peak in $I(Q)$ and $S(Q)$ for Z60 sharpens after annealing, as shown in Figure 2b,d, suggesting the development of atomic ordering. Furthermore, some small diffraction peaks are recognizable in the annealed Z60. These diffraction peaks can be clearly observed in the difference curve of $I(Q)$ before and after annealing, and the difference also appears in $G(r)$, as shown in Figure 2f. Comparing with the diffraction patterns of Zr_2Cu (bct) and ZrCu (fcc), calculated by the Rietveld simulation using RIETAN-FP [11], the diffraction peak positions measured approximately correspond to their both diffraction patterns. Considering the precipitation temperature of ZrCu, more than 988K [12], the crystalline phase precipitated in Z60 is expected to be Zr_2Cu since the annealing temperature was less than T_g (697 K). The volume fraction of this crystalline phase is predicted to be a few percent, and, hence, the crystalline phase with such a small volume fraction would not affect the macroscopic deformation behavior.

Figure 3 shows the Transmission Electron Microscope (TEM) images of the annealed Z60. Ambiguous fringe contrast related to nanocrystallization can be seen in the glassy amorphous matrix, as shown in Figure 3a. In addition, many close-range crystal-like orderings could be recognized in Figure 3b, which are typically marked by white circles and shown in the magnified images in Figure 3c,d, although the structure in general remains amorphous. This is a similar feature of the microstructure to the previous works [1,2]. Furthermore, relatively large crystal grains with about a few hundred nanometers were also observed slightly, which may contribute to diffraction peaks shown in Figure 2b.

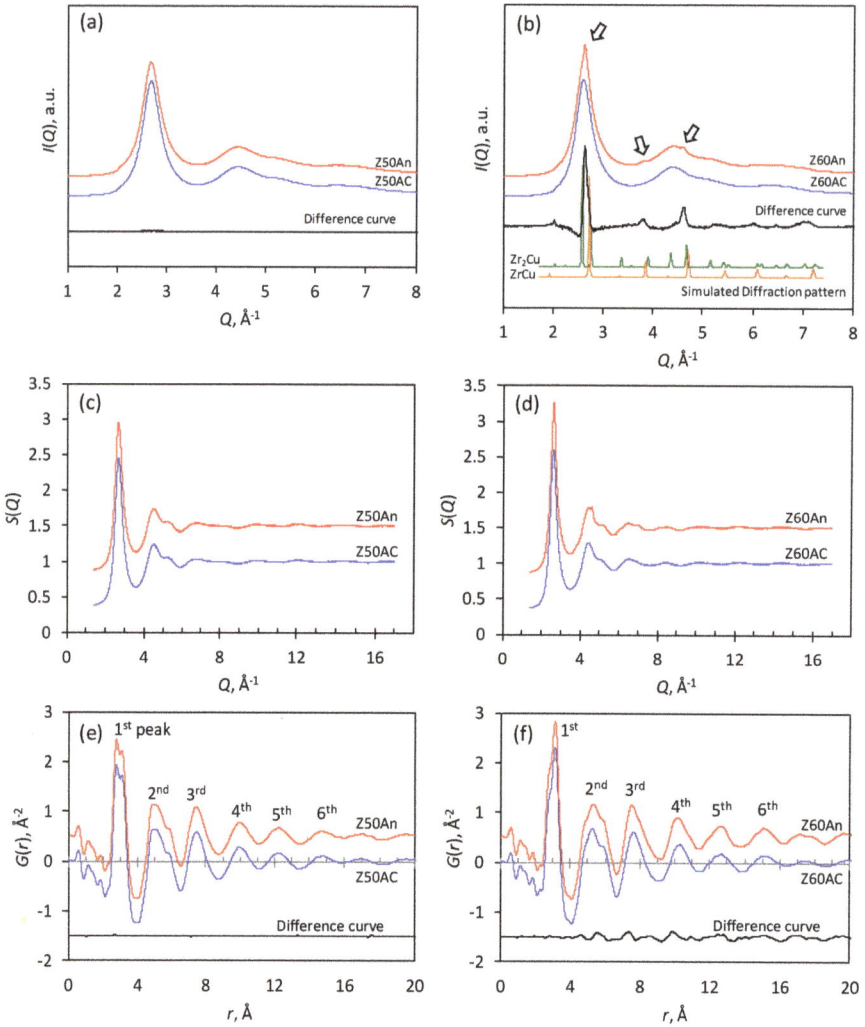

Figure 2. The intensity functions $I(Q)$ of (**a**) $Z_{50}Cu_{40}Al_{10}$ (Z50) and (**b**) $Z_{60}Cu_{30}Al_{10}$ (Z60) BMGs before (blue line) and after (red line) annealing and difference curve between them. "AC" and "An" after sample name denotes "as-cast" and "annealed", respectively. Arrows in (**b**) indicate the distinct diffraction peaks. For comparison, diffraction peak patterns of ZrCu and Zr_2Cu obtained by Rietveld simulation are shown in (**b**). (**c**) and (**d**) show the structure function $S(Q)$ of Z50 and Z60 before (blue line) and after (red line) annealing, respectively. (**e**) and (**f**) show the atomic pair distribution functions $G(r)$ before (blue line) and after (red line) annealing and difference curve between them for Z50 and Z60, respectively. In figures (**c**) to (**f**), $S(Q)$ and $G(r)$ after annealing are intentionally offset by +0.5 for easy comparison.

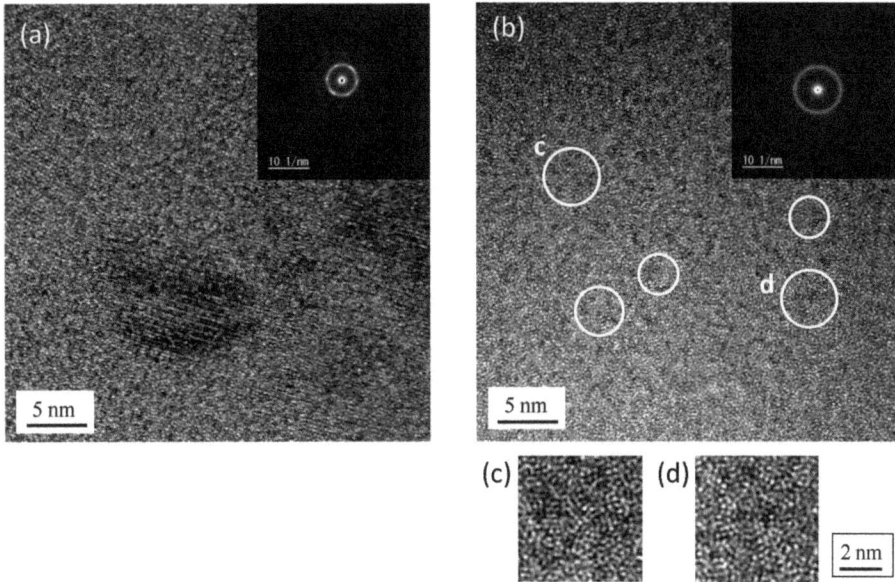

Figure 3. TEM images of the annealed Z60. Ambiguous fringe contrast related to nanocrystallization can be seen in (**a**), and random close-range crystal-like orderings could be recognized as typically shown by white circles in (**b**). Images in the circles, c and d, in (**b**) are magnified in (**c**) and (**d**), respectively. Electron diffraction in the inset of (**a**) shows the diffraction spots and that in the inset of (**b**) indicates the halo ring showing amorphous structure.

3.2. Tensile Deformation Behavior

Figure 4 shows a comparison of the strain changes measured by the strain gauge and QSM for Z50 and Z60. The macroscopic Young's modulus, E_M of Z50 measured by the strain gauge is increased from 87 GPa to 96 GPa by annealing, which is a typical trend caused by structural relaxation [2]. The macroscopic Poisson's ratios, v_M before and after annealing are 0.34 and 0.36, respectively. A decrease in the Poisson's ratio with a few % to 10% is commonly observed after annealing [13–15], but cannot be found in this result due to less accuracy. In contrast, E_M for Z60 is constant at 86 GPa, regardless of whether annealing was performed. The macroscopic Poisson's ratios v_M shows slightly decreasing from 0.37 to 0.36 by annealing, which is a typical trend caused by structural relaxation.

The microscopic deformations measured by QSM are derived from a shift in the first peak of $Q(r)$, fitted by the Voigt function for an initial state of the applied loading, providing the microscopic Young's modulus, E_Q and Poisson's ratio, v_Q. The Young's modulus E_Q of the as-cast Z50 is 103 GPa that is larger than E_M (=87 GPa), and slightly increases to 106 GPa after annealing. The Poisson's ratio v_Q before annealing is 0.32 that is smaller than the macroscopic Poisson's ratio v_M (=0.34), and decreases to 0.29 after annealing. On the other hand, E_Q of the as-cast Z60 is 93 GPa that is larger than E_M (=86 GPa), and increases to 100 GPa after annealing, which is a different feature from the macroscopic Young's modulus showing a constant level before and after annealing. The microscopic Poisson's ratio v_Q of the as-cast Z60 is 0.30 that is smaller than v_M (=0.37), and slightly decreases to 0.29 after annealing. As described above, the microscopic deformation obtained by QSM exhibits different trends from the macroscopic deformation, which can be also found for the Zr–Al–Ni–Cu BMG reported by Sato *et al.* [5]. Furthermore, the relation between macroscopic and microscopic elastic moduli for the hypoeutectic metallic glass is different from that for the eutectic metallic glass.

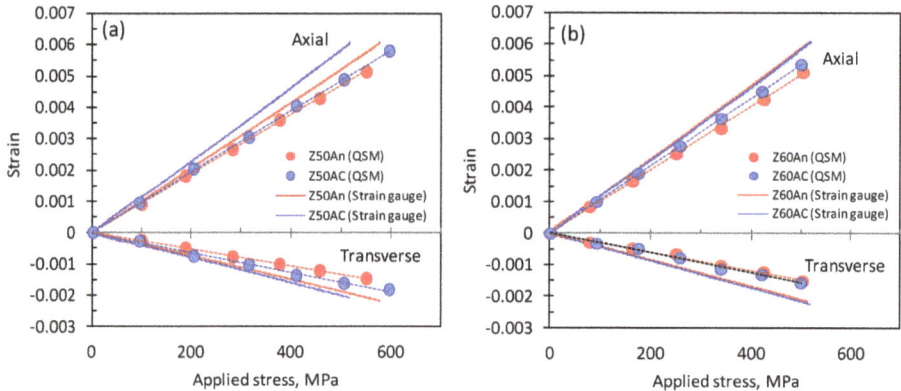

Figure 4. Stress-strain relations of (**a**) Z50 and (**b**) Z60 before (blue) and after (red) annealing, derived from macroscopic strains measured by the strain gauge and from microscopic strains determined by QSM. Average error bars are from ± 1.3 to $\pm 2.4 \times 10^{-4}$.

3.3. Comparison of Young's Modulus with Different Scales

To understand the microscopic elastic behavior accurately, the microscopic Young's moduli, E_D were assessed by a shift in each peak of $G(r)$, and they are plotted in Figure 5, as a function of r. The comparison of the Young's moduli and the Poisson's ratios measured by the strain gauge, QSM and DSM, is presented in Table 1. The Poisson's ratios were determined from the stress-strain relations in the axial and transverse directions, measured by each method.

Figure 5a shows the microscopic Young's modulus E_D as a function of r for Z50, compared with E_Q and E_M. The Young's modulus of the first peak at the lowest value of r is larger than the Young's moduli determined at larger values of r beyond the second peak. This is a typical trend for the metallic glasses as shown in previous studies [3–5]. The local strain at the lowest value of r represents the local structural deformation between first nearest-neighbor atoms, which depends on the inherent stiffness of atomic bonds. In contrast, the strains at larger values of r exhibit the average structural deformation including a redistribution of the free-volume. Here, the Young's modulus, E_{D_A} derived from the average structural deformation was calculated by averaging values in a flat region of E_D, and those before and after annealing were calculated to be 96 GPa and 100 GPa, respectively. These values are approximately 7 GPa smaller than the microscopic Young's modulus E_Q obtained by QSM. The difference between E_{D_A} and E_Q can be explained by considering E_Q to be increased by an influence of the local structural deformation at the lowest value of r. Moreover, a slight increase in E_{D_A} by annealing would be caused by hardening due to structural relaxation. In contrast, E_{D_A} before annealing is 9 GPa larger than E_M (=87 GPa) measured by the strain gauge, whereas E_M approaches E_{D_A} after annealing, *i.e.*, E_{D_A} = 100 GPa and E_M = 96 GPa.

Figure 5b shows a change in the microscopic Young's modulus, E_D as a function of r for Z60, compared with E_Q and E_M. The trend of a change in E_D is similar to that for Z50 shown in Figure 5a. However, a rise in E_D at the lowest value of r is smaller than that for Z50, suggesting the inherent stiffness of atomic bonds for Z60 to be smaller. The Young's modulus of the average structural deformation E_{D_A} before annealing is larger than the macroscopic Young's modulus E_M, *i.e.*, E_{D_A} = 91 GPa and E_M = 86 GPa. Furthermore, E_Q is 93 GPa that is slightly larger than E_{D_A}, but the Young's moduli for all scales tend to be comparable for the as-cast Z60. The Young's moduli E_M and E_{D_A} are almost unchanged by annealing, while E_Q is clearly increased. In addition, the annealed Z60 shows a specific trend that the microscopic Young's modulus E_D determined at the second peak of $G(r)$ increases after annealing.

The Poisson's ratio also shows a characteristic trend that the macroscopic Poisson's ratio ν_M is larger than the microscopic Poisson's ratios ν_Q and ν_D, which is independent of the sample condition. Moreover, any drastic changes cannot be observed before and after annealing.

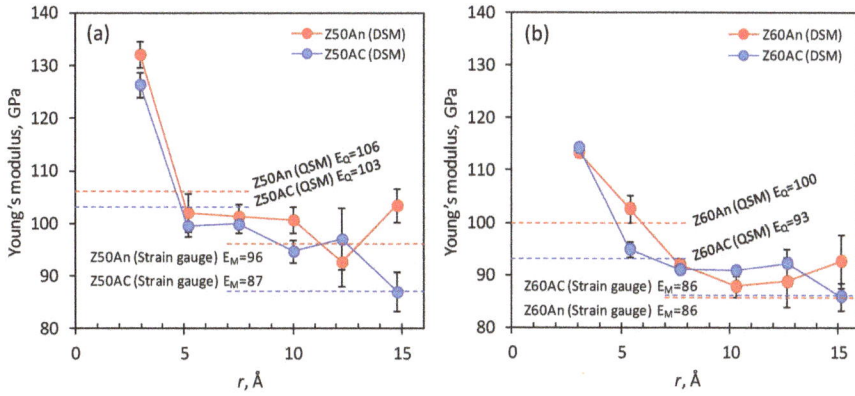

Figure 5. Microscopic Young's moduli of (**a**) Z50 and (**b**) Z60 before (blue) and after (red) annealing, which are derived from DSM, compared with Young's moduli evaluated by QSM and the strain gauge.

Table 1. Young's moduli and Poisson's ratios of Z50 and Z60 evaluated by reciprocal-space (Q-space) method (QSM), direct-space method (DSM) and the strain gauge. "AC" and "An" denote "as-cast" and "annealed", respectively.

Elastic Modulus	Z50		Z60	
	AC	An	AC	An
E_Q (QSM), GPa	103 ± 1	106 ± 1	93 ± 0	100 ± 1
ν_Q (QSM)	0.32	0.29	0.30	0.29
E_{D_A} (DSM Average), GPa	96 ± 5	100 ± 4	91 ± 3	90 ± 2
ν_{D_A} (DSM Average)	0.28	0.27	0.27	0.24
E_M (Strain gauge), GPa	87 ± 0	96 ± 0	86 ± 0	86 ± 0
ν_M (Strain gauge)	0.34	0.36	0.37	0.36

4. Discussion

4.1. Strain-Scale Observed by Each Technique

First of all, let us classify strain-scales observed by QSM, DSM and strain gauge. The strain-scale determined by QSM indicates an average deformation of nanoscale structures with less crystal periodicity involving crystal-like orderings and a glassy amorphous structure or none at all. On the other hand, the strain scale determined by DSM provides an average microscopic deformation of all composed structures including crystalline phases. In contrast, the strain-scale determined by the strain gauge is a macroscopic deformation of the specimen.

4.2. Deformation Model of Eutectic Z50-BMG

The microstructural model suggested by Ichitsubo *et al.* [16] provides an idea of the deformation model for the metallic glass. Figure 6a shows the schematic illustration of the simplified microstructural model of Z50 before and after annealing. The microstructure of a metallic glass is known to be a heterogeneous structure composed of strongly-bonded regions (SBRs) with low mobility and high density (low free-volume fraction), surrounded by weakly-bonded regions (WBRs) with high mobility and low density (high free-volume fraction). It is further suggested that an icosahedral atomic configuration generally exists in SBR in Zr-based metallic glasses [17]. Therefore, $G(r)$ in Figure 2e involves the structural information of both regions, *i.e.*, SBR and WBR, weighted by each volume fraction. Since the volume fraction of SBRs for the Zr-based BMG is typically larger than that of WBRs [18], the microscopic strain determined by DSM should be dominated by SBRs. Consequently, E_{D_A} in Table 1 predominantly represents the microscopic Young's modulus of SBR. On the other hand, the macroscopic deformation measured by the strain gauge must be a total deformation including both WBRs and SBRs. As the elastic constant of WBR is softer than that of SBR, the macroscopic Young's modulus E_M depends on the volume fraction of each region [18]. In particular, E_M can approach the microscopic Young's modulus E_{D_A} as the volume fraction of SBRs increases. In the present study, E_M is significantly smaller than E_{D_A} before annealing, suggesting relatively inhomogeneous microstructure with high volume fraction of WBRs. After annealing, in contrast, reduction in the free-volume in WBRs due to structural relaxation increases the volume fraction of SBRs, constructing relatively homogeneous microstructure. Therefore, E_M approaches E_{D_A} by annealing since the elastic homogeneity is developed with an increase in the volume fraction of SBRs.

The relation between macroscopic and microscopic Poisson's ratios can be explained by the same mechanism of the Young's modulus with different Poisson's ratios between WBR and SBR. Ichitsubo *et al.* suggests that the Poisson's ratio of WBR is higher than that of SBR [18], explaining the present results showing the macroscopic Poisson's ratio to be larger than the microscopic Poisson's ratio. However, we cannot see any changes in the Poisson's ratio by annealing, suggesting that the measurement resolution might be insufficient for observing its changes

The elastic homogeneity inhibits generation of a shear transformation zone, leading to decrease in its toughness or ductility and plasticity. Therefore, degradation of deformability of the eutectic Z50 due to structural relaxation would be originated from homogenization of the glassy nanostructure by reduction in the volume fraction of WBRs.

4.3. Deformation Model of Hypoeutectic Z60-BMG

Figure 6b shows the schematic illustration of the simplified microstructural model of Z60 before and after annealing. Since the crystal-like orderings are partially recognized in the annealed Z60, the as-cast Z60 might originally have some crystal-like ordered SBRs which can be nuclei of nanoclusters after annealing. This is supported by the fact showing the first peaks of $I(Q)$ and $S(Q)$ for the as-cast Z60 in Figure 2b, and d is slightly sharper compared with those for the as-cast Z50. In addition, the as-cast Z60 has relatively homogeneous microstructure with large volume fraction of SBRs since E_M is originally close to E_{D_A}.

It is known that the intensity damping of $G(r)$ is correlated to damping of the structural coherence [19]. Accordingly, a discrete $G(r)$ for the crystal-like ordered regions with the size of a few nanometers or less decreases immediately with an increase of r. Therefore, an increase in E_D at the second peak of $G(r)$ for the annealed Z60 would be affected by nanocluster formation after annealing. A similar trend appears in E_Q to be increased after annealing since E_Q could be affected by the change in the Young's modulus E_D at the second peak of $G(r)$. In contrast, the average deformation determined at larger values of r above the third peak of $G(r)$ represents the microscopic Young's modulus of the average microstructure (E_{D_A} in Table 1), and is almost unchanged by annealing. Therefore, an increase in E_D at the second peak of $G(r)$ after annealing suggests that there are locally harder regions such as nanoclusters in relatively softer amorphous glassy matrix. As described above, the annealed Z60 exhibits structural inhomogeneity owing to nanocluster formation, likewise elastic homogeneity would be decreased after complete structural relaxation. However, the result that E_M and E_{D_A} are almost unchanged by annealing suggests that the elastic properties of the annealed SBRs are almost unchanged from that of the as-cast SBRs, and that WBRs surrounding nanoclusters would play a role of damper to minimize an influence of nanoclusters on the macroscopic elastic constant. In addition, nanocluster formation in SBRs after annealing might improve deformability of glassy structure by branching and pinning the shear band around the nanoclusters [20]. This would be one reason why the hypoeutectic Z60 inhibits degradation of mechanical properties after complete structural relaxation.

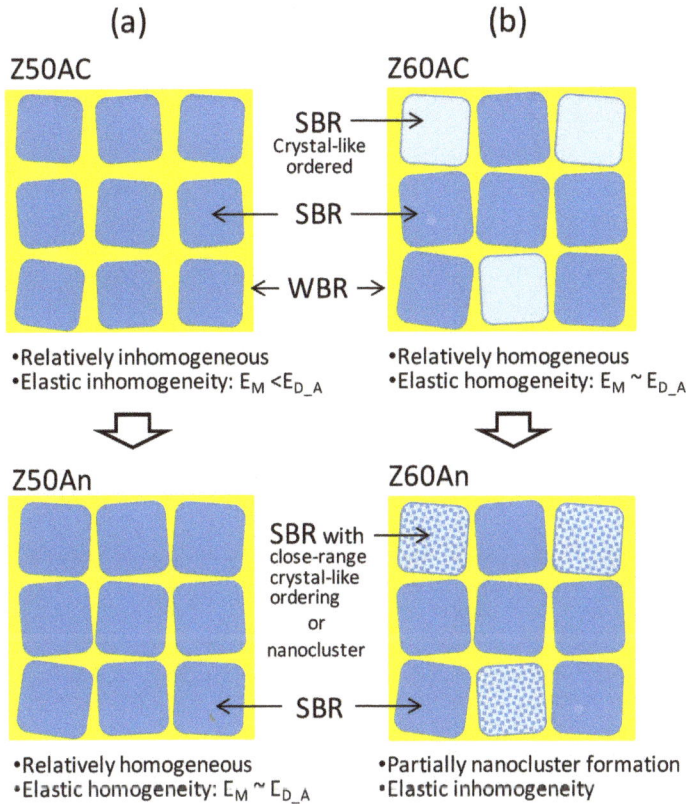

Figure 6. Schematic illustration of the simplified microstructural models of (**a**) Z50 and (**b**) Z60 before and after annealing. The weakly-bonded regions (WBR) shown by the yellow color surround the strongly-bonded regions (SBR) shown by the blue color.

5. Conclusions

In this study, the difference of the elastic deformation behaviors between eutectic $Zr_{50}Cu_{40}Al_{10}$ and hypoeutectic $Zr_{60}Cu_{30}Al_{10}$ BMGs was investigated by comparing strains with different scales obtained by the Direct-space method (DSM), Reciprocal-space (Q-space) method (QSM) and the strain gauge method. The eutectic $Zr_{50}Cu_{40}Al_{10}$ BMG obtains more homogeneous microstructure by free-volume annihilation after annealing, improving resistance to deformation but degrading ductility because of a decrease in the volume fraction of WBRs (weakly-bonded regions) with relatively high mobility. On the other hand, the as-cast hypoeutectic $Zr_{60}Cu_{30}Al_{10}$ BMG originally has homogeneous nanostructure with high volume fraction of SBRs (strongly-bonded regions) but loses structural and elastic homogeneities because of nanocluster formation partially after annealing. Such structural changes by annealing might develop unique mechanical properties showing no degradations of ductility and toughness for the structural-relaxed hypoeutectic $Zr_{60}Cu_{30}Al_{10}$ BMGs.

Acknowledgments: This work has been supported by a Grant-in-Aid of the Ministry of Education, Sports, Culture, Science and Technology, Japan, Scientific Research (A) (No. 23246109). The synchrotron radiation experiment was performed at the SPring-8 with the approval of Japan Synchrotron Radiation Research Institute (JASRI) as Proposal No. 2013B3724. The authors wish to acknowledge the experimental assistance of T. Shobu and A. Shiro at Japan Atomic Energy Agency (JAEA), and K. Shimizu at Tokyo City University. The authors would also like to acknowledge N. Igawa at JAEA for his beneficial assistance.

Author Contributions: H.S., M.I, S.S and J.S. conceived and designed the experiments; T.W. and A.M. contributed experimental instruments; H.S., R.Y., S.T. and J.S. performed the experiments and analyzed the data; all authors contributed to the interpretation of the data; H.S. wrote the paper.

Conflicts of Interest: The authors declare no conflict of interest.

References

1. Yokoyama, Y.; Yamasaki, T.; Nishijima, M.; Inoue, A. Drastic Increase in the Toughness of Structural Relaxed Hypoeutectic $Zr_{59}Cu_{31}Al_{10}$ Bulk Glassy Alloy. *Mater. Trans.* **2007**, *48*, 1276–1281. [CrossRef]
2. Yokoyama, Y.; Yamasaki, T.; Liaw, P.K.; Inoue, A. Study of the structural relaxation-induced embrittlement of hypoeutectic Zr–Cu–Al ternary bulk glassy alloys. *Acta Mater.* **2008**, *56*, 6097–6108. [CrossRef]
3. Poulsen, H.F.; Wert, J.A.; Neuefeind, J.; Honkimaki, V.; Daymond, M. Measuring strain distributions in amorphous materials. *Nat. Mater.* **2005**, *4*, 33–36. [CrossRef]
4. Hufnagel, T.C.; Ott, R.T. Structural aspects of elastic deformation of a metallic glass. *Phys. Rev. B* **2006**. [CrossRef]
5. Sato, S.; Suzuki, H.; Shobu, T.; Imafuku, M.; Tsuchiya, Y.; Wagatsuma, K.; Kato, H.; Setyawan, A.D.; Saida, J. Atomic-scale characterization of elastic deformation of Zr-Based metallic glass under tensile stress. *Mater. Trans.* **2010**, *51*, 1381–1385. [CrossRef]
6. Liss, K.-D.; Qu, D.D.; Yan, K.; Reid, M. Variability of Poisson's ratio and enhanced ductility in amorphous metal. *Adv. Eng. Mater.* **2013**, *15*, 347–351. [CrossRef]
7. Watanuki, T.; Machida, A.; Ikeda, T.; Ohmura, A.; Kaneko, H.; Aoki, K.; Sato, T.J.; Tsai, A.P. Development of a single-crystal X-ray diffraction system for hydrostatic-pressure and low-temperature structural measurement and its application to the phase study of quasicrystals. *Philos. Mag.* **2007**, *87*, 2905–2911. [CrossRef]
8. Fujihisa, H. Recent progress in the power X-ray diffraction image analysis program PIP. *Rev. High Press. Sci. Technol.* **2005**, *15*, 29–35. [CrossRef]
9. Juhás, P.; Davis, T.; Farrow, C.L.; Billinge, S.J.L. PDFgetX3: A rapid and highly automatable program for processing powder diffraction data into total scattering pair distribution functions. *J. Appl. Crystallogr.* **2013**, *46*, 560–566. [CrossRef]
10. Ishii, A.; Hori, F.; Iwase, A.; Fukumoto, Y; Yokoyama, Y.; Konno, T.J. Relaxation of free volume in $Zr_{50}Cu_{40}Al_{10}$ bulk metallic glasses studied by positron annihilation measurements. *Mater. Trans.* **2008**, *49*, 1975–1978. [CrossRef]
11. Izumi, F.; Momma, K. Three-dimensional visualization in powder diffraction. *Solid State Phenom.* **2007**, *130*, 15–20. [CrossRef]
12. Zeng, K.J.; Hamalainen, M. A new thermodynamic description of the Cu–Zr system. *J. Phase Equilib.* **1994**, *15*, 577–586. [CrossRef]
13. Louzguine-Luzgin, D.V.; Fukuhara, M.; Inoue, A. Specific volume and elastic properties of glassy, icosahedral quasicrystalline and crystalline phases in Zr–Ni–Cu–Al–Pd alloy. *Acta Mater.* **2007**, *55*, 1009–1015. [CrossRef]
14. Kumar, G.; Rector, D.; Conner, R.D.; Schroers, J. Embrittlement of Zr-based bulk metallic glasses. *Acta Mater.* **2009**, *57*, 3572–3583. [CrossRef]
15. Ngai, K.L.; Wang, L.-M.; Liu, R.; Wang, W.H. Microscopic dynamics perspective on the relationship between Poisson's ratio and ductility of metallic glass. *J. Chem. Phys.* **2014**. [CrossRef]
16. Ichitsubo, T.; Matsubara, E.; Yamamoto, T.; Chen, H.S.; Nishiyama, N.; Saida, J.; Anazawa, K. Microstructure of fragile metallic glasses inferred from ultrasound-accelerated crystallization in Pd-based metallic glasses. *Phys. Rev. Lett.* **2005**. [CrossRef] [PubMed]
17. Matsubara, E.; Ichitsubo, T.; Itoh, K.; Fukunaga, T.; Saida, J.; Nishiyama, N.; Kato, H.; Inoue, A. Heating rate dependence of T_g and T_x in Zr-based BMGs with characteristic structures. *J. Alloys Compds.* **2009**, *483*, 8–13. [CrossRef]

18. Ichitsubo, T.; Kato, H.; Matsubara, E.; Biwa, S.; Hosokawa, S.; Matsuda, K.; Uchiyama, H.; Baron, A.Q.R. Static heterogeneity in metallic glasses and its correlation to physical properties. *J. Non-Cryst. Solids* **2011**, *357*, 494–500. [CrossRef]

19. Kodama, K.; Iikubo, S.; Taguchi, T.; Shamoto, S. Finite size effects of nanoparticles on the atomic pair distribution functions. *Acta Crystallogr.* **2006**, *A62*, 444–453. [CrossRef] [PubMed]

20. Saida, J.; Setyawan, A.D.; Kato, H.; Inoue, A. Nanoscale multistep shear band formation by deformation-induced nanocrystallization in Zr–Al–Ni–Pd bulk metallic glass. *Appl. Phys. Lett.* **2005**. [CrossRef]

metals

MDPI

Article

Deformation in Metallic Glasses Studied by Synchrotron X-Ray Diffraction

Takeshi Egami [1,2,3,*], Yang Tong [1,4,†] and Wojciech Dmowski [1,†]

1   Department of Materials Science and Engineering, Joint-Institute for Neutron Sciences,
     University of Tennessee, Knoxville, TN 37996, USA; yangtong@um.cityu.edu.hk (Y.T.);
     wdmowski@utk.edu (W.D.)
2   Department of Physics and Astronomy, University of Tennessee, Knoxville, TN 37996, USA
3   Oak Ridge National Laboratory, Materials Science and Technology Division, Oak Ridge, TN 37831, USA
4   Department of Mechanical and Biomedical Engineering, City University of Hong Kong, Hong Kong, China
*   Correspondence: egami@utk.edu; Tel.: +1-865-574-5165; Fax: +1-865-576-8631
†   These authors contributed equally to this work.

Academic Editor: Klaus-Dieter Liss
Received: 10 December 2015; Accepted: 7 January 2016; Published: 11 January 2016

Abstract: High mechanical strength is one of the superior properties of metallic glasses which render them promising as a structural material. However, understanding the process of mechanical deformation in strongly disordered matter, such as metallic glass, is exceedingly difficult because even an effort to describe the structure qualitatively is hampered by the absence of crystalline periodicity. In spite of such challenges, we demonstrate that high-energy synchrotron X-ray diffraction measurement under stress, using a two-dimensional detector coupled with the anisotropic pair-density function (PDF) analysis, has greatly facilitated the effort of unraveling complex atomic rearrangements involved in the elastic, anelastic, and plastic deformation of metallic glasses. Even though PDF only provides information on the correlation between two atoms and not on many-body correlations, which are often necessary in elucidating various properties, by using stress as means of exciting the system we can garner rich information on the nature of the atomic structure and local atomic rearrangements during deformation in glasses.

Keywords: metallic glasses; mechanical deformation; anisotropic PDF analysis; high-energy X-ray diffraction

---

1. Introduction

Glass generally is a symbol of something extremely fragile. Indeed, conventional glasses, mostly oxide glasses, shatter helplessly upon impact. However, a relatively new family member of glasses, metallic glass, is stronger and tougher than oxide glasses, and even compares favorably to most of crystalline metallic materials in their mechanical properties. For this reason they are promising as structural materials [1], and beginning to be used in watches and mobile phones. However, it is not easy to understand why they are strong and how they mechanically fail. Crystalline materials lattice defects, such as dislocations, can be readily defined as deviations from lattice periodicity, and crystalline materials fail because of the motion of these defects. In glasses, defects cannot be uniquely defined because of the extensive disorder in their structures. Nevertheless, phenomenologically structural defects, such as the shear-transformation-zones (STZs), have been postulated and have facilitated elucidation of mechanical properties [2,3]. However, atomistic details of STZs remain elusive.

Diffraction measurements, by their nature, provide only the information on two-atom positional correlation. However, most physical properties depend on more collective atomic correlations, which diffraction measurements cannot directly assess. Luckily, in crystalline materials, because of lattice symmetry and periodicity, it is possible to construct an accurate three-dimensional model out of two-body correlation alone. However, we are unable to do so with liquids and glasses, and it is not possible because of the extensive structural disorder. The structures of liquids and glasses are usually expressed in terms of the atomic pair-distribution function (PDF), $g(r)$, which describe the probability of finding two atoms separated by the distance, r. PDF has an advantage that it can be determined through the Fourier-transformation of the structure function, $S(Q)$, which can be directly measured by diffraction experiments [4,5]. However, all PDFs, more or less, look alike, and they are not strongly discriminatory in determining the structure, allowing high degeneracy of similar structures.

On the other hand, if the system is perturbed by a field, such as a stress field, the response often reveals the nature of the structure more informatively. Usually, the symmetry is broken by the field, so that response can be accurately detected by symmetry discrimination. In this article we discuss how the measurement of response of metallic glasses to the applied stress contribute to the understanding of the subtle nature of the glassy system. In our view, it constitutes one of the triumphs of synchrotron radiation research in interrogating the nature of metallic materials.

2. Synchrotron X-Ray Diffraction Measurement under Stress

2.1. Anisotropic PDF Analysis

The structure of a macroscopically isotropic glass or liquid is usually described by the atomic pair-density function (PDF) defined by:

$$\rho_0 g(r) = \frac{1}{4\pi r^2 N} \sum_{i,j} \delta\left(r - |r_{ij}|\right) \tag{1}$$

where ρ_0 is the atomic number density, N the number of atoms in the system, and $\delta(x)$ is the delta function [4,5]. PDF is related to the structure function $S(Q)$ by:

$$\rho_0 g(r) = \frac{1}{2\pi^2 r} \int_0^\infty [S(Q) - 1] \sin(Qr) \, Q \, dQ \tag{2}$$

where Q is the scattering vector (= $4\pi\sin\Theta/\lambda$, Θ is the diffraction angle and λ is the wavelength of the probe). $S(Q)$ can be determined by X-ray, neutron, or electron diffraction measurement after correction for geometry, absorption and background [4,5]. Theoretically, the integration in Equation (2) should be carried out to infinity. However, in reality, because $\sin\Theta$ is equal or less than unity the maximum value of Q that can be attained, Q_{Max}, is less than $4\pi/\lambda$. Therefore we have to use high-energy X-ray to access a wide range of Q. Synchrotron radiation is an ideal probe for such a purpose, and X-rays with energy higher than 100 keV are routinely used for PDF measurement. Furthermore, by using a two-dimensional X-ray detector $S(Q)$ can be measured over a wide range of Q at once, making the diffraction measurement very fast [6].

The PDF analysis usually assumes that the sample is isotropic, so $S(Q)$ depends only on the magnitude of Q, not the direction. However, when a solid is subjected to shear or uniaxial stress the sample is no longer isotropic, and both $g(r)$ and $S(Q)$ depend on the direction of r and Q. In this case, we use the spherical harmonics expansion:

$$g(r) = \sum_{\ell,m} g_\ell^m(r) Y_\ell^m(r/r) \tag{3}$$

$$S(Q) = \sum_{\ell,m} S_\ell^m(Q) Y_\ell^m(Q/Q) \tag{4}$$

which are connected through:

$$\rho_0 g_\ell^m (r) = \frac{i^\ell}{2\pi^2} \int_0^\infty S_\ell^m (Q) \, J_\ell \, (Qr) \, Q^2 dQ \tag{5}$$

where $J_\ell(x)$ is the spherical Bessel function [5,7]. For the isotropic ($\ell = 0$) component, $S_0{}^0(Q)$, Equation (5) is reduced to Equation (1), because $J_0(x) = \sin x/x$. The anisotropic structure function, $S_\ell{}^m(Q)$ can be determined from $S(Q)$ by:

$$S_\ell^m (Q) = \iint S (\mathbf{Q}) \, Y_\ell^m (\mathbf{Q}/Q) \, d\Omega \tag{6}$$

where $d\Omega = d(\cos\theta)d\varphi$, making use of the orthonormal properties of the spherical harmonics. In the case of a sample deformed under a uniaxial stress applied along the z-axis, with the X-ray beam along the y-axis, the data from the two-dimensional detector takes the form of $S(Q, \theta)$. In most cases, the only relevant terms in Equation (4) are $S_0{}^0(Q)$ (isotropic structure function) and the anisotropic term, $S_2{}^0(Q)$, which are obtained by:

$$S_0^0 (Q) = \int_{-1}^{1} S (Q, \theta) \, d\cos\theta \tag{7}$$

$$S_2^0 (Q) = \frac{1}{2} \int_{-1}^{1} S (Q, \theta) \left[3\cos^2\theta - 1 \right] d\cos\theta \tag{8}$$

If the deformation is affine, that is, the strain is uniform, and the displacement of the i-th atom is given by a single strain tensor:

$$\Delta r_i^\alpha = \varepsilon^{\alpha\beta} r_i^\beta \tag{9}$$

Then, it can be shown that $S_2{}^0(Q)$ s given by:

$$S_{2,affine}^0 (Q) = -\varepsilon \left(\frac{1}{5} \right)^{1/2} \frac{2 \, (1 + \nu)}{3} Q \frac{d}{dQ} S_0^0 (Q) \tag{10}$$

In the same way the anisotropic PDF for affine deformation is given by:

$$g_{2,affine}^0 (r) = -\varepsilon \left(\frac{1}{5} \right)^{1/2} \frac{2 \, (1 + \nu)}{3} r \frac{d}{dr} g_0^0 (r) \tag{11}$$

Therefore, the measurement of $S_2{}^0(Q)$ does not provide any new information. In general, however, deformation in glasses is heterogeneous at the atomic level, and the real anisotropic $S(Q)$ and PDF deviate from Equations (10) and (11). In such a case, the measurement does provide new and unique information. Below we describe how the determination of the anisotropic PDF of a metallic glass sample deformed by thermomechanical creep by a diffraction experiment using high-energy X-ray from a synchrotron radiation source facilitated understanding of the microscopic process of deformation.

2.2. Procedure of Determining Anisotropic PDF

The anisotropy in $S(Q)$ is often small, and care has to be exercised in order to measure it accurately. In particular, the sensitivity of a two-dimensional (2D) detector is always slightly inhomogeneous and anisotropic. If we did not compensate for this anisotropy, the data could become very distorted. We make it a standard practice to repeat measurements at two sample orientations, rotated by 90 degrees to each other. In this section, we discuss the procedure of such a measurement in the case of samples processed by thermomechanical (creep) deformation.

2.2.1. Sample Preparation

1) Prepare a cylindrical or rectangular sample of bulk metallic glass of chosen composition. Make sure that the two end surfaces are smooth and parallel to each other.

2) Subject the sample to uniaxial stress, and raise the temperature to T_a, which has to be below the glass transition temperature, T_g. One has to be careful to keep the stress level below the yield stress at that temperature, because the yield stress is strongly temperature dependent, particularly near T_g.
3) Cool the sample down to room temperature with the stress still applied.
4) Remove the stress, and cut a sample thin enough for X-ray diffraction (~0.5 mm) from the crept sample. The cut should be made parallel to the stress axis so that, in transmission geometry, the scattering vector Q probes the direction parallel and normal to the applied stress.
5) The reference sample annealed in the same conditions (same temperature and time) should be prepared with a similar thickness. In general, the thickness is chosen in reference to absorption of X-rays ($t\mu < 0.2$, t is the thickness and μ is the absorption coefficient) and samples containing heavy elements should be thinner.

2.2.2. Diffraction Measurement

6) Set-up the beam line for a high-energy PDF experiment with a 2D detector. The detector should be placed at a distance that covers enough Q space for reliable normalization and Fourier transformation, up to 18–25 Å^{-1}.
7) Place the sample with the plane normal to the beam and the stress axis aligned vertically to the ground.
8) Turn on the X-ray and take data for a sufficiently long time, typically 5–15 min (Data Set #1). The data should be in the form of $I(Q, \theta)$, where θ is the azimuthal angle. The counting time and number of exposures depend on specifics of the detector and X-ray beam intensity. Ideally we want to have ~10^6 counts per inverse Å after integration over θ, at high Q.
9) Turn the sample by 90° around the beam so that the stress axis is horizontal, and repeat the measurement (Data Set #2). This step is useful because the sensitivity and background of the 2D detector are usually not isotropic. It is helpful to have a sample holder that can rotate the sample by 90° without changing its distance to the detector.
10) Rotate Data Set #2 by 90° and subtract it from Data Set #1, to remove the effect of the efficiency, and the flat field anisotropy of the 2D detector.
11) Determine $S(Q, \theta)$ from the raw data, $I(Q, \theta)$, by correcting for absorption, background, Compton intensity and the atomic scattering factor, $<f(Q)>^2$, and normalizing to unity at $Q \to \infty$.

2.2.3. Data Processing

12) Ideally the isotropic and anisotropic $S(Q)$ is obtained by:

$$S_0^0(Q) = \frac{1}{2}\int_0^\pi [S(Q,\theta) + S(Q,-\theta)]\sin\theta d\theta \tag{12}$$

$$S_2^0(Q) = \frac{1}{4}\int_0^\pi [S(Q,\theta) + S(Q,-\theta)]\left[3\cos^2\theta - 1\right]\sin\theta d\theta \tag{13}$$

However, as we noted, we make measurements with two sample orientations to compensate for the anisotropy of detector efficiency, and obtain two sets of data, Data Sets #1 and #2. The method of obtaining $S_2^0(Q)$ from the two sets of data is as follows.

13) The anisotropic part of Data Set #1 has the form:

$$S_{aniso1}(Q,\theta) = \frac{S_2^0(Q)}{2}\left(3\cos^2\theta - 1\right) \tag{14}$$

On the other hand, for Data Set #2, the stress is along the *x*-axis, so that:

$$S_{aniso2}(Q,\theta) = \frac{S_2^0(Q)}{2}\left(3\sin^2\theta - 1\right) \tag{15}$$

Thus, the difference is:

$$\begin{aligned}\Delta S(Q,\theta) &= S_{aniso1}(Q,\theta) - S_{aniso2}(Q,\theta)\\ &= \frac{3S_2^0(Q)}{2}\left(\cos^2\theta - \sin^2\theta\right) = 2S_{aniso1}(Q,\theta) - \frac{S_2^0(Q)}{2}\end{aligned} \tag{16}$$

Therefore:

$$S_{aniso1}(Q,\theta) = \frac{\Delta S(Q,\theta)}{2} + \frac{S_2^0(Q)}{4} \tag{17}$$

Then:

$$S_2^0(Q) = \frac{1}{8}\int_0^\pi \left[\Delta S(Q,\theta) + \Delta S(Q,-\theta)\right]\left[3\cos^2\theta - 1\right]\sin\theta d\theta \tag{18}$$

because the second term in Equation (17) integrates out to zero using this integration. Usually, $S_2^0(Q)$ is much smaller than $S_0^0(Q)$, and thus requires a higher precision to determine. By taking the difference in Equation (16), the background and the Compton scattering intensity are automatically removed, making the result more accurate.

14) Carry out the Bessel transformation to obtain the isotropic and anisotropic PDF:

$$\rho_0 g_0^0(r) = \frac{1}{2\pi^2}\int_0^\infty S_0^0(Q)\frac{\sin(Qr)}{Qr}Q^2 dQ \tag{19}$$

$$\rho_0 g_2^0(r) = -\frac{1}{2\pi^2}\int_0^\infty S_2^0(Q)\left[\left(3 - Q^2 r^2\right)\frac{\sin(Qr)}{Q^3 r^3} - \frac{3\cos(Qr)}{Q^2 r^2}\right]J_\ell(Qr)Q^2 dQ \tag{20}$$

Note that the range of *Q* is limited to $Q_{\max} = 4\pi/\lambda$. Therefore, one has to be careful not to introduce termination errors [4,5].

15) Calculate the anisotropic PDF for affine deformation using Equation (11).

16) Sometimes it is not possible to rotate the sample by 90°, for instance when we carry out an *in situ* measurement using a heavy mechanical testing machine. In such a case, we use an isotropic sample, such as a metallic glass, well annealed and relaxed at a temperature close to T_g, to determine the $S_2^0(Q)$ of the detector using the step 12), and subtract the detector $S_2^0(Q)$ from the result.

3. Results

Here, we show typical results obtained for the sample treated with thermomechanical creep to demonstrate how the anisotropic PDF can be determined. Metallic glass $Zr_{55}Cu_{30}Ni_5Al_{10}$ ($T_g = 707$ K [8]) with the dimension of 2.5 × 2.5 × 5 (mm) was subjected to compressive stress of 1.0 GPa for 60 min at $T = 623$ K, which is substantially below the glass transition temperature. Using this process the sample undergoes thermomechanical creep deformation and becomes anelastically distorted. The structure is no longer isotropic, but retains the memory of anelastic deformation [7,9]. After cooling the sample to room temperature, a piece with the dimension of 2 × 0.5 × 2 (mm) was cut out of the sample. The X-ray diffraction measurement was carried out at the 6-ID and 1-ID beam lines of the Advanced Photon Source (APS), Argonne National Laboratory. The incident X-ray energy was set to 100 keV. The beam size was 300 × 300 (µm) and a Perkin-Elmer detector or a GE 2D detector with the pixel size 200 × 200 (µm) and 2048 × 2048 pixels was used. The distance between the sample and the detector was 35 cm, which allowed $S(Q)$ to be determined up to $Q = 22$ Å^{-1}.

Figure 1a shows the raw data from the 2D detector. In this figure the diffraction ring is actually distorted elliptically, but it is difficult to recognize it by eye. If one takes the difference between Data

Set #1 and Data Set #2 the distortion is more clearly seen, as shown in Figure 1b. Both data sets were corrected for absorption, Compton scattering, and sample-independent background from the set-up to obtain $S(Q, \theta)$. Then the isotropic and anisotropic $S(Q)$ were determined using Equations (17) and (18). The step size for integration was $\Delta Q = 0.023$ Å^{-1}, $\Delta\theta = 0.052$ radian.

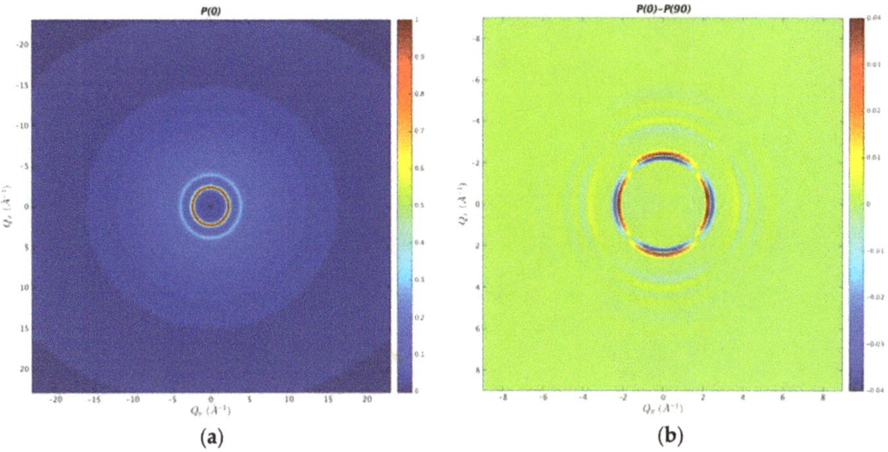

Figure 1. (**a**) High-energy X-ray diffraction pattern from metallic glass $Zr_{55}Cu_{30}Ni_5Al_{10}$, colors indicating intensity; (**b**) Difference between Data Set #1 (the stress axis vertical) and Data Set #2 (stress axis horizontal).

$S_0{}^0(Q)$ and $S_2{}^0(Q)$ thus obtained are shown in Figures 2 and 3. In Figure 3, $S_2{}^0(Q)$ is compared to $S_{2,affine}^0(Q)$. They are clearly different due to extensive non-affine deformation and atomic rearrangement as discussed below. The isotropic and anisotropic PDFs, obtained by Equations (19) and (20), are shown in Figures 4 and 5. Again, the anisotropic PDF is compared to the affine anisotropic PDF in Figure 5.

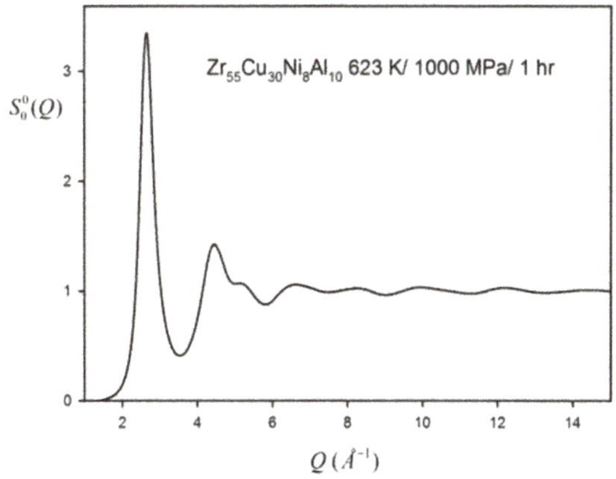

Figure 2. Isotropic structure function $S_0{}^0(Q)$ for a sample creep deformed at 623 K under 1 GPa for 1 h.

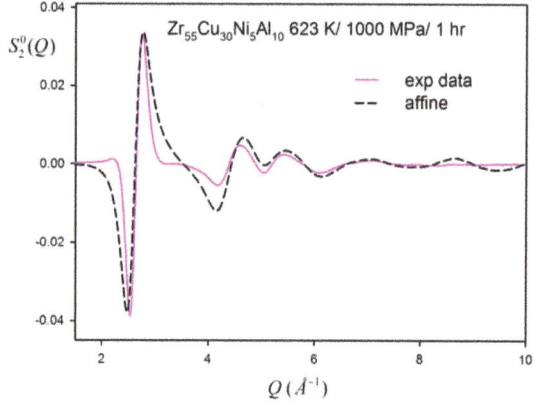

Figure 3. Anisotropic structure function $S_2{}^0(Q)$. Line in pink shows $S_{2,affine}^0 (Q)$.

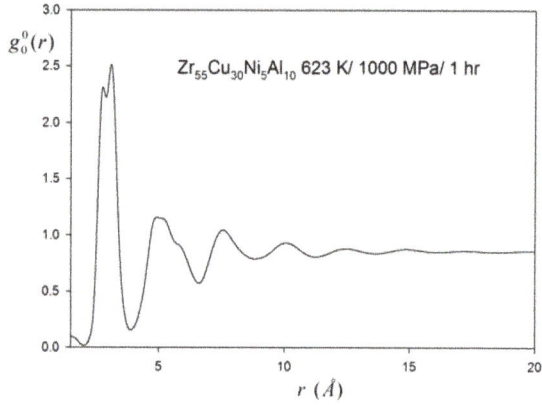

Figure 4. Isotropic PDF, $g_0{}^0(r)$.

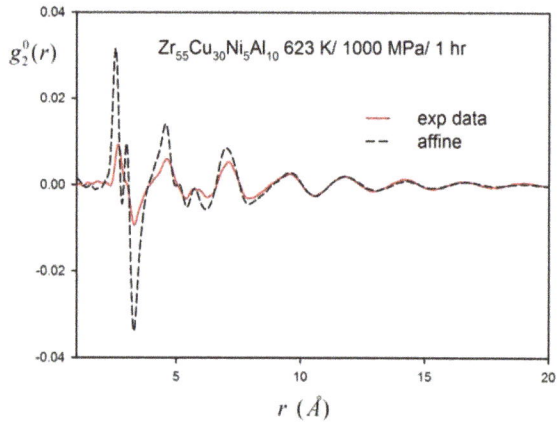

Figure 5. Anisotropic PDF $g_2{}^0(r)$ compared to the affine PDF (red curve).

It is constructive to compare the results with those obtained for a sample elastically deformed at room temperature with the applied stress well below the yield stress. In this case, in order to deform the sample, heavy mechanical testing equipment is involved, which makes it impossible to rotate the sample by 90°. Thus, we followed the procedure (16) to eliminate the effect of anisotropy in the detector. To test this procedure, we prepared two samples out of the metallic glass sample processed by creep treatment. One was a thin plate piece cut out of the sample with the plane parallel to the applied uniaxial stress direction, and the other was cut perpendicular to the stress. As shown in Figure 6, the one parallel to the stress shows a strong $S_2{}^0(Q)$ component, whereas the one perpendicular to the stress does not.

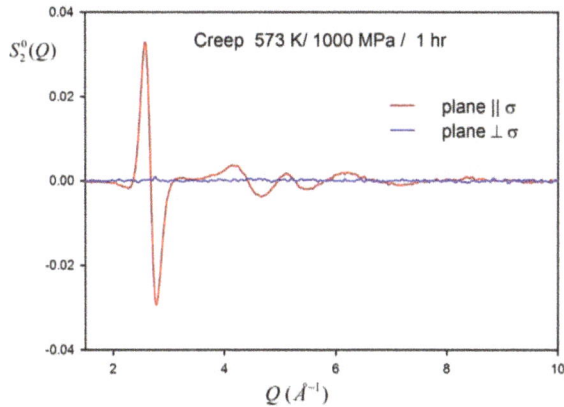

Figure 6. The $S_2{}^0(Q)$ determined for two cross-sections of the sample after the creep treatment: With the plane parallel to axial stress and plane perpendicular to the uniaxial stress.

During RT deformation , because the applied stress is below the yield stress, metallic glasses respond to stress elastically following Hook's law. However, at the atomic level, the structure changes by breaking and forming of atomic bonds, even below the yield stress [10,11]. Compared to the case of creep deformation shown in Figure 5, the anisotropic PDF $g_2{}^0(r)$ under applied stress is much closer to the affine anisotropic PDF because the deformation is mostly elastic as shown in Figure 7. However, there are significant differences in the first peak area (2–4 Å) of $g_2{}^0(r)$, which reflect the intrinsically anelastic nature of a glass [11,12]. Figures 5 and 7 cannot be directly compared, because the result in Figure 7 was obtained while the sample was still under stress, whereas that in Figure 5 was obtained after releasing the stress. To allow closer examination of the deviations from the affine PDF, the difference between the affine PDF and the data PDF, $\Delta g_2{}^0(r)$, is shown in Figure 8 for the applied stress of 400 and 1000 MPa. The difference is not limited to the first peak area, but extends to 6 Å or beyond. A part of this relaxation reflects the non-affine elastic strain because of the disorder in the structure [13,14]. However, such an effect is rather small [12], and much of it originates from bond cutting and forming [7], which occurs even in the nominally elastic regime [10–12].

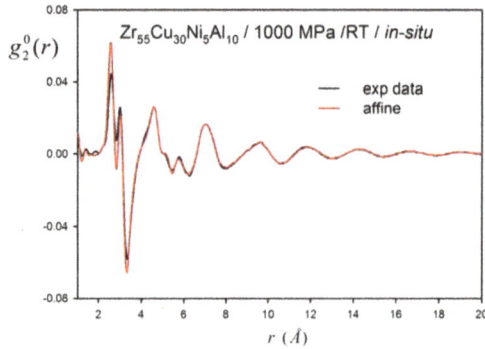

Figure 7. The anisotropic PDF, $g_2{}^0(r)$, determined during *in situ* experiment at 1000 MPa.

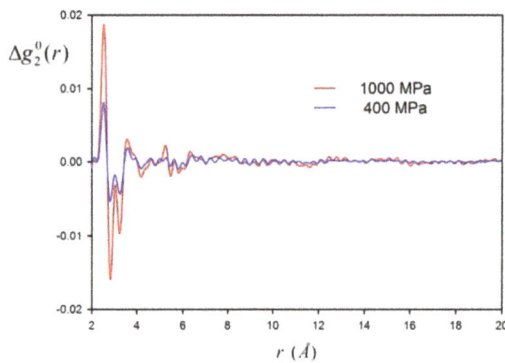

Figure 8. The difference PDF, $\Delta g_2{}^0(r) = g_2{}^0(\text{affine}) - g_2{}^0(\text{data})$, for the applied stress of 400 and 1000 MPa.

4. Discussion

As shown in Figures 3 and 5 the actual anisotropic $S(Q)$ and PDF are significantly deviated from the one for affine deformation. One way to interpret this deviation is to introduce local strain, which is dependent on r, by fitting the following equation to the data:

$$g_2^0(r) = -\varepsilon(r) \left(\frac{1}{5}\right)^{1/2} \frac{2(1+\nu)}{3} r \frac{\mathrm{d}}{\mathrm{d}r} g_0^0(r) \qquad (21)$$

Figure 9 shows the r-dependent local strain, $\varepsilon(r)$, normalized by $\varepsilon(\infty)$ for creep deformation (blue) and elastic deformation (red). A similar analysis was made earlier using the $S(Q)$ data with Q parallel ($\theta = 0$) and perpendicular ($\theta = 90°$) to the stress axis [15–17]. However, in [15–17] the strain was determined from the PDF obtained using Equation (19) rather than Equation (20), which introduced some errors. Note that Equation (19) is only valid for isotropic systems, and once the system becomes anisotropic, we have to use the spherical harmonics expansion shown above. The correct procedure with Equation (20) is now widely used [18–20].

For creep deformation (blue symbols in Figure 9), the local strain $\varepsilon(r)$ is small at short distances and increases with r, saturating to a constant, $\varepsilon(\infty)$, which is equal to the macroscopic recoverable strain. The total creep strain includes both anelastic strain (recoverable) and plastic strain (unrecoverable). Plastic strain leaves no signature on the structure other than some rejuvenation, and $\varepsilon(\infty)$ corresponds only to anelastic strain. The difference, $\Delta\varepsilon(r) = \varepsilon(\infty) - \varepsilon(r)$, represents screening, or the local strain modification, through atomic rearrangements. The sample that underwent creep was treated thermomechanically at a temperature (623 K), which is below T_g, but high enough to activate anelastic deformation. While the sample was held at this temperature under the stress, anelastic local atomic rearrangements occurred to relax the applied stress. When the sample was cooled down and the stress was removed, the regions in which local atomic rearrangements took place were strained in the opposite direction. The stress produced by these strained regions is balanced by the long-range stress, which corresponds to $G\varepsilon(\infty)$, where G is the shear elastic modulus. Consequently, the actual strain at the nearest neighbor is only 30% of the total strain at a long range $\varepsilon(\infty)$ as seen here. The strain is reduced compared to $\varepsilon(\infty)$ up to 10 Å, indicating that the region of atomic rearrangements extend as far as the fourth neighbor shell.

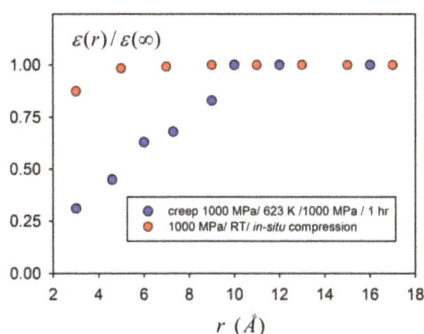

Figure 9. *r*-dependent local strain determined from Equation (21) for the sample creep deformed at 623 K under 1 GPa for 1 h (blue), and for the sample under load of 1 GPa at room temperature (red).

In comparison, for elastic deformation (red symbols in Figure 9) the strain is nearly 100%. This is mainly because, in this case, the diffraction measurement was made with the applied stress on, so that much of the strain is truly elastic. Nevertheless, $\varepsilon(r)/\varepsilon(\infty)$ is less than unity at the nearest neighbor distance, due to non-collinear atomic displacement [13] and anelasticity [10,11]. It is noteworthy that the local strain modification, $\Delta\varepsilon(r) = \varepsilon(\infty) - \varepsilon(r)$, is much more short-range (only up to 4 Å or so, including only the nearest neighbor shell) for elastic deformation than for creep (up to 10 Å).

In crystals, the topology of atomic connectivity is well defined because of lattice periodicity. Lattice defects, such as dislocations and vacancies, are defined as local deviations from the periodic lattice. In glasses, on the other hand, the topology of atomic connectivity is open, in a sense that it is easily changed by thermal excitation or applied stress, through local bond breaking and forming. Much of the deviations from the affine deformation observed for the anisotropic PDF reflect such local bond rearrangements. The local stresses these rearrangements create are balanced by the long-range stress produced by the long-range strain $\varepsilon(\infty)$, to make the total stress zero.

These local atomic rearrangements are phenomenologically described in terms of the shear-transformation-zones (STZs), which facilitated the elucidation of mechanical properties [2,3]. Whereas the atomic level details of STZ still remain elusive, a variety of computer simulations, such as References [3] and [24], made the outline of STZs clearer. The reported size of STZ varies greatly, from 20 to few hundred atoms [21–24]. However, according to our recent work [25,26], the size of STZ is actually much smaller, involving only about five atoms at the saddle point of the potential energy landscape and about 17 atoms after relaxation from the saddle point to the nearest minimum. Atoms involved in STZ form a small cluster with the size extending only to 2–3 atomic distances.

Such a small size of STZ is more consistent with the spatial extension of the atomic rearrangement for the apparently elastic deformation seen in Figures 8 and 9. Therefore, it appears that in the elastic regime the applied stress activates only individual STZs, so that atomic rearrangements induced by stress are limited to immediate locality, not much beyond the nearest neighbors. When the stress is removed, atomic rearrangements in the opposite direction take place. Thus, macroscopically the system appears elastic, and the original dimension is restored when the stress is removed. However, microscopically it is anelastic, in a sense that local atomic rearrangements take place and the local topology of atomic connectivity is altered during deformation. A proof of such an anelastic nature of deformation below the yield stress is the fact that the system can be rejuvenated even during apparently elastic deformation [27].

On the other hand, it was shown that activation of STZ can induce a cascade or avalanche of other STZ actions [26]. The large extension of the zone of atomic rearrangement during creep, as shown in Figures 5 and 9 indicates such avalanche of STZ activity is taking place during thermomechanical creep. Whereas, at room temperature, STZ avalanche occurs only for rapidly quenched unstable samples [26]. Figure 9 suggests that at elevated temperatures it may happen even in well-annealed stable sample. The sample is then cooled down to room temperature for diffraction measurement. When the stress is removed, the original sample dimension is not recovered because some local deformations are anelastically frozen-in. Much of the frozen-in strain is recovered when the sample is relaxed without stress, but a part of deformation becomes plastic by percolation and is never recovered.

5. Conclusions

By their nature, diffraction measurements provide only the information regarding two-atom positional correlation. However, most physical properties depend on more collective atomic correlations, which diffraction measurements cannot directly assess. This problem is exacerbated in liquids and glasses because of strong disorder. Nevertheless, by perturbing the system and looking at the response, it is possible to learn more about the nature of the system. As an example, we discussed the structural response of metallic glasses to applied stress. Applied stress breaks the symmetry of the system, so it becomes possible to carry out measurements with high accuracy by focusing on the emergent symmetry component induced by stress.

Our analysis indicates that anelastic atomic rearrangements at room temperature induced by applied stress below the yield stress are limited to immediate neighborhood of atoms, most likely representing activation of a single STZ. However, when the stress is applied at a temperature close to T_g atomic rearrangements are extensive and spatially more extended, and the sample shows creep deformation. Such deformation must be a consequence of a cascade or avalanche of multiple STZ actions, resulting in macroscopically anelastic behavior. As illustrated in the results above, the study of the structural change due to applied stress using synchrotron X-ray diffraction leads to revelation of atomistic details of the deformation mechanism in metallic glasses when the result are analyzed using the anisotropic PDF method.

Acknowledgments: This work was supported by the US Department of Energy, Office of Science, Basic Energy Sciences, Materials Science and Engineering Division.

Author Contributions: This paper was written by T.E. with the aid by T.Y. and W.D. who provided data and information on experimental details.

Conflicts of Interest: The authors declare no conflict of interest.

References

1. Greer, A.L. Metallic glasses. *Science* **1995**, *267*, 1947–1953. [CrossRef] [PubMed]
2. Argon, A.S. Mechanisms of inelastic deformation in metallic glasses. *J. Phys. Chem. Solids* **1982**, *43*, 945–961. [CrossRef]
3. Falk, M.L.; Langer, J.S. Dynamics of viscoplastic deformation in amorphous solids. *Phys. Rev. E* **1998**, *57*, 7192–7205. [CrossRef]
4. Warren, B.E. *X-ray Diffraction*; Dover Publications: New York, NY, USA, 1969.

5. Egami, T.; Billinge, S.J.L. *Underneath the Bragg Peaks: Structural Analysis of Complex Materials*, 2nd ed.; Pergamon Materials Series, Elsevier: Amsterdam, The Netherland, 2012.

6. Chupas, P.J.; Qiu, X.; Hanson, J.C.; Lee, P.L.; Grey, C.P.; Billinge, S.J.L. Rapid-acquisition pair distribution function (RA-PDF) analysis. *J. Appl. Cryst.* **2003**, *36*, 1342–1347. [CrossRef]

7. Suzuki, Y.; Haimovich, J.; Egami, T. Bond-orientational anisotropy in metallic glasses observed by X-ray diffraction. *Phys. Rev. B* **1987**, *35*, 2162–2168. [CrossRef]

8. Yokoyama, Y.; Inoue, K.; Fukaura, K. Pseudo Float melting state in ladle arc-melt-type furnace for preparing crystalline inclusion-free bulk amorphous alloy. *Mater. Trans.* **2002**, *43*, 2316–2320. [CrossRef]

9. Dmowski, W.; Egami, T. Observation of structural anisotropy in metallic glasses induced by mechanical deformation. *J. Mater. Res.* **2007**, *22*, 412–418. [CrossRef]

10. Suzuki, Y.; Egami, T. Shear deformation of glassy metals: Breakdown of Cauchy relationship and anelasticity. *J. Non-Cryst. Solids* **1985**, *75*, 361–366. [CrossRef]

11. Dmowski, W.; Iwashita, T.; Chuang, C.-P.; Almer, J.; Egami, T. Elastic heterogeneity in metallic glasses. *Phys. Rev. Lett.* **2010**. [CrossRef] [PubMed]

12. Egami, T.; Iwashita, T.; Dmowski, W. Mechanical properties of metallic glasses. *Metals* **2013**, *3*, 77–113. [CrossRef]

13. Waire, D.; Ashby, M.F.; Logan, J.; Weis, J. On the use of pair potentials to calculate the properties of amorphous metals. *Acta Metall.* **1971**, *19*, 779–788. [CrossRef]

14. Maloney, C.E.; Lemaître, A. Amorphous systems in athermal, quasistatic shear. *Phys. Rev. E* **2006**. [CrossRef] [PubMed]

15. Poulsen, H.F.; Wert, J.A.; Neuefeind, J.; Honkimäki, V.; Daymond, M. Measuring strain distributions in amorphous materials. *Nat. Mater.* **2005**, *4*, 33–36. [CrossRef]

16. Hufnagel, T.C.; Ott, R.T.; Almer, J. Structural aspects of elastic deformation of a metallic glass. *Phys. Rev. B* **2006**. [CrossRef]

17. Wang, X.D.; Bednarcik, J.; Saksl, K.; Franz, H.; Cao, Q.P.; Jiang, J.Z. Tensile behavior of bulk metallic glasses by *in situ* X-ray diffraction. *Appl. Phys. Lett.* **2007**. [CrossRef]

18. Mattern, N.; Bednarcik, J.; Pauly, S.; Wang, G.; Das, J.; Eckert, J. Structural evolution of Cu-Zr metallic glasses under tension. *Acta Mater.* **2009**, *57*, 4133–4139. [CrossRef]

19. Vempati, U.K.; Valavala, P.K.; Falk, M.L.; Almer, J.; Hufnagel, T.C. Length-scale dependence of elastic strain from scattering measurements in metallic glasses. *Phys. Rev. B* **2012**. [CrossRef]

20. Qu, D.D.; Liss, K.-D.; Sun, Y.J.; Reid, M.; Almer, J.D.; Yan, K.; Wang, Y.B.; Liao, X.Z.; Shen, J. Structural origins for the high plasticity of a Zr-Cu-Ni-Al bulk metallic glass. *Acta Mater.* **2013**, *61*, 321–330. [CrossRef]

21. Schuh, C.A.; Lund, A.C.; Nieh, T.G. New regime of homogeneous flow in the deformation map of metallic glasses: Elevated temperature nanoindentation experiments and mechanistic modeling. *Acta Mater.* **2004**, *52*, 5879–5891. [CrossRef]

22. Pan, D.; Inoue, A.; Sakurai, T.; Chen, M.W. Experimental characterization of shear transformation zones for plastic flow of bulk metallic glasses. *Proc. Natl. Acad. Sci. USA* **2008**, *105*, 14769–14772. [CrossRef] [PubMed]

23. Ju, J.D.; Jang, D.; Nwankpa, A.; Atzmon, M. An atomically quantized hierarchy of shear transformation zones in a metallic glass. *J. Appl. Phys.* **2011**. [CrossRef]

24. Choi, I.-C.; Zhao, Y.; Yoo, B.-G.; Kim, Y.-J.; Suh, J.-Y.; Ramamurty, U.; Jiang, J. Estimation of the shear transformation zone size in a bulk metallic glass through statistical analysis of the first pop-in stresses during spherical nanoindentation. *Scr. Mater.* **2012**, *66*, 923–926. [CrossRef]

25. Fan, Y.; Iwashita, T.; Egami, T. How thermally activated deformation starts in metallic glass. *Nat. Commun.* **2014**. [CrossRef] [PubMed]

26. Fan, Y.; Iwashita, T.; Egami, T. Crossover from Localized to Cascade Relaxations in Metallic Glasses. *Phys. Rev. Lett.* **2015**. [CrossRef] [PubMed]

27. Tong, Y.; Iwashita, T.; Dmowski, W.; Bei, H.; Yokoyama, Y.; Egami, T. Structural Rejuvenation in Bulk Metallic Glasses. *Acta Mater.* **2015**, *86*, 240–248. [CrossRef]

metals

MDPI

Article

How Can Synchrotron Radiation Techniques Be Applied for Detecting Microstructures in Amorphous Alloys?

Gu-Qing Guo, Shi-Yang Wu, Sheng Luo and Liang Yang *

College of Materials Science and Technology, Nanjing University of Aeronautics and Astronautics, Nanjing 210016, China; guoguqing@nuaa.edu.cn (G.-Q.G.); shiyangwu0914@gmail.com (S.-Y.W.); nuaaluosheng@163.com (S.L.)

* Author to whom correspondence should be addressed; yangliang@nuaa.edu.cn; Tel.: +86-25-52112903; Fax: +86-25-52112626.

Academic Editor: Klaus-Dieter Liss
Received: 23 September 2015; Accepted: 2 November 2015; Published: 4 November 2015

Abstract: In this work, how synchrotron radiation techniques can be applied for detecting the microstructure in metallic glass (MG) is studied. The unit cells are the basic structural units in crystals, though it has been suggested that the co-existence of various clusters may be the universal structural feature in MG. Therefore, it is a challenge to detect microstructures of MG even at the short-range scale by directly using synchrotron radiation techniques, such as X-ray diffraction and X-ray absorption methods. Here, a feasible scheme is developed where some state-of-the-art synchrotron radiation-based experiments can be combined with simulations to investigate the microstructure in MG. By studying a typical MG composition ($Zr_{70}Pd_{30}$), it is found that various clusters do co-exist in its microstructure, and icosahedral-like clusters are the popular structural units. This is the structural origin where there is precipitation of an icosahedral quasicrystalline phase prior to phase transformation from glass to crystal when heating $Zr_{70}Pd_{30}$ MG.

Keywords: metallic glasses; extended X-ray absorption fine structure; X-ray diffraction; reverse Monte Carlo simulation; microstructure

1. Introduction

The atomic structure of metallic glass (MG) is a long-standing issue and has been attracting great interest since the 1960s [1–7] because of its unique properties and forming ability which is strongly related to its atomic structure. The microstructure of MG is rather complex, and the three-dimensional structural picture in this class of alloys is far from being established. Fortunately, thus far, several structural models have been proposed theoretically, enhancing the understanding of the glass-forming mechanisms in binary MG by building and stacking clusters in space to reveal their short-range and medium-range orderings [5,8–10]. However, there is a challenge in probing the atomic-scale structure by using conventional experimental techniques. As an advanced experimental platform, a synchrotron radiation facility can provide a series of state-of-the-art techniques for detecting the microstructure of various materials, especial for some amorphous materials [11]. In the previous work, synchrotron radiation methods such as X-ray diffraction (XRD) and extended X-ray absorption fine structure (EXAFS) and neutron diffraction have been applied for studying MG [12–14]. Nevertheless, because XRD and EXAFS can only provide the average atomic distributions and the average surroundings of each kind of atom (element-specific), respectively, whether performing these synchrotron radiation experiments can directly reveal the complex microstructure of MG is a controversial issue [15,16].

Metals **2015**, *5*, 2048–2057

In this work, a feasible scheme for addressing this issue is developed by performing a series of state-of-the-art synchrotron radiation-based experiments combined with simulations to investigate the microstructures of amorphous alloys [17,18]. $Zr_{70}Pd_{30}$ binary alloy is selected as the research prototype due to the following reasons: (1) compared with multicomponent alloys, the $Zr_{70}Pd_{30}$ alloy has a relatively simple composition, enhancing the reliability of structural results; (2) an icosahedral quasicrystalline primary phase (I-phase) was detected when heating the $Zr_{70}Pd_{30}$ MG [19], and it was suggested that icosahedral clusters are the basic building blocks in the microstructure of this glassy alloy [20]. Therefore, as a contrast, the icosahedral cluster can be used as the initial structural model to directly fit EXAFS signals.

2. Experimental Section

The $Zr_{70}Pd_{30}$ binary ingot was prepared by arc-melting high-purity metals (99.9% Zr and 99.9% Pd) [21,22]. Amorphous ribbons with a cross-section of 0.04×2 mm^2 were produced from the ingot via single-roller melt spinning at a wheel surface velocity of 40 m/s in purified Ar atmosphere. Firstly, X-ray diffraction (Cu K_α, radiation) and high-resolution electron microscopy measurements were performed to confirm the amorphous state of the as-prepared sample. Subsequently, room temperature X-ray diffraction (XRD) measurement was performed using a high-energy synchrotron radiation monochromatic beam (about 100 KeV) on beam line BW5 in Hasylab, Germany [23]. Two-dimension diffraction data was collected by a Mar345 image plate, and then was integrated to Q-space (Q is the wave vector transfer) after subtracting the corresponding background by using program Fit2D [24]. The output data was normalized by software PDFgetX to obtain structure factor S(Q) according to the Faber-Ziman equation [25]. Furthermore, extended X-ray absorption fine structure (EXAFS) measurements for Zr and Pd K-edge were carried out using transmission mode at beam lines BL14W1, in the Shanghai Synchrotron Radiation Facility of China and U7C, in the National Synchrotron Radiation Laboratory (NSRL) of China. These EXAFS raw data were normalized via a standard data-reduced procedure, employing the Visual Processing in EXAFS Researches (VIPER) [26].

In this work, two methods were applied for detecting the microstructure of $Zr_{70}Pd_{30}$ MG, based on the normalized EXAFS and XRD data. The first method is that we directly fit the Zr and the Pd K-edge EXAFS signals with the Zr- and the Pd-centered icosahedral cluster models simultaneously, using the software VIPER. According to our previous work [14], these two icosahedral models could be obtained by extracting some typical clusters from ZrPd binary crystalline phases, such as the Zr_2Pd_1 tetragonal phase. The second method is that rather than fitting the EXAFS signals directly, we simulated all the EXAFS and XRD data simultaneously under the framework of reverse Monte-Carlo (RMC) [27], because the RMC-simulation technique is an efficient iterative method for building a structural model in disordered systems with detailed structural information that agree quantitatively with experimental data (such as synchrotron radiation-based XRD and EXAFS, and neutron-diffraction data) [28].

3. Results and Discussion

When directly fitting the EXAFS signals with the Zr- and Pd-centered icosahedral cluster models, fixing and unfixing the coordination numbers (CNs) of the cluster models were both tried. Fixing the CNs of the cluster model to fit the EXAFS signal is usually adopted when the short-range ordering (local structure) of the measured sample resembles the cluster model. Because it was suggested that the icosahedral clusters are the building blocks in ZrPd binary MG, the icosahedral quasicrystalline primary phase could appear during annealing [19]. Thus, it seems reasonable to fit the EXAFS signals by fixing the CNs of the icosahedral cluster models. Figure 1 shows the Zr and the Pd K-edge EXAFS signals ($\kappa^3 \cdot \chi(\kappa)$), as well as their fitted data.

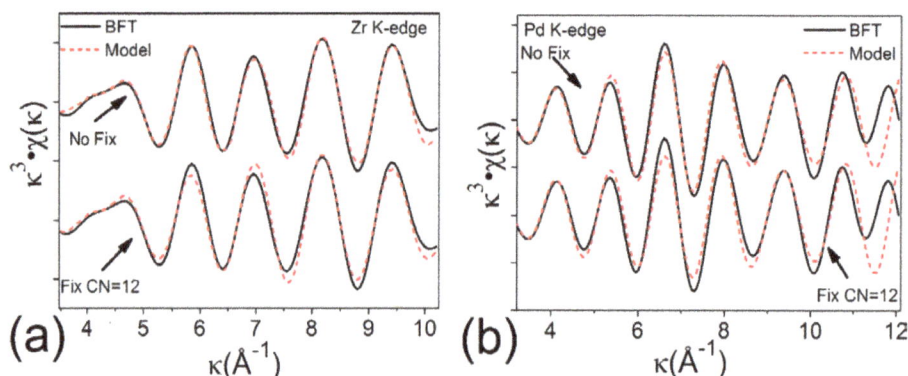

Figure 1. The fitted EXAFS signals of (**a**) Zr K-edge; and (**b**) Pd K-edge. The experimental and fitted data are plotted with solid and dashed lines, respectively. Here κ is the wave vector, and the original EXAFS signal ($\chi(\kappa)$) is weighted by κ^3. The icosahedral cluster model is the initial fitting model. Here, fixing and unfixing the CN of icosahedral cluster models were both tried.

These fitted Zr and Pd K-edge EXAFS curves are consistent with their experimental counterparts. All the fitted data are listed in Table 1, including the CNs around Zr or Pd centers, atomic distances (*R*), energy shifts, and the relative displacement of atoms. These fitted data are some physical parameters that not only can reflect the short-range orderings in the microstructure of MG, but also can be applied for evaluating the success of the fitting itself. As shown in Table 1, when the total CNs around Zr or Pd centers are fixed, the energy shifts have abnormal values larger than 20 eV, which are seldom observed if the structural model can fit the experimental data properly. On the other hand, when the total CNs around Zr or Pd centers are not fixed, there are abnormal CN values far from that of the icosahedral cluster (CN = 12). For instance, the CN of Zr centers (6.0 + 2.9 = 8.9) is much smaller than 12, while that of Pd centers (2.8 + 17.1 = 19.9) is much larger than 12. The abnormal values of these physical parameters indicate that it is not proper to fit the EXAFS signals of MG directly by using the icosahedral cluster model, whether the CN of this model is fixed or not.

Table 1. Atomic structural information obtained from the EXAFS fitting, including coordination numbers (CNs) around Zr or Pd centers, atomic distances (*R*), energy shifts, and the relative displacement of atoms.

Absorption Edge	Fitting Condition	R (Å) ± 0.02		CN ± 0.1		σ^2 (Å) ± 0.001		E_0 Shift (eV) ± 0.01	
		ZrZr	ZrPd	ZrZr	ZrPd	ZrZr	ZrPd	ZrZr	ZrPd
Zr K-edge	CN unfixed	3.10	3.00	6.0	2.9	0.038	0.014	1.20	−11.80
	CN fixed	3.10	3.01	8.8	3.2	0.045	0.015	24.01	−23.34
-	-	PdPd	PdZr	PdPd	PdZr	PdPd	PdZr	PdPd	PdZr
Pd K-edge	CN unfixed	2.74	2.95	2.8	17.1	0.013	0.068	−9.72	−4.62
	CN fixed	2.71	2.90	2.7	9.3	0.015	0.078	−21.72	−22.58

Figure 2a shows the original two-dimensional X-ray diffraction pattern of $Zr_{70}Pd_{30}$ and Figure 2b is the RMC-simulated structural model. Figure 2c–f show the simulated *S*(*Q*) curve, the *G*(*r*) curve, the Zr and Pd K-edge EXAFS spectra, as well as their corresponding experimental data. As shown in Figure 2a,c, *i.e.*, the two-dimensional diffraction pattern and the one-dimensional diffraction data, except for a bright hole (sharp peak) located at about 2.5–3.5 Å^{-1} and some other halos (broad peaks) ranging from 4–6 Å^{-1} and 6–8 Å^{-1}, there is no circle (sharp Bragg peak). Therefore, the full amorphous structure in this sample is confirmed. The good matching between all the experiment-simulation pairs confirms success of the RMC simulation. The simulated structural model shown in Figure 2b contains

Metals **2015**, *5*, 2048–2057

a mass of position-determined atoms, so that atomic- and cluster-level structural information can be deduced accordingly. On the atomic scale, the nearest atomic-pair distances are listed in Table 2. Such data are compared with the sum of their Goldschmidt atomic radii (SGAR) and those extracted from the corresponding crystalline compounds, such as the Zr_2Pd_1 tetragonal phase obtained by annealing treatment on the $Zr_{70}Pd_{30}$ MG [19]. Atomic-pair distances in the Zr_2Pd_1 I-phase are not provided here due to the lack of such values. It is found that the Zr–Pd bond apparently differs from both values of the SGAR and the tetragonal phase so that its length is shortened by 0.1–0.2 Å. This strongly suggests that a strong interaction between Zr and Pd atoms occurs in the $Zr_{70}Pd_{30}$ binary glassy alloy, which is consistent with the previous work [29], in particular the findings of the d-band bonding changes with the transition metals [30,31].

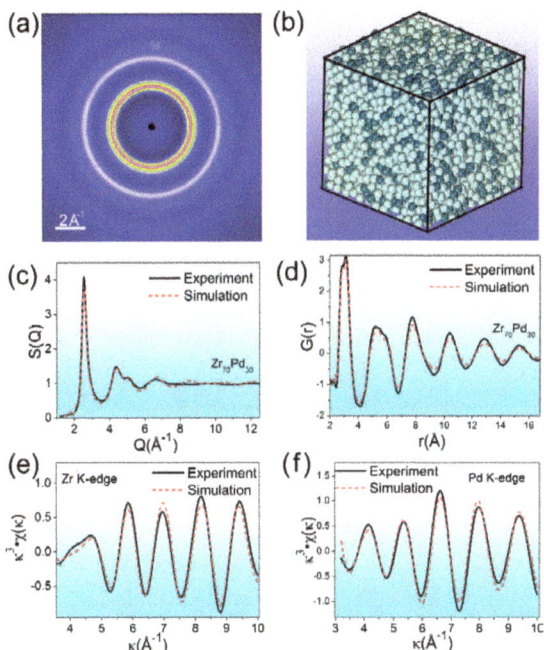

Figure 2. (a) The original two-dimensional X-ray diffraction pattern of $Zr_{70}Pd_{30}$ and (b) the RMC-simulated structural model. (c), (d), (e), and (f) are the RMC-simulated $S(Q)$, $G(r)$, Zr K-edge, and Pd K-edge EXAFS signals, as well as their corresponding experimental data. The experimental and the simulated data are plotted with solid and dashed lines, respectively.

Table 2. The nearest atomic-pair distances deduced from the RMC-simulated structural models of $Zr_{70}Pd_{30}$ MG, the Goldschmidt atomic radii (SGAR) values, and the corresponding Zr_2Pd_1 crystalline alloy.

Atomic Pairs	Atomic-Pair Distances (Å) $\pm$ 0.02		
	$Zr_{70}Pd_{30}$ MG	SGAR	Crystalline Alloy
Zr–Zr	3.20	3.20	3.23
Zr–Pd	2.90	2.97	3.06
Pd–Pd	2.76	2.74	3.31
Pd–Zr	2.90	2.97	3.06

Furthermore, Voronoi tessellation [32] is carried out to analyze the RMC-simulated structural model, from which various indexed Voronoi clusters (VCs) can be obtained. Distributions of the major VCs centered with Zr or Pd atoms are plotted in Figure 3. In the present work, Zr atoms should be regarded as the solvents while Pd atoms should be the solutes, considering their concentrations. It is found that there are hundreds of types of VCs, and the most popular solvent-centered and solute-centered VCs are indexed as <0,2,8,4>, <0,1,10,2>, <0,3,6,4>, and <0,2,8,2>, <0,3,6,3>, <0,2,8,1>, respectively. This is consistent with the co-existence of various clusters presented in the theoretical work [5]. In each <n3,n4,n5,n6> indexed VC, ni denotes the number (n) of the i-fold rotation symmetry, and indicates the number of shell atoms connected with other i shell atoms. In addition, the CN of the center atom in a <n3,n4,n5,n6> VC can be deduced because Σni stands for the number of the shell atoms around the center. For instance, the center atom in a <0,2,8,2> VC has a CN of 12. As shown in Figure 3, the solvent and the solute have CNs ranging from 12–15 and 10–13, respectively. This indicates that the average CN around the solvent is obviously larger than its counterpart around the solute. According to the efficient cluster-packing model [33], the optimal CN relates to the size ratio between the center atom and the shell atoms in clusters. In our case, it is obvious that the Zr atom has a larger radius than that of the Pd atom. Therefore, it is reasonable that the Zr centers have more near neighbors.

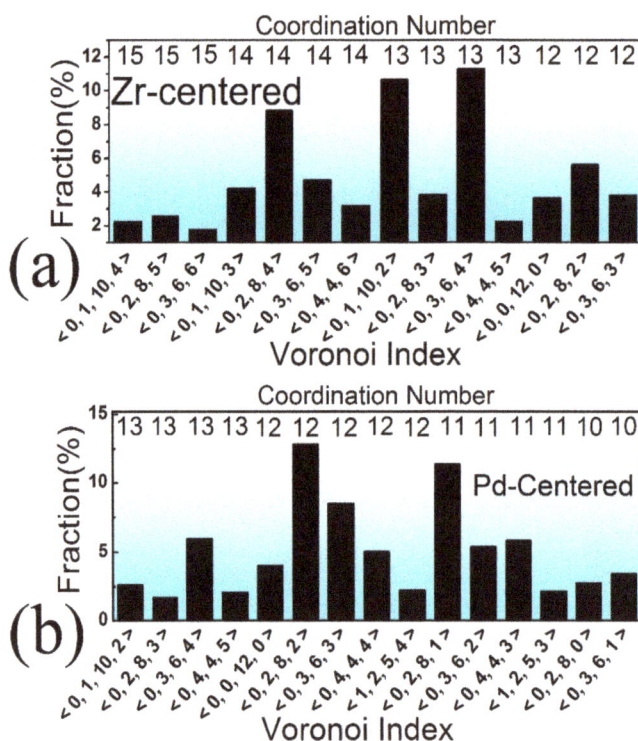

Figure 3. Distribution of the major Voronoi clusters centered with (**a**) Zr and (**b**) Pd atoms. Only those whose fractions are larger than 1.5% are selected. The CN value denotes the number of shell atoms of each VC, *i.e.*, the CN around the center atom.

It has been pointed out that the ideal icosahedral or icosahedral-like VCs are indexed as <0,0,12,0>, <0,2,8,2>, <0,3,6,3>, <0,4,4,4>, <0,1,10,2>, and so on [32]. Configurations of these icosahedral or icosahedral-like VCs are plotted in Figure 4a–e. The icosahedron deduced from the Zr_2Pd_1 I-phase is also shown in Figure 4f, which is extremely similar with that shown in Figure 4a.

Metals **2015**, *5*, 2048–2057

It is observed in Figure 3 that icosahedral or icosahedral-like VCs have relatively high fractions. In particular, it is worth noting that the icosahedral-like VCs with a CN of 12 (such as <0,2,8,2> and <0,3,6,3>) are some popular structural units centered with Pd atoms. According to the relationship between the R value (the size ratio between the center and the shell atoms, where the optimal R value is the so-called R^*) and the CN [29], the estimated R value is close to the R^* corresponding to a CN of 12, which is in agreement with our result mentioned above.

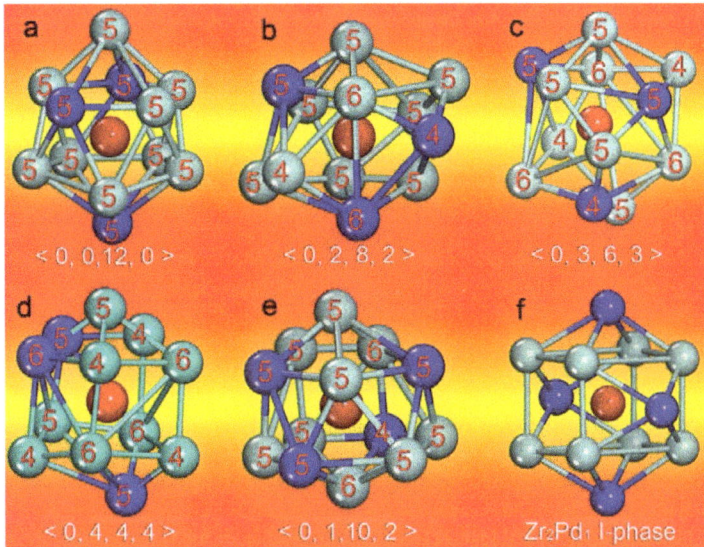

Figure 4. Configurations of the icosahedral or icosahedral-like VCs extracted from the RMC-simulated structural model, including: (**a**) <0,0,12,0>; (**b**) <0,2,8,2>; (**c**) <0,3,6,3>; (**d**) <0,4,4,4>; and (**e**) <0,1,10,2>. The number labeled on each shell atom stands for the number of its neighbor (connected) shell atoms, and also indicates the *i*-fold rotation symmetry. The icosahedron deduced from the Zr_2Pd_1 I phase is shown in (**f**), which is extremely similar with that shown in (**a**).

Here we can explain why there is precipitation of an icosahedral quasicrystalline phase prior to phase transformation from glass to crystal when heating $Zr_{70}Pd_{30}$ MG. We have revealed that there are popular icosahedral or icosahedral-like VCs in the microstructure of $Zr_{70}Pd_{30}$ MG. There is not a large configuration discrepancy between VCs in the $Zr_{70}Pd_{30}$ MG and those in the Zr_2Pd_1 I-phase, leading to their relatively small energy barrier of phase transformation. Therefore, the Zr_2Pd_1 I-phase is easy to form by rearranging and stacking those icosahedral or icosahedral-like VCs with quasi-periodicity during annealing. When adequate energy is provided to overcome the energy barrier during annealing, both the Zr_2Pd_1 I-phase and the residual amorphous phase will transform into the Zr_2Pd_1 tetragonal phase, as observed experimentally [19].

4. Conclusions

In summary, how synchrotron radiation techniques can be applied for detecting the microstructure in MG is studied. It is found that fitting the EXAFS signal of MG with a structural model directly cannot provide reliable structural information. A feasible scheme for investigating the microstructure of amorphous alloys is required. Combining synchrotron radiation-based experiments with simulations is tried in this work. It is revealed that the co-existence of various clusters is the intrinsic nature in the amorphous structure, and some icosahedral or icosahedral-like VCs are the popular structural units. This leads to their relatively small energy barrier for the amorphous-to-quasicrystal phase

Metals **2015**, *5*, 2048–2057

transformation, and can explain why there is precipitation of an icosahedral quasicrystalline phase prior to phase transformation from glass to crystal when annealing the $Zr_{70}Pd_{30}$ MG.

Acknowledgments: The authors would like to thank the HASYLAB in Germany, the Shanghai Synchrotron Radiation Facility in China, and the National Synchrotron Radiation Laboratory of China for the use of the advanced synchrotron radiation facilities. Financial support from the National Natural Science Foundation of China (Grant No. U1332112 and 51471088), the Fundamental Research Funds for the Central Universities (Grant No. NE2015004), the Funding for Outstanding Doctoral Dissertation in NUAA (Grant No. BCXJ12-08), the Funding of Jiangsu Innovation Program for Graduate Education (Grant No. CXLX13-152), and the project funded by the Priority Academic Program Development (PAPD) of Jiangsu Higher Education Institutions are gratefully acknowledged.

Author Contributions: G.-Q.G. performed simulation work upon the experimental data. L.Y. performed analysis of this work and wrote this article. S.-Y.W. and S.L. contributed to the experimental research work.

Conflicts of Interest: The authors declare no conflict of interest.

References

1. Cohen, M.H.; Turnbull, D. Metastability of amorphous structures. *Nature* **1964**, *203*, 964. [CrossRef]
2. Finney, J.L. Modeling the structures of amorphous metals and alloys. *Nature* **1977**, *266*, 309–314. [CrossRef]
3. Doye, J.P.K.; Wales, D.J. The structure and stability of atomic liquids: From clusters to bulk. *Science* **1996**, *271*, 484–487. [CrossRef]
4. Tan, H.; Zhang, Y.; Ma, D.; Feng, Y.P.; Li, Y. Optimum glass formation at off-eutectic composition and its relation to skewed eutectic coupled zone in the La based La-Al-(Cu,Ni) pseudo ternary system. *Acta Mater.* **2003**, *51*, 4551–4561. [CrossRef]
5. Sheng, H.W.; Luo, W.K.; Alamgir, F.M.; Bai, J.M.; Ma, E. Atomic packing and short-to-medium-range order in metallic glasses. *Nature* **2006**, *439*, 419–425. [CrossRef] [PubMed]
6. Hirata, A.; Guan, P.F.; Fujita, T.; Hirotsu, Y.; Inoue, A.; Yavari, A.R.; Sakurai, T.; Chen, M.W. Direct observation of local atomic order in a metallic glass. *Nat. Mater.* **2011**, *10*, 28–33. [CrossRef] [PubMed]
7. Wu, Z.W.; Li, M.Z.; Wang, W.H.; Liu, K.X. Hidden topological order and its correlation with glass-forming ability in metallic glasses. *Nat. Commun.* **2015**. [CrossRef] [PubMed]
8. Bernal, J.D. Geometrical approach to the structure of liquids. *Nature* **1959**, *183*, 141–147.
9. Gaskell, P.H. A new structural model for transition metal-metalloid glasses. *Nature* **1978**, *276*, 484–485. [CrossRef]
10. Miracle, D.B. A structural model for metallic glasses. *Nat. Mater.* **2004**, *3*, 697–702. [CrossRef] [PubMed]
11. Pfeiffer, F.; Weitkamp, T.; Bunk, O.; David, C. Phase retrieval and differential phase-contrast imaging with low-brilliance X-ray sources. *Nat. Phys.* **2006**, *2*, 258–261. [CrossRef]
12. Lefebvre, S.; Quivy, A.; Bigot, J.; Calvayrac, Y.; Bellissent, R. A neutron diffraction determination of short-range order in a $Ni_{63.7}Zr_{36.3}$ glass. *J. Phys. F* **1985**, *15*, L99–L103. [CrossRef]
13. Luo, W.K.; Sheng, H.W.; Alamgir, F.M.; Bai, J.M.; He, J.H.; Ma, E. Icosahedral Short-Range Order in Amorphous Alloys. *Phys. Rev. Lett.* **2004**. [CrossRef]
14. Yang, L.; Xia, J.H.; Wang, Q.; Dong, C.; Chen, L.Y.; Ou, X.; Liu, J.F.; Jiang, J.Z.; Klementiev, K.; Saksl, K.; *et al.* Design of Cu_8Zr_5-based bulk metallic glasses. *Appl. Phys. Lett.* **2006**. [CrossRef]
15. Sadoc, J.F.; Dixmier, J. Structural investigation of amorphous CoP and NiP alloys by combined X-ray and neutron scattering. *Mater. Sci. Eng.* **1978**, *23*, 187–192. [CrossRef]
16. Saksl, K.; Franz, H.; Jovari, P.; Klementiev, K.; Welter, E.; Ehnes, A.; Saida, J.; Inoue, A.; Jiang, J.Z. Evidence of icosahedral short-range order in $Zr_{70}Cu_{30}$ and $Zr_{70}Cu_{29}Pd_1$ metallic glasses. *Appl. Phys. Lett.* **2003**, *8333*, 3924–3926. [CrossRef]
17. Yang, L.; Guo, G.Q. Structural origin of the high glass-forming ability in Gd doped bulk metallic glasses. *Appl. Phys. Lett.* **2010**. [CrossRef]
18. Yang, L.; Guo, G.Q.; Chen, L.Y.; Wei, S.H.; Jiang, J.Z.; Wang, X.D. Atomic structure in Al-doped multicomponent bulk metallic glass. *Scr. Mater.* **2010**, *63*, 879–882. [CrossRef]
19. Saida, J.; Kasai, M.; Matsubara, E.; Inoue, A. Stability of glassy state in Zr-based glassy alloys correlated with nano icosahedral phase formation. *Ann. Chim. Sci. Mater.* **2002**, *27*, 77–89. [CrossRef]

20. Saida, J.; Matsushita, M.; Inoue, A. Direct observation of icosahedral cluster in $Zr_{70}Pd_{30}$ binary glassy alloy. *Appl. Phys. Lett.* **2002**, *79*, 412–414. [CrossRef]
21. Xu, D.H.; Duan, G.; Johnson, W.L. Unusual glass-forming ability of bulk amorphous alloys based on ordinary metal copper. *Phys. Rev. Lett.* **2004**. [CrossRef]
22. Takagi, T.; Ohkubo, T.; Hirotsu, Y.; Murty, B.S.; Hono, K.; Shindo, D. Local structure of amorphous $Zr_{70}Pd_{30}$ alloy studied by electron diffraction. *Appl. Phys. Lett.* **2001**, *79*, 485–487. [CrossRef]
23. Liss, K.D.; Bartels, A.; Schreyer, A.; Clemens, H. High energy X-rays: A tool for advanced bulk investigations in materials science and physics. *Textures Microstruct.* **2003**, *35*, 219–252. [CrossRef]
24. Hammersley, A.P.; Svensson, S.O.; Hanfland, M.; Fitch, A.N.; Häusermann, D. Two-dimensional detector software: From real detector to idealised image or two-theta scan. *High Press. Res.* **1996**, *14*, 235–248. [CrossRef]
25. Faber, T.E.; Ziman, J.M. A theory of the electrical properties of liquid metals. *Philos. Mag.* **1965**, *11*, 153–173. [CrossRef]
26. Klementev, K.V. Extraction of the fine structure from X-ray absorption spectra. *J. Phys. D* **2001**, *34*, 209–217. [CrossRef]
27. Yang, L.; Guo, G.Q.; Chen, L.Y.; Huang, C.L.; Ge, T.; Chen, D.; Liaw, P.K.; Saksl, K.; Ren, Y.; Zeng, Q.S.; *et al.* Atomic-Scale Mechanisms of the Glass-Forming Ability in Metallic Glasses. *Phys. Rev. Lett.* **2012**. [CrossRef]
28. McGreevy, R.L.; Pusztai, L. Reverse Monte Carlo Simulation: A new technique for the determination of disordered structures. *Mol. Simul.* **1988**, *1*, 359–367. [CrossRef]
29. Yang, L.; Guo, G.Q.; Jiang, J.Z.; Chen, L.Y.; Wei, S.H. "Soft" atoms in $Zr_{70}Pd_{30}$ metal-metal amorphous alloy. *Scr. Mater.* **2010**, *63*, 883–886. [CrossRef]
30. Takahara, Y.; Narita, N. Local electronic structures and chemical bonds in Zr-based metallic glasses. *Mater. Trans.* **2004**, *45*, 1172–1176. [CrossRef]
31. Huang, L.; Wang, C.Z.; Hao, S.G.; Kramer, M.J.; Ho, K.M. Atomic size and chemical effects on the local order of Zr_2M (M = Co, Ni, Cu, and Ag) binary liquids. *Phys. Rev. B* **2010**. [CrossRef]
32. Wang, S.Y.; Kramer, M.J.; Xu, M.; Wu, S.; Hao, S.G.; Sordelet, D.J.; Ho, K.M.; Wang, C.Z. Experimental and *ab initio* molecular dynamics simulation studies of liquid $Al_{60}Cu_{40}$ alloy. *Phys. Rev. B* **2009**. [CrossRef]
33. Miracle, D.B.; Sanders, W.S.; Senkov, O.N. The influence of efficient atomic packing on the constitution of metallic glasses. *Philos. Mag.* **2003**, *83*, 2409–2428. [CrossRef]

metals

MDPI

Article

Detecting Structural Features in Metallic Glass via Synchrotron Radiation Experiments Combined with Simulations

Gu-Qing Guo, Shi-Yang Wu, Sheng Luo and Liang Yang *

College of Materials Science and Technology, Nanjing University of Aeronautics and Astronautics,
Nanjing 210016, China; guoguqing@nuaa.edu.cn (G.-Q.G.); shiyangwu0914@gmail.com (S.-Y.W.);
nuaaluosheng@163.com (S.L.)
* Author to whom correspondence should be addressed; yangliang@nuaa.edu.cn; Tel.: +86-25-52112903;
  Fax: +86-25-52112626.

Academic Editor: Klaus-Dieter Liss
Received: 23 September 2015; Accepted: 3 November 2015; Published: 9 November 2015

Abstract: Revealing the essential structural features of metallic glasses (MGs) will enhance the understanding of glass-forming mechanisms. In this work, a feasible scheme is provided where we performed the state-of-the-art synchrotron-radiation based experiments combined with simulations to investigate the microstructures of ZrCu amorphous compositions. It is revealed that in order to stabilize the amorphous state and optimize the topological and chemical distribution, besides the icosahedral or icosahedral-like clusters, other types of clusters also participate in the formation of the microstructure in MGs. This cluster-level co-existing feature may be popular in this class of glassy materials.

Keywords: metallic glasses; extended X-ray absorption fine structure; X-ray diffraction; reverse Monte Carlo simulation; microstructure

1. Introduction

Metallic glasses (MGs) have drawn intense interest due to their unique properties since the discovery of the first glassy alloy with the composition of $Au_{75}Si_{25}$ in 1960 [1]. Thus far, vast efforts have been devoted to developing alloys with high glass-forming abilities (GFAs), which may be applied as potential engineering materials. Understanding the glass-formation mechanisms to guide the preparation of alloy materials with high GFAs is desired, thus a number of rules, principles, and criteria have been presented to address this long-standing issue [2–6].

It has been realized that the formation of glassy alloys is strongly influenced by their microstructures, and various clusters should be the basic units forming the atomic structures of MGs. Thus far, many structural models have been proposed by studying clusters theoretically [7–10], enhancing the understanding of the glass-forming mechanisms in binary MGs by building and stacking clusters in space to form their short-range and medium-range orderings. In addition, besides some conventional experimental techniques, synchrotron radiation methods such as X-ray diffraction (XRD) and extended X-ray absorption fine structure (EXAFS) and neutron diffraction also have been applied for studying the microstructures of MGs [11–13]. Therefore, investigating the microstructures of MGs and glass-forming mechanisms by combining some advanced experimental and theoretical methods is expected.

In this work, a feasible scheme is provided where we performed a series of state-of-the-art synchrotron radiation-based experiments (XRD and EXAFS) combined with calculations (simulations) to investigate the microstructures of amorphous alloys. $Zr_{70}Cu_{30}$ and $Zr_{54}Cu_{46}$ binary compositions are selected as the study objects for the following reasons: (1) ZrCu is a typical simple (binary) alloy system for investigating glass formation, which has attracted intensive interest recently [14–16]; (2) the ZrCu binary system has a broad composition range-enable formation of amorphous alloys (30–80 at. % for the Zr component) [17]. Solute-centered clusters are regarded as the building blocks and are helpful for forming the amorphous structure in alloys [9,10]. However, when the concentration of Zr is comparable to that of Cu, a glassy structure also can be formed, while in this case, it is hard to say whether Zr or Cu atoms are the solutes. Considering the concentrations of Zr and Cu in $Zr_{70}Cu_{30}$ and $Zr_{54}Cu_{46}$, Cu could be regarded as the solute in the former while neither Zr nor Cu could be the solutes in the latter. Revealing the underlying structural forming mechanisms in MGs by studying these two ZrCu amorphous compositions is appropriate.

2. Experimental Section

The alloy ingots were prepared by arc melting the mixture of Zr (99.9 wt. %) and Cu (99.9 wt. %) elements in Ti-gettered high-purity argon atmosphere. The ingots were melted at least five times in order to ensure their compositional homogeneity. The corresponding amorphous ribbons were fabricated by melt-spinning, producing a cross-section of $0.04 \times 2 \ mm^2$.

To obtain high-resolution radial distribution functions for amorphous alloys, getting XRD data with a relatively large Q (wave vector transfer) value is required, which is expressed as:

$$Q = \frac{4\pi \sin \theta}{\lambda} \tag{1}$$

Therefore, the synchrotron radiation-based high-energy (about 100 keV) X-ray diffraction measurements were performed for two samples at the beam line, BW5, of Hasylab in Germany [18–20]. The samples measured at room temperature in transmission mode were illuminated for about 200 s by a well-collimated incident beam with a $0.8 \ mm^2$ cross-section. The sample-to-image plate distance was set to be about 500 mm, so that raw diffraction patterns with Q values up to about 20 Å^{-1} were measured. The setup layout for this measurement is plotted in Figure 1. Because the proper penetrating depth here is about 1 mm while the ribbon depth is only about 40 microns, the ribbons were cut into very small pieces and filled in the capillaries. The two-dimensional diffraction data for both empty capillaries and capillaries filled with MG pieces were recorded using a Mar345 image plate [19,20]. After subtracting the background of empty capillaries, the diffraction data of samples could be obtained. The two-dimensional diffraction data were integrated into one-dimensional data by using the program Fit2D [21]. The integrated data were corrected for polarization, sample absorption, fluorescence contribution, inelastic scattering, and so on. Then the total structural factor $S(Q)$ was obtained by using the Faber-Ziman equation, employing the software PDFgetX [22].

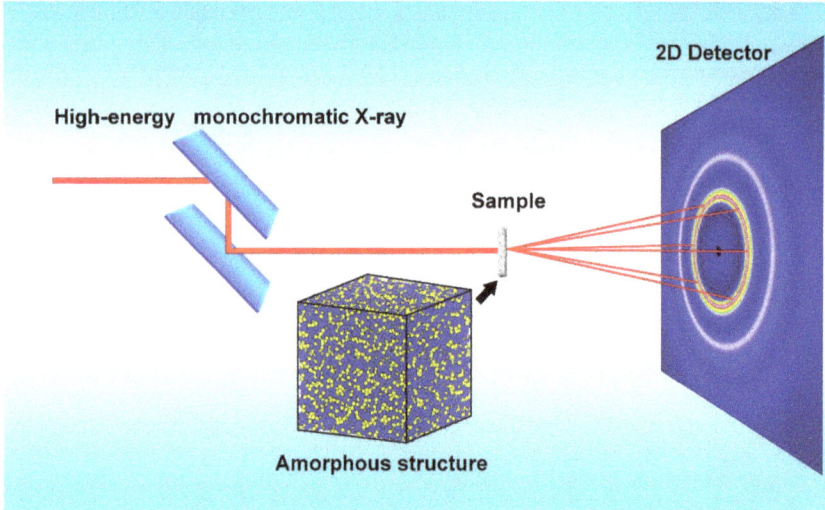

Figure 1. The setup layout of synchrotron radiation-based high-energy (about 100 keV) X-ray diffraction measuring amorphous alloys. By combining XRD and EXAFS data with simulations we can obtain an amorphous structural model, as shown in this figure.

Subsequently, because the Zr and Cu K-edge absorption energies are so different (8.979 keV and 17.998 keV) that their absorption signals could not interfere with each other, and both Zr and Cu have relatively high concentrations here, the transmission mode was adopted for both Zr and Cu K-edges that allowed their EXAFS spectra to be measured at the beam lines BL14W1, in the Shanghai Synchrotron Radiation Facility of China, and U7C, in the National Synchrotron Radiation Laboratory (NSRL) of China. The calculated proper depths of our samples required for Zr and Cu K-edge EXAFS measurements are about 20 and 45 microns, respectively. Therefore, the ribbon samples were polished until their depths were about 20 microns when measuring the Cu K-edge signals. The EXAFS spectra were normalized via a standard data-reduced procedure [23], employing the software Visual Processing in EXAFS Researches (VIPER) [24].

In order to obtain the atomic structural information as reliably as possible, both the normalized diffraction and EXAFS data were simulated simultaneously under the framework of reverse Monte-Carlo (RMC) [25]. The RMC simulation technique is an iterative method for building a structural model in disordered systems with detailed structural information that agrees quantitatively with experimental data (such as the synchrotron radiation-based XRD, EXAFS, and neutron-diffraction data) [26]. In this work, synchrotron radiation-based XRD and EXAFS data were simulated via the RMC method, using the software RMCA [27]. The initial cubic boxes built contain 40,000 randomly distributed Zr and Cu, according to the Zr_xCu_{100-x} (x = 70 and 54) compositions. During RMC simulation, atoms move randomly within a determined time interval. The experimental data are compared to the simulation with the iterative calculation [28],

$$\delta^2 = \frac{1}{\varepsilon^2}\sum_n (S_m(Q_n) - S_{\exp}(Q_n))^2 + \frac{1}{\varepsilon_{Cu}^2}\sum_n (\chi_{m,El}(k_n) - \chi_{\exp,El}(k_n))^2 \qquad (2)$$

where δ^2 represents the deviation between the experimental and simulation data, ε parameters regulate the weight of the data set given in the fitting procedure, E_i denotes Cu or Zr elements, and $S(Q)$ and $\chi(k)$ parameters are the XRD structural factor and the EXAFS signal, respectively. The subscripts "m" and "exp" represent the simulations and the experiments, respectively. Once simulation and experimental data converge, the simulation is stopped, and all the atoms are "frozen" in the cubic box. The result is an atomic structural model available for further analyses. Because all the XRD and EXAFS should be fitted well with all the corresponding theoretical counterparts calculated from the same structural model, such constraint confirms the reliability RMC of the simulation.

Additionally, the simulated structural models were further analyzed by the Voronoi tessellation method [23,25]. According to the Voronoi original algorithm, each convex Voronoi polyhedron (VP) can be built by connecting the perpendicular bisectors between a center atom and all of its neighboring atoms. Each VP may be indexed as $<n3,n4,n5,n6, \ldots >$, where ni denotes the number of i-edged faces on the surface of this polyhedron. Each VP should be embedded in a corresponding convex Voronoi cluster (VC), which is made up of one center atom and its neighboring shell atoms [25,29]. Thus, Σni also stands for the number of the shell atoms in one VC, *i.e.*, the coordination number (CN) of the center atom. The Voronoi algorithm also requires that all the VCs should be closed structural units, which can be accomplished by piling up a set of Delaunay tetrahedrons with the shared vertex at an atom (the center atom of the VC) [30]. This is done so their surfaces are only made up of triangular faces, *i.e.*, they could be regarded as deltahedra [10]. Euler's formula is defined by

$$V - E + F = 2 \tag{3}$$

where V, E, and F stand for the number of vertexes, edges, and faces of VCs, respectively. Because each vertex of VCs should be occupied by one atom, V also denotes the CN of VCs. V, F, and E should satisfy the following equations

$$3E = 2F = \sum i \times ni \tag{4}$$

$$V = 0.5F + 2 \tag{5}$$

obviously, $\sum i \times ni$ must be divisible by 6.

3. Results and Discussion

Figure 2a,c show the two-dimensional X-ray diffraction patterns, the structural factor, $S(Q)$, and the total pair distribution function, $G(r)$. $S(Q)$ and $G(r)$ curves can be deduced from the two-dimensional diffraction pattern. The amorphous nature of these two ZrCu samples can be confirmed because there are no circle lines or dots in the two-dimensional diffraction patterns and no sharp Bragg peaks behind the first strong peak in the $S(Q)$ curves, and these features are usually found in the diffraction data of polycrystals or single crystals [31]. For both $S(Q)$ and $G(r)$ curves, there are differences in their first-, second- and third-shell distributions (SDs) in terms of the peak intensity, position, and width. In particular, obvious differences between their second SDs in $S(Q)$ and their first SDs in $G(r)$ indicate that clear structural changes between these two samples do exist because the split of the second SD in $S(Q)$ reflects the chemical short-range information in MGs [32] and the first SD in $G(r)$ relates to local structural information.

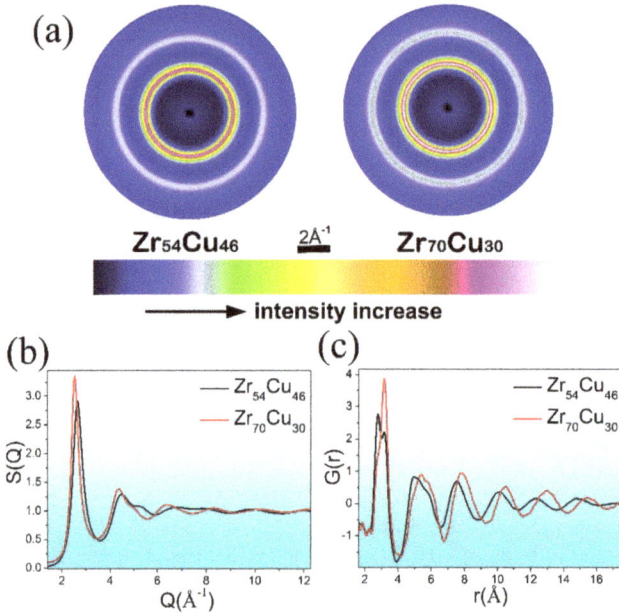

Figure 2. (a) The two-dimensional X-ray diffraction patterns of both ZrCu amorphous samples; and the deduced data: (b) the structural factor ($S(Q)$); and (c) the total pair distribution function ($G(r)$). To highlight the $S(Q)$ difference between $Zr_{70}Cu_{30}$ and $Zr_{54}Cu_{46}$, the Q region here was shortened to about 12 Å^{-1}.

By reverse Fourier transforming the EXAFS signal into real space, the radial distribution function (RDF) could be obtained whose peak area and peak position relate to CN and atomic-pair distance information, respectively. Zr and Cu K-edge RDFs are shown in Figure 3a,b, respectively. As expected, the peak shapes are obviously different between these two selected samples. This also indicates that a difference in the microstructure between them exists. In particular, it is interesting that a peak split appears in both the Zr and Cu K-edge RDFs of $Zr_{70}Cu_{30}$ while no peak split is shown in those of $Zr_{54}Cu_{46}$. In crystal alloys, their local structures are usually formed by distributing the nearest atoms at several relatively localized positions around center atoms, resulting in a split of their first-shell main peak in their EXAFS RDFs. However, in amorphous alloys, the nearest atoms are relatively randomly distributed around the center atoms, resulting in a continuous distribution (Gaussian distribution) in the RDF. In other words, a sole first-shell main peak without an obvious split in the RDF usually appears, denoting the local structural information [33]. This indicates that the neighbor atoms around the center atoms have relatively localized positions in $Zr_{70}Cu_{30}$ [34].

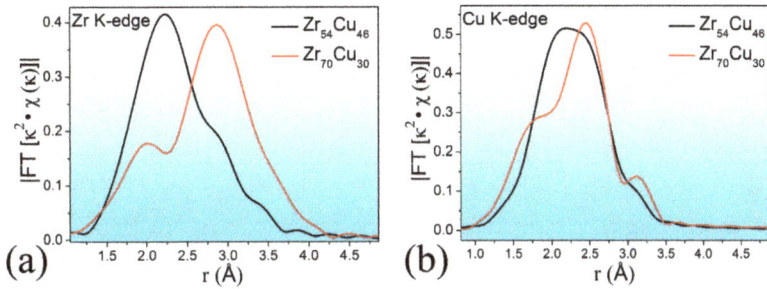

Figure 3. The radial distribution functions (RDFs) obtained by reverse Fourier transforming the EXAFS signal into real space, including: (**a**) Zr K-edge and (**b**) Cu K-edge.

Figure 4a–c show the XRD and EXAFS experimental data, as well as their corresponding RMC simulated curves. To ensure the proper interpretation of all the structural information during EXAFS normalization, the EXAFS data for Zr and Cu K-edge signals were both weighted by κ^3 values. This does not reduce the reliability of RMC simulation because the simulated Zr and Cu K-edge EXAFS spectra also were strictly weighted by κ^3 values so that no systematic errors could be generated from these κ-weight normalizations [35].

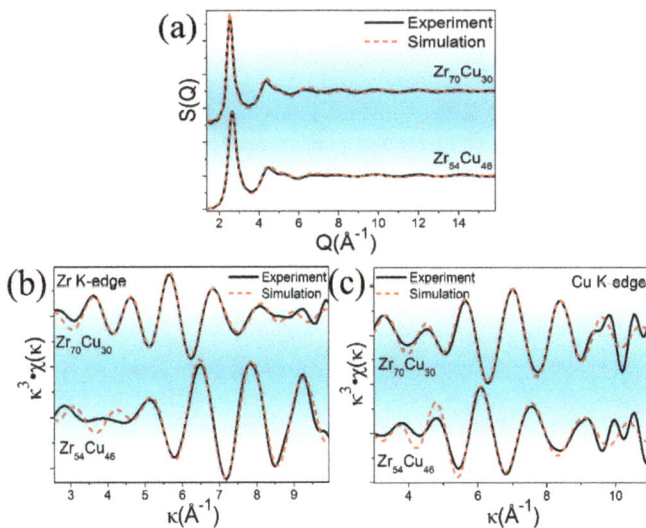

Figure 4. XRD and EXAFS experimental data as well as their corresponding simulated curves, including (**a**) $S(Q)$; (**b**) Zr K-edge; and (**c**) Cu K-edge EXAFS data. The experimental and simulated data are plotted with solid and dashed lines, respectively. Both experimental and simulated Zr and Cu K-edge signals were weighted by κ^3. To highlight the $S(Q)$ difference between the experimental and the simulated data for both samples, the Q region here was shortened to about 16 Å^{-1}.

The good matching between all the experiment/simulation pairs confirms the success of the RMC simulations. Based on the simulated structural models, atomic-level structural information can be deduced. The CN values around Zr and Cu center atoms, as well as all kinds of atomic-pair distances, are listed in Table 1. We could find no obvious difference in the atomic-pair distances between $Zr_{70}Cu_{30}$ and $Zr_{54}Cu_{46}$, and all the CN values are reasonable, considering the concentrations of Zr and Cu in these two compositions. It seems that such atomic-level structural parameters can barely provide any unique information for amorphous alloys.

Table 1. Atomic structural information obtained from the RMC simulation, including: coordination numbers (CNs) around Zr or Cu centers and atomic-pair distances (R).

Centers	Atomic Pairs	CN ± 0.1		R (Å) ± 0.02	
		$Zr_{54}Cu_{46}$	$Zr_{70}Cu_{30}$	$Zr_{54}Cu_{46}$	$Zr_{70}Cu_{30}$
Zr atom	Zr–Zr	7.5	9.7	3.15	3.19
	Zr–Cu	5.4	3.5	2.96	2.96
Cu atom	Cu–Cu	4.1	2.5	2.66	2.65
	Cu–Zr	6.4	8.6	2.96	2.96

Via the Voronoi tessellation, all the VCs whose distributions are plotted in Figure 5 can be extracted. Because it has been revealed that various clusters may co-exist in the microstructure of MGs [10,16], it is no surprise that we can deduce hundreds of types of VCs. Here, only those whose fractions are larger than 1.5% are selected and shown in Figure 5. For the Zr- centered VCs, it is found that there are some popular VCs (fractions are larger than 5%) in both ZrCu samples, such as <0,2,8,2>, <0,3,6,3>, and <0,1,10,2>. These VCs have been validated to be typical icosahedral or icosahedral-like VCs in previous work [25]. For instance, like the ideal icosahedron (<0,0,12,0>), <0,2,8,2> and <0,3,6,3> are distorted icosahedra, having the same CN value of 12. For another example, if we add one atom on the shell of the <0,0,12,0> VC, it changes into <0,1,10,2>. The high weights of these icosahedral or icosahedral-like VCs in both $Zr_{70}Cu_{30}$ and $Zr_{54}Cu_{46}$ MG compositions indicate that icosahedral or icosahedral-like VCs are the preferred building blocks for forming the microstructure of glassy alloys. This is consistent with the viewpoint presented in previous work that icosahedral or icosahedral-like VCs ease the formation of the amorphous structure in alloys [36,37].

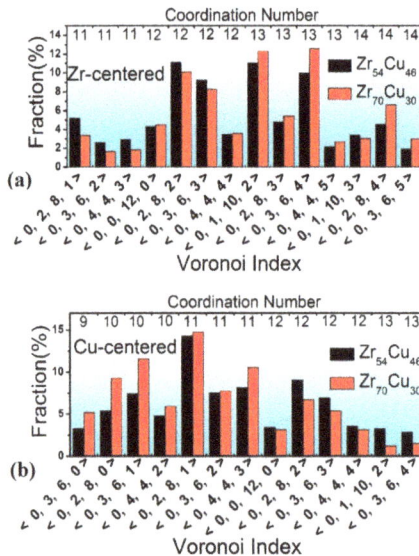

Figure 5. Distribution of the major Voronoi clusters, including: (**a**) Zr-centered VCs and (**b**) Cu-centered VCs. Only those whose fractions are larger than 1.5% are selected. The CN value denotes the number of shell atoms of the corresponding VC, *i.e.*, the CN around the center atom.

However, other types of clusters also exist here. For instance, we notice that one non-icosahedral VC indexed as <0,3,6,4> has a relatively large fraction in both samples. According to the efficient cluster-packing model [38], the optimal CN relates to the size ratio between the center atom and the shell atoms in clusters. According to the concentrations of Zr and Cu components in both samples, we can estimate that the optimal CNs around the Zr centers in $Zr_{54}Cu_{46}$ and $Zr_{70}Cu_{30}$ are 12.8 and 13.6, respectively. These CNs are a little bit larger than 12 (CN of the ideal icosahedron). Therefore, besides some distorted icosahdral VCs with CNs of 12, such as <0,2,8,2> and <0,3,6,3>, other VCs with CNs of 13 (such as <0,1,10,2> and <0,3,6,4>) and even 14, which also are popular clusters, the average CN around Zr centers equals the estimated optimal CN value. In addition, it is worth noting that there is a tendency for $Zr_{70}Cu_{30}$ to have relatively high (low) fractions of VCs with large (small) CNs compared with the $Zr_{54}Cu_{46}$ composition. This is because the optimal CN of $Zr_{70}Cu_{30}$ (13.6) is larger than that of $Zr_{54}Cu_{46}$ (12.8) due to the increase of the size ratio between the center atom and the shell atoms when replacing Cu atoms with Zr atoms (Zr atoms are larger than Cu ones).

Concerning the Cu centers, there are some popular VCs with relatively high weights in both ZrCu compositions, such as <0,2,8,1>, <0,2,8,2>, <0,3,6,3>, <0,4,4,3>, <0,3,6,2>, <0,3,6,1>, and so on. Although <0,2,8,1>, <0,2,8,2>, and <0,3,6,3> may be regarded as icosahedral-like VCs (for instance, <0,2,8,1> could be formed by removing one shell atom from <0,0,12,0>), others with CNs of 11 or 10 have no icosahedral-like features at all. This is because the optimal CNs around Cu centers in $Zr_{54}Cu_{46}$ and $Zr_{70}Cu_{30}$ are calculated to be 11.3 and 10.9, respectively. These CNs are a little bit smaller than 12. In addition, there also is a tendency for $Zr_{54}Cu_{46}$ and $Zr_{70}Cu_{30}$ to have relatively high (low) fractions of Cu-centered VCs with large (small) CNs. This also is because of the relatively high Zr concentration in $Zr_{70}Cu_{30}$.

From the results and discussion mentioned above, we can conclude that icosahedral or icosahedral-like VCs are the favorite structural units in the microstructure of MGs, because stacking such clusters with abundant five-fold rotation symmetrical features [29] can result in the exclusion of structural periodicity, which is required in crystals. Nevertheless, the microstructures of MGs could not be formed only by stacking icosahedral or icosahedral-like VCs because of the following reasons: (1) it has been revealed that fractal features are popular in amorphous alloys and that icosahedral clusters fail to be packed to fill space [39]; (2) the optimal CN relates to the size ratio between the center atom and the shell atoms of VCs. This value usually does not equal 12 (the CN of a standard icosahedron). Therefore, besides icosahedral or icosahedral-like VCs, some other clusters with no icosahedral-like features (usually with CN not equal to 12) also should exist to fill space. That should be the structural nature of amorphous alloys.

In addition, although icosahedral or icosahedral-like clusters co-existing with other types of clusters is revealed to be the structural nature in both $Zr_{70}Cu_{30}$ and $Zr_{54}Cu_{46}$, we have found that there are some obvious differences in the distributions of Zr- and Cu-centered VCs, as shown in Figure 5a,b. A tendency could be observed which shows that $Zr_{70}Cu_{30}$ has the higher (lower) fractions of Zr- (Cu-) centered VCs with large CN values compared to $Zr_{54}Cu_{46}$, which is due to the higher concentration of Zr having a relatively large Goldschmidt atomic radius of 1.60 Å. This strongly relates to the atomic-scale differences between $Zr_{70}Cu_{30}$ and $Zr_{54}Cu_{46}$ shown in Figure 2, Figure 3. For different systems or compositions enabling the formation of glassy alloys, there must be some atomic- and cluster-level structural differences due to the different concentrations of all the containing elements with different atomic sizes. Nevertheless, icosahedral or icosahedral-like clusters are the preferred structural building blocks. To fill in space efficiently, they should be densely packed with the help of other non-icosahedral clusters, leading to the formation of the microstructure in various glass formers.

Because ZrCu is a typical binary alloy system, it enables the formation of MG, revealing the structural mechanisms for its glass formation that have drawn intense interest that a number of theoretical or experimental works have been published [14–16,40–43]. Therefore, it is necessary to compare the present work with previous work in terms of studying methods and structural information. For instance, in a previous report, the authors did RMC simulation upon neutron diffraction data and compared it with Molecular Dynamics (MD) simulations. They concluded that in the ZrCu system, the basic structural units correspond to "Superclusters", which is different from our conclusion that "icosahedral-like VCs and other types of clusters co-exist in glassy structure". This difference is probably because the Voronoi tessellation applied in our work can give more detailed cluster-level structural information. For another instance, in another article, the authors did MD simulations and concluded that a string-like backbone network formed by icosahedral clusters is the amorphous structural basis, which is similar to our conclusion. This indicates that theoretical methods such as MD also can provide detailed structural information on amorphous alloys.

In addition, in order to compare the microstructures between different alloy systems, a ZrCu binary and a ZrCuAl ternary MG compositions also are studied, as described in the Appendix section.

4. Conclusions

In summary, a feasible scheme for investigating the microstructure of amorphous alloys is provided by combining synchrotron radiation–based experiments with simulations. It is revealed that although there are some distribution differences of clusters (local structures) between $Zr_{70}Cu_{30}$ and $Zr_{54}Cu_{46}$ MG compositions, icosahedral or icosahedral-like VCs are preferred structural units in both samples. It is further revealed that in order to increase the cluster packing efficiency (space filling) and obtain the optimal CN of clusters corresponding to the chemical distribution of the center and shell atoms, icosahedral-like VCs co-existing with other types of clusters should be the structural nature in amorphous alloys. This work will enhance the understanding of glass-forming mechanisms at the atomic- and cluster-level structural aspect.

Acknowledgments: The authors would like to thank the HASYLAB in Germany, the Shanghai Synchrotron Radiation Facility in China, and the National Synchrotron Radiation Laboratory of China for the use of the advanced synchrotron radiation facilities. Financial support from the National Natural Science Foundation of China (Grant No. U1332112 and 51471088), the Fundamental Research Funds for the Central Universities (Grant No. NE2015004), the Funding for Outstanding Doctoral Dissertation in NUAA (Grant No. BCXJ12-08), the Funding of Jiangsu Innovation Program for Graduate Education (Grant No. CXLX13-152), and the project funded by the Priority Academic Program Development (PAPD) of Jiangsu Higher Education Institutions are gratefully acknowledged.

Author Contributions: Gu-qing Guo performed simulation work with the experimental data. Lang Yang performed the analysis of this work and wrote this article. Shi-yang Wu and Sheng Luo contributed to the experimental research work.

Conflicts of Interest: The authors declare no conflict of interest.

Appendix

Appendix A. Introduction

In order to compare the microstructures between different alloy systems, the $Zr_{48}Cu_{45}Al_7$ ternary and $Zr_{50}Cu_{50}$ binary metallic glass (MG) compositions are studied. They have large differences in physical properties such as glass-forming ability (GFA). In detail, since the critical casting size is an important indicator of GFA, we can say that $Zr_{48}Cu_{45}Al_7$ has a higher GFA than $Zr_{50}Cu_{50}$, considering the fact that $Zr_{50}Cu_{50}$ has a critical casting size not more than 2 mm in diameter [44], while such value is enhanced up to 6 mm in $Zr_{48}Cu_{45}Al_7$ [45].

The experimental and the simulation methods are the same with those described in the manuscript. Here we show their atomic-level and cluster-level microstructure below, and will discuss the relationship between the microstructure and GFA. More information refers to our previous work [46].

Appendix B. Atomic-Level Structural Information

From the reverse Monte-Carlo (RMC) simulated atomic model, the coordination number (CN) and the nearest interatomic distances can be deduced, as listed in Table A1. The cut-off distances are set to be 3.90, 3.80, 3.80, 3.65, 3.65, and 3.5 Å for Zr–Zr, Zr–Cu, Zr–Al, Cu–Cu, Cu–Al, and

Al–Al atomic pairs, respectively, which are consistent with the maximum ranges of the first shell distribution in partial $G(r)$s. For comparison, interatomic distances are also obtained by summing their Goldschmidt atomic radii. Little difference in the length of Zr–Zr, Zr–Cu, and Cu–Cu pairs can be found. However, a large deviation from the interatomic distances calculated based on the Goldschmidt atomic radii is observed, with values of 0.27, 0.14, and 0.30 Å for Zr–Al, Cu–Al, and Al–Al couples, respectively, *i.e.*, Al–M (M = Zr, Cu, and Al) pairs are greatly shortened in this composition. A similar bond-shortening phenomenon was ever detected in Al-based amorphous alloys [28]. Additionally, in previous computational work on ZrCuAl, strong bonding of Al–Cu was also suggested [47]. The origin of the shortened Al–M pair distance should be the strong bonding effect between Al and it neighbor atoms.

Table A1. Atomic structure information, including average CNs of M (M = Zr, Cu and Al) atoms obtained by RMC simulation; D_{RMC} and D_{SGAR} are the interatomic distances (atomic bonds) calculated from the RMC model and the sum of the Goldschmidt atomic radii (SGAR), respectively.

Atomic Pair	CN	D_{RMC} (Å)	D_{SGAR} (Å)
Zr–Zr	6.13	3.18	3.20
Zr–Cu	5.28	2.85	2.88
Zr–Al	0.82	2.76	3.03
Cu–Zr	5.62	-	-
Cu–Cu	4.74	2.58	2.56
Cu–Al	0.6	2.57	2.71
Al–Zr	5.94	-	-
Al–Cu	4.05	-	-
Al–Al	0.36	2.56	2.86

Appendix C. Cluster-Level Structural Information

The distribution of major Voronoi clusters (VCs) centered with Zr, Cu, and Al atoms in $Zr_{48}Cu_{45}Al_7$ and $Zr_{50}Cu_{50}$ is plotted in Figure A1a–c, respectively. Except for little difference for each VC fraction, the similar VC distribution tendency indicates that the surrounding of Zr and Cu atoms is highly alike in both alloys. As shown in Figure A1a, dominant VCs indexed as <0,1,10,2>, <0,3,6,4>, <0,2,8,2>, <0,3,6,3>, and <0,2,8,1> account for a total fraction of about 50%, and most of them may be regarded as icosahedral-like (distorted or irregular icosahedra) clusters [25]. We notice that the ideal icosahedron indexed as <0,0,12,0> only possesses a weight of 6%. Its low fraction may result from the large size gap between Zr and Cu atoms, which may retard the formation of such a regular cluster whose shell atoms are symmetrically packed around the center atom. Concerning Cu-centered VCs, it is remarkable that the fraction of <0,2,8,1> is twice larger than any other fraction, which is quite consistent with a recent report [48]. However, the weight of <0,0,12,0> is rather low, which matches well with the previous work [49]. In short, icosahedral-like Zr- and Cu-centered clusters are the main building blocks in both ZrCu and ZrCuAl samples. Additionally, it is interesting that four Al-centered major clusters indexed as <0,3,6,0>, <0,3,6,1>, <0,2,8,0>, and <0,2,8,1> are deduced with a total weight over 60%, as plotted in Figure A1c. These four VCs are relatively smaller clusters with a CN range of 9–11. We notice that the shortened Al atomic radius is 1.28 Å, one half of the Al-Al bond length (in Table A1). The atomic size ratio R^* between the Al solute and its neighbors is thus calculated at 0.87 and may lead to an optimal cluster with a CN of 10–11 [38], which is consistent with the CN range of these four Al-centered major VCs.

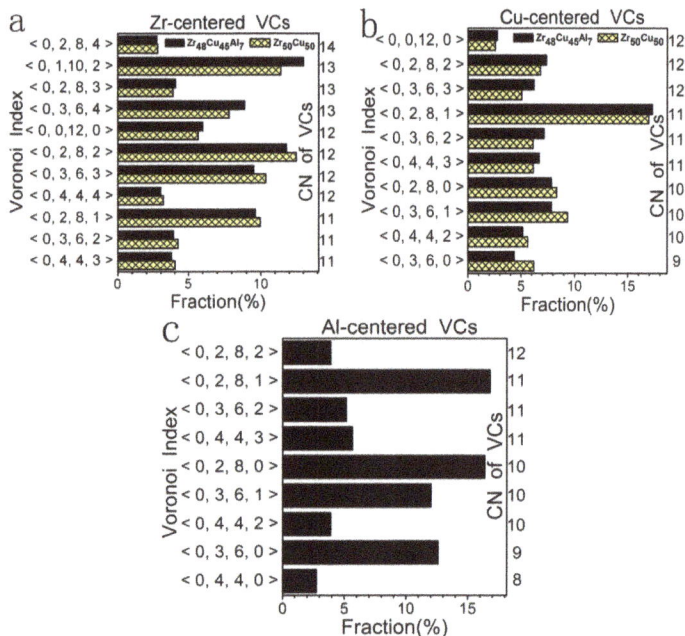

Figure A1. Distribution of major VCs, centered with (**a**) Zr; (**b**) Cu; and (**c**) Al atoms. Note only VCs possessing a weight over 2.5% are selected.

Appendix D. Relationship between Microstructure and GFA

To reveal the relationship between the local structure and GFA in ZrCuAl, further analysis upon the deduced structural features caused by the Al addition is required. Since structure heredity occurs between molten liquid and solid MG [50], we may discuss the local structure in a molten liquid state based on the above results. Inside Al-centered clusters, Al solute atoms are prone to bonding with Zr and Cu solvent atoms. Depending on the strong interatomic bonding, Cu and Zr neighbors may be tightly connected by Al centers, resulting in shortened pair distances. Therefore, the mobility of atoms is sharply decreased, which may lead to the increase of viscosity in the molten state and ease of glass formation [51]. During the quench, rearrangements of Zr, Cu, and Al atoms by changing or breaking Al-connected bonds are largely retarded. This tendency may help preserve the local structure of the liquid to solid state, which may contribute to the increase of GFA in the ZrCuAl alloy. Considering the cluster scale for $Zr_{50}Cu_{50}$ composition, icosahedral-like Zr- and Cu-centered clusters with CNs of 11–13 should be the major building blocks. The cluster packing efficiency in the $Zr_{50}Cu_{50}$ alloy may be not very high because it was suggested that voids could be created due to the incomplete filling in space only by icosahedral-like clusters [52]. However, in the corresponding Al-doped ZrCu amorphous alloy, some of the Al atoms may be regarded as glue atoms [13,53], which may occupy the interstices around Zr- and Cu-centered icosahedral-like clusters to connect and fix them. These Al glue atoms are usually surrounded with fewer neighbors, resulting in smaller Al-centered VCs, such as the deduced <0,2,8,0>, <0,3,6,0>, and <0,3,6,1> with CNs of 9–10. The space may be filled with VCs with various shapes and volumes and the cluster-packing efficiency is accordingly increased. Such cluster-dense packing cause by Al atoms also may reduce the mobility of most atoms and clusters in molten liquid. Therefore, the crystallization is avoided, and the amorphous alloy with enhanced glass-forming ability was obtained.

References

1. Klement, W.; Willens, R.H.; Duwez, P. Non-crystalline structure in solidified gold-silicon alloys. *Nature* **1960**, *187*, 869–870. [CrossRef]

2. Turnbull, D. Under what conditions can a glass be formed? *Contemp. Phys.* **1969**, *10*, 473–488. [CrossRef]

3. Greer, A.L. Confusion by Design. *Nature* **1993**, *366*, 303–304. [CrossRef]

4. Inoue, A. Stabilization of metallic supercooled liquid and bulk amorphous alloys. *Acta Mater.* **2000**, *48*, 279–306. [CrossRef]

5. Lu, Z.P.; Tan, H.; Li, Y.; Ng, S.C. The correlation between reduced glass transition temperature and glass forming ability of bulk metallic glasses. *Scr. Mater.* **2000**, *42*, 667–673. [CrossRef]

6. Lu, Z.P.; Liu, C.T. Glass formation criterion for various glass-forming systems. *Phys. Rev. Lett.* **2003**. [CrossRef]

7. Bernal, J.D. A geometrical approach to the structure of liquids. *Nature* **1959**, *183*, 141–147.

8. Gaskell, P.H. A new structural model for transition metal-metalloid glasses. *Nature* **1978**, *276*, 484–485. [CrossRef]

9. Miracle, D.B. A structural model for metallic glasses. *Nat. Mater.* **2004**, *3*, 697–702. [CrossRef] [PubMed]

10. Sheng, H.W.; Luo, W.K.; Alamgir, F.M.; Bai, J.M.; Ma, E. Atomic packing and short-to-medium-range order in metallic glasses. *Nature* **2006**, *439*, 419–425. [CrossRef] [PubMed]

11. Lefebvre, S.; Quivy, A.; Bigot, J.; Calvayrac, Y.; Bellissent, R. A neutron diffraction determination of short-range order in a $Ni_{63.7}Zr_{36.3}$ glass. *J. Phys. F* **1985**, *15*, L99–L103. [CrossRef]

12. Luo, W.K.; Sheng, H.W.; Alamgir, F.M.; Bai, J.M.; He, J.H.; Ma, E. Icosahedral Short-Range Order in Amorphous Alloys. *Phys. Rev. Lett.* **2004**. [CrossRef]

13. Yang, L.; Xia, J.H.; Wang, Q.; Dong, C.; Chen, L.Y.; Ou, X.; Liu, J.F.; Jiang, J.Z.; Klementiev, K.; Saksl, K.; *et al.* Design of Cu_8Zr_5-based bulk metallic glasses. *Appl. Phys. Lett.* **2006**. [CrossRef]

14. Wang, X.D.; Yin, S.; Cao, Q.P.; Jiang, J.Z.; Franz, H.; Jin, Z.H. Atomic structure of binary $Cu_{64.5}Zr_{35.5}$ bulk metallic glass. *Appl. Phys. Lett.* **2008**. [CrossRef]

15. Li, M.; Wang, C.Z.; Hao, S.G.; Kramer, M.J.; Ho, K.M. Structural heterogeneity and medium-range order in Zr_xCu_{100-x} metallic glasses. *Phys. Rev. B* **2009**. [CrossRef]

16. Yang, L.; Guo, G.Q.; Chen, L.Y.; Huang, C.L.; Ge, T.; Chen, D.; Liaw, P.K.; Saksl, K.; Ren, Y.; Zeng, Q.S.; *et al.* Atomic-Scale Mechanisms of the Glass-Forming Ability in Metallic Glasses. *Phys. Rev. Lett.* **2012**. [CrossRef]

17. Saida, J.; Kasai, M.; Matsubara, E.; Inoue, A. Stability of glassy state in Zr-based glassy alloys correlated with nano icosahedral phase formation. *Ann. Chim. Sci. Mater.* **2002**, *27*, 77–89. [CrossRef]

18. Liss, K.D.; Bartels, A.; Schreyer, A.; Clemens, H. High energy X-rays: A tool for advanced bulk investigations in materials science and physics. *Textures Microstruct.* **2003**, *35*, 219–252. [CrossRef]

19. Qu, D.D.; Liss, K.D.; Yan, K.; Reid, M.; Almer, J.D.; Wang, Y.B.; Liao, X.Z.; Shen, J. On the atomic anisotropy of thermal expansion in bulk metallic glass. *Adv. Eng. Mater.* **2011**, *13*, 861–864. [CrossRef]

20. Liss, K.D.; Qu, D.D.; Yan, K.; Reid, M. Variability of Poisson's Ratio and Enhanced Ductility in Amorphous Metal. *Adv. Eng. Mater.* **2013**, *15*, 347–351. [CrossRef]

21. Hammersley, A.P.; Svensson, S.O.; Hanfland, M.; Fitch, A.N.; Häusermann, D. Two-dimensional detector software: From real detector to idealised image or two-theta scan. *High Press. Res.* **1996**, *14*, 235–248. [CrossRef]

22. Qiu, X.; Thompson, J.W. PDFgetX2: A GUI-driven program to obtain the pair distribution function from X-ray powder diffraction data. *J. Appl. Crystallogr.* **2004**, *37*, 110–116. [CrossRef]

23. Guo, G.Q.; Yang, L. Structural mechanisms of the microalloying-induced high glass forming abilities in metallic glasses. *Intermetallics* **2015**, *65*, 66–74. [CrossRef]

24. Klementev, K.V. Extraction of the fine structure from X-ray absorption spectra. *J. Phys. D* **2001**, *34*, 209–217. [CrossRef]

25. Wang, S.Y.; Kramer, M.J.; Xu, M.; Wu, S.; Hao, S.G.; Sordelet, D.J.; Ho, K.M.; Wang, C.Z. Experimental and *ab initio* molecular dynamics simulation studies of liquid $Al_{60}Cu_{40}$ alloy. *Phys. Rev. B* **2009**, *79*, 144205–144209. [CrossRef]

26. McGreevy, R.L.; Pusztai, L. Reverse Monte Carlo Simulation: A new technique for the determination of disordered structures. *Mol. Simul.* **1988**, *1*, 359–367. [CrossRef]

27. McGreevy, R.L. Reverse Monte Carlo modelling. *J. Phys. Condens. Matt.* **2001**, *13*, R877–R913. [CrossRef]

28. Saksl, K.; Jovari, P.; Franz, H.; Zeng, Q.S.; Liu, J.F.; Jiang, J.Z. Atomic structure of $Al_{89}La_6Ni_5$ metallic glass. *J. Phys. Condens. Matter* **2006**, *18*, 7579–7592. [CrossRef] [PubMed]

29. Yang, L.; Guo, G.Q. Preferred clusters in metallic glasses. *Chin. Phys. B* **2010**, *12*. [CrossRef]

30. Medvedev, N.N. The algorithm for three-dimensional Voronoi polyhedra. *J. Comput. Phys.* **1986**, *67*, 223–229. [CrossRef]

31. Zeng, Q.S.; Sheng, H.W.; Ding, Y.; Wang, L.; Yang, W.G.; Jiang, J.Z.; Mao, W.L.; Mao, H.K. Long-range topological order in metallic glass. *Science* **2011**, *332*, 1404–1406. [CrossRef] [PubMed]

32. Mattern, N.; Kuhn, U.; Hermann, H.; Ehrenberg, H.; Neuefeind, J.; Eckert, J. Short-range order of $Zr_{62-x}Ti_xAl_{10}Cu_{20}Ni_8$ bulk metallic glasses. *Acta Mater.* **2002**, *50*, 305–314. [CrossRef]

33. Yang, L.; Guo, G.Q.; Zhang, G.Q.; Chen, L.Y. Structural origin of the high glass-forming ability in Y-doped bulk metallic glasses. *J. Mater. Res.* **2010**, *25*, 1701–1705. [CrossRef]

34. Guo, G.Q.; Yang, L.; Huang, C.L.; Chen, D.; Chen, L.Y. Structural origin of the different glass-forming abilities in ZrCu and ZrNi metallic glasses. *J. Mater. Res.* **2011**, *26*, 2098–2102.

35. Keen, D.A.; McGreevy, R.L. Structural modelling of glasses using reverse Monte Carlo simulation. *Nature* **1990**, *344*, 423–425. [CrossRef]

36. Saksl, K.; Franz, H.; Jovari, P.; Klementiev, K.; Welter, E.; Ehnes, A.; Saida, J.; Inoue, A.; Jiang, J.Z. Evidence of icosahedral short-range order in $Zr_{70}Cu_{30}$ and $Zr_{70}Cu_{29}Pd_1$ metallic glasses. *Appl. Phys. Lett.* **2003**, *8333*, 3924–3926. [CrossRef]

37. Saida, J.; Matsushita, M.; Inoue, A. Direct observation of icosahedral cluster in $Zr_{70}Pd_{30}$ binary glassy alloy. *Appl. Phys. Lett.* **2002**, *79*, 412–414. [CrossRef]

38. Miracle, D.B.; Sanders, W.S.; Senkov, O.N. The influence of efficient atomic packing on the constitution of metallic glasses. *Philos. Mag.* **2003**, *83*, 2409–2428. [CrossRef]

39. Ma, D.; Stoica, A.D.; Wang, X.L. Power-law scaling and fractal nature of medium-range order in metallic glasses. *Nature Mater.* **2009**, *8*, 30–34. [CrossRef] [PubMed]

40. Antonowicz, J.; Pietnoczka, A.; Pękała, K.; Latuch, J.; Evangelakis, G.A. Local atomic order, electronic structure and electron transport properties of Cu-Zr metallic glasses. *J. Appl. Phys.* **2014**. [CrossRef]

41. Almyras, G.A.; Papageorgiou, D.G.; Lekka, C.E.; Mattern, N.; Eckert, J.; Evangelakis, G.A. Atomic cluster arrangements in Reverse Monte Carlo and Molecular Dynamics structural models of binary Cu-Zr Metallic Glasses. *Intermetallics* **2011**, *19*, 657–661. [CrossRef]

42. Bokas, G.B.; Lagogianni, A.E.; Almyras, G.A.; Lekka, Ch.E.; Papageorgiou, D.G.; Evangelakis, G.A. On the role of Icosahedral-like clusters in the solidification and the mechanical response of Cu-Zr metallic glasses by Molecular Dynamics simulations and Density Functional Theory computations. *Intermetallics* **2013**, *43*, 138–141. [CrossRef]

43. Antonowicz, J.; Pietnoczka, A.; Drobiazg, T.; Almyras, G.A.; Papageorgiou, D.G.; Evangelakis, G.A. Icosahedral order in Cu-Zr amorphous alloys studied by means of X-ray absorption fine structure and molecular dynamics simulations. *Philos. Mag.* **2012**, *92*, 1865–1875. [CrossRef]

44. Zhu, Z.W.; Zhang, H.F.; Sun, W.S.; Ding, B.Z.; Hu, Z.Q. Processing of bulk metallic glasses with high strength and large compressive plasticity in $Cu_{50}Zr_{50}$. *Scr. Mater.* **2006**, *54*, 1145–1149. [CrossRef]

45. Xu, D.H.; Duan, G.; Johnson, W.L. Unusual glass-forming ability of bulk amorphous alloys based on ordinary metal copper. *Phys. Rev. Lett.* **2004**. [CrossRef]

46. Yang, L.; Guo, G.Q.; Chen, L.Y.; Wei, S.H.; Jiang, J.Z.; Wang, X.D. Atomic structure in Al-doped multicomponent bulk metallic glass. *Scr. Mater.* **2010**, *63*, 879–882. [CrossRef]

47. Cheng, Y.Q.; Ma, E.; Sheng, H.W. Atomic level structure in multicomponent bulk metallic glass. *Phys. Rev. Lett.* **2009**. [CrossRef]

48. Fujita, T.; Konno, K.; Zhang, W.; Kumar, V.; Matsuura, M.; Inoue, A.; Sakurai, T.; Chen, M.W. Atomic-scale heterogeneity of a multicomponent bulk metallic glass with excellent glass forming ability. *Phys. Rev. Lett.* **2009**. [CrossRef]

49. Wang, X.D.; Jiang, Q.K.; Cao, Q.P.; Bednarcik, J.; Franz, H.; Jiang, J.Z. Atomic structure and glass forming ability of $Cu_{46}Zr_{46}Al_8$ bulk metallic glass. *J. Appl. Phys.* **2008**. [CrossRef]

50. Dyre, J.C. Glasses: Heirs of liquid treasures. *Nat. Mater.* **2004**, *3*, 749–750. [CrossRef] [PubMed]

51. Mukherjee, S.; Schroers, J.; Zhou, Z.; Johnson, W.L.; Rhim, W.K. Viscosity and specific volume of bulk metallic glass-forming alloys and their correlation with glass forming ability. *Acta Mater.* **2004**, *52*, 3689–3695. [CrossRef]

52. Doye, J.P.K.; Wales, D.J.; Simdyankin, S.I. Global optimization and the energy landscapes of Dzugutov clusters. *Faraday Discuss.* **2001**, *118*, 159–170. [CrossRef] [PubMed]

53. Xia, J.H.; Qiang, J.B.; Wang, Y.M.; Wang, Q.; Dong, C. Ternary bulk metallic glasses formed by minor alloying of Cu_8Zr_5 icosahedron. *Appl. Phys. Lett.* **2006**. [CrossRef]

metals

MDPI

Article

Characterization of Deformation Behavior of Individual Grains in Polycrystalline Cu-Al-Mn Superelastic Alloy Using White X-ray Microbeam Diffraction

Eui Pyo Kwon [1,*], Shigeo Sato [2], Shun Fujieda [3], Kozo Shinoda [3], Ryosuke Kainuma [4], Kentaro Kajiwara [5], Masugu Sato [5] and Shigeru Suzuki [3,*]

[1] Convergence Components & Agricultural Machinery Application Group, Korea Institute of Industrial Technology, Gimje 54325, Korea
[2] Graduate School of Science and Engineering, Ibaraki University, Hitachi 316-8511, Japan; shigeo.sato.ar@vc.ibaraki.ac.jp
[3] Institute of Multidisciplinary Research for Advanced Materials, Tohoku University, Sendai 980-8577, Japan; fujieda@tagen.tohoku.ac.jp (S.F.); shinoda@tagen.tohoku.ac.jp (K.S.)
[4] Department of Materials Science, Graduate School of Engineering, Tohoku University, Sendai 980-8579, Japan; kainuma@material.tohoku.ac.jp
[5] Japan Synchrotron Radiation Research Institute, SPring-8, Hyogo 679-5198, Japan; kajiwara@spring8.or.jp (K.K.); msato@spring8.or.jp (M.S.)
* Authors to whom correspondence should be addressed; ackep@kitech.re.kr (E.P.K.); ssuzuki@tagen.tohoku.ac.jp (S.S.); Tel.: +82-63-920-1286 (E.P.K.); Fax: +82-63-920-1280 (E.P.K.); Tel./Fax: +81-22-217-5177 (S.S.).

Academic Editors: Klaus-Dieter Liss and Hugo F. Lopez
Received: 3 September 2015; Accepted: 29 September 2015; Published: 9 October 2015

Abstract: White X-ray microbeam diffraction was applied to investigate the microscopic deformation behavior of individual grains in a Cu-Al-Mn superelastic alloy. Strain/stresses were measured *in situ* at different positions in several grains having different orientations during a tensile test. The results indicated inhomogeneous stress distribution, both at the granular and intragranular scale. Strain/stress evolution showed reversible phenomena during the superelastic behavior of the tensile sample, probably because of the reversible martensitic transformation. However, strain recovery of the sample was incomplete due to the residual martensite, which results in the formation of local compressive residual stresses at grain boundary regions.

Keywords: white X-ray microbeam diffraction; Cu-Al-Mn alloys; superelasticity; microscopic stresses; martensitic transformation

1. Introduction

Characterization of the microscopic deformation behavior in individual grains and their interactions are important for understanding the overall deformation behavior and mechanical properties of polycrystalline materials. In particular, microstructural characterization would be important for shape memory and superelastic alloys, as their unique functional properties originate from microstructural changes caused by reversible martensitic transformation. Deformation induced martensitic transformation upon loading, and subsequent reverse transformation upon unloading give rise to the phenomenon of superelasticity. For example, Cu-Al-Mn shape memory alloys that undergo a β_1 (bcc) $\leftrightarrow \beta_1'$ (monoclinic) martensitic transformation exhibit superelasticity. Their recoverable strain is significantly improved when microstructural parameters, such as crystallographic orientation, grain size, and constituent phases, are optimized [1,2]. For example, Sutou *et al.* reported that superelastic strain was enhanced with increasing grain size relative to specimen thickness, and >5% superelastic strain was obtained in coarse grain samples, even in the absence of texture [3].

In order to understand the relationship between the microstructure and macroscopic properties of superelastic alloys, experiments have recently focused on measuring deformation behavior at the microscopic level using modern experimental tools such as X-ray diffraction based on synchrotron radiation. Berveiller *et al.* used the synchrotron diffraction technique to investigate microscopic deformation behavior in individual grains of Cu-based superelastic alloys [4]. It was found that deformation of the alloy was accompanied by lattice rotation in grains and splitting of grains into sub-domains, which were attributed to formation of the martensite phase during tensile loading. Upon unloading, the initial lattice orientation was recovered and the sub-domains merged, indicating that the lattice rotation and sub-domain formation are reversible phenomena. In our previous investigation on an Fe-Mn-Si-Cr shape memory alloy [5], white X-ray microbeam diffraction was employed to examine microscopic stress evolution in individual austenite grains during shape memory behavior. A recently developed imaging technique and its combined use with an energy dispersive detector at the BL28B2 beamline of SPring-8 allowed us to measure strain/stress on the local area of interest. It was found that after tensile deformation of the alloy, large compressive stress developed due to the formation of martensite. When subsequent recovery annealing was performed, it almost disappeared due to reverse martensitic transformation. In addition, the magnitude of compressive stress depended on the grain orientation, as the martensitic transformation depended strongly on orientation. These experimental findings suggest that internal stress is an important factor that determines the shape memory effect. This is in good agreement with the study by Tomota *et al.*, which demonstrated that internal stress plays an important role in promoting the reversible motion of Shockley partial dislocation, thereby improving the shape memory effect [6]. Microscopic stress analysis of superelastic alloys may also help us to expand our understanding of their macroscopic deformation behavior. Although there have been many microstructural analyses on Cu-based superelastic alloys, to the best of our knowledge, microscopic stress evolution of the alloys has not yet been reported.

In the present work, we performed the first *in situ* observation of microstructure and strain/stress evolution during tensile deformation of superelastic Cu-Al-Mn alloys using white X-ray microbeam diffraction. The strain/stress measurements were performed at different positions within a single grain and in several grains having different orientations to investigate the heterogeneous deformation behavior and orientation dependence of stress evolution.

2. Experimental Section

2.1. Materials

The materials used in this study were polycrystalline Cu-18% Al-11.5% Mn (atomic%) alloy sheets. A sample with a very large average grain size of about 400 μm was used, in which the size was obtained by controlling the solution-heat treatment temperature and time [3]. For the tensile test, a small tensile sample with a gauge size of 3 mm length and 1 mm width was prepared. To use the sample in white X-ray microbeam diffraction experiments, the thickness (t) was reduced to about 200 μm by electropolishing in a solution consisting of 20% H_2SO_4, 47% H_3PO_4, and 33% distilled water at room temperature. Such thickness is much lower than the average grain size (400 μm), thus preventing the overlap of diffraction patterns generated from multiple grains. Figure 1 shows the strain-stress curve obtained during the superelastic behavior of the sample. When the sample is subjected to tensile loading by 8% strain and subsequent unloading, it shows good superelastic strain (ε_{SE}) of 5.8% and residual strain (ε_R) of 0.4%. White X-ray microbeam diffraction experiments were performed during a tensile cycle, *i.e.*, before loading → loading to 8% strain → unloading.

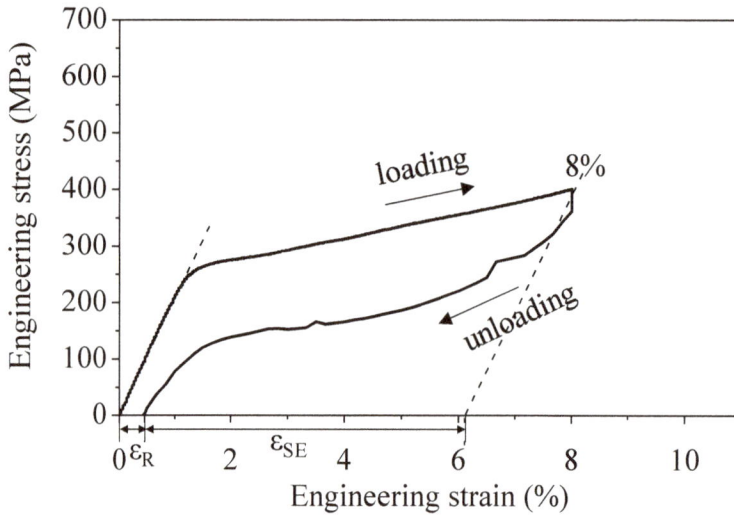

Figure 1. Stress-strain curve of a Cu-Al-Mn tensile sample.

2.2. White X-ray Microbeam Diffraction Experiments

White X-ray diffraction experiments were conducted at SPring-8 on the beamline BL28B2 (Japan Synchrotron Radiation Research Institute: JASRI, Hyogo, Japan), where a high-energy white X-ray microbeam was available. Figure 2 shows a schematic diagram of the tensile sample used for X-ray microbeam diffraction experiments. The diffraction experiments were carried out on a measurement area of 1.5 mm × 0.75 mm in the sample gauge region. The beam size was controlled to 15 μm × 15 μm using the incident slit to illuminate a local area within the respective grains.

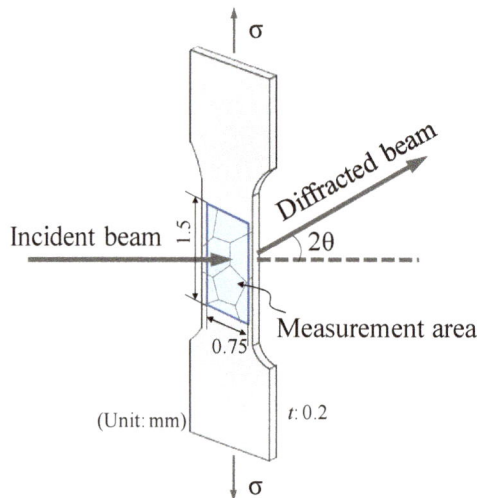

Figure 2. Schematic of tensile sample used for X-ray microbeam diffraction experiments.

Prior to the strain/stress measurements, the grain image of the sample was obtained by performing X-ray microbeam diffraction experiments in scanning mode. Details of the visualization method are given in [7]. Then, the measurement positions were chosen using the obtained grain image. Figure 3a shows the grain image of the undeformed gauge region of the tensile sample, which clearly depicts the presence of grain boundaries (GB) that appear as a bright contrast in the microstructure. Stress-induced martensite (SM) formed in the deformed microstructure also appears as a bright contrast, as will be discussed in the next section. It should be noted that the grain image highly resembles the orientation image (Figure 3b) obtained by electron backscattered diffraction (EBSD) in terms of the GB structure. The EBSD image shows the grain orientation in loading direction. The combined microstructural analysis and EBSD orientation image enabled us to measure strain/stress on the grains with known orientation. The strain/stress measurements were performed on several positions in some grains, as denoted by the white dots in Figure 3a.

To examine the stress evolution behavior in the grains with different orientations, the stresses were measured in five grains labeled as G1–G5 in the EBSD image of Figure 3b. These grains have specific crystallographic orientations, which are marked in the inverse pole figure of Figure 4a. They have different Schmid factor values for the martensitic transformation, as shown in Figure 4b. Note that the Schmid factor is much higher in the G3 and G4 grains than in the others. We therefore expect the martensitic transformation to be favored in the G3 and G4 grains. The transformation strain (%) induced by the martensitic transformation is shown in the stereo-triangle of Figure 4c [1]. The transformation strain is the largest in the (104) orientation, meaning that tensile deformation along the (104) direction would result in greater elongation than that obtained by deformation along the other directions. The G3 and G4 orient close to (104), and therefore would elongate largely by tensile deformation due to the favored martensitic transformation.

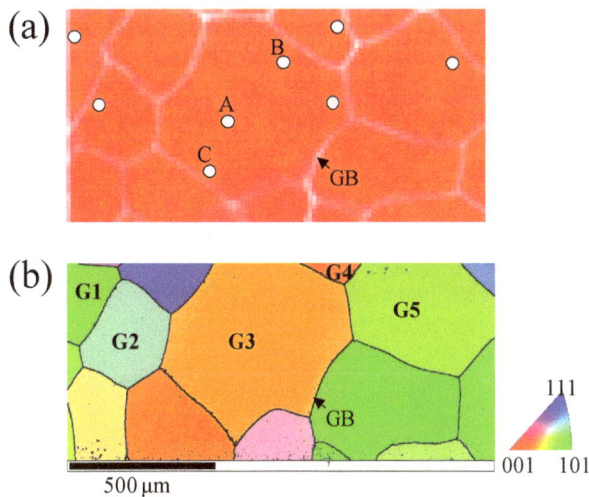

Figure 3. Images showing the microstructure of Cu-Al-Mn alloy. The measurement area shown in Figure 2 was imaged by (**a**) white X-ray microbeam diffraction and (**b**) electron backscattered diffraction (EBSD). White dots in (**a**) denote strain/stress measurement points. GB indicates grain boundary.

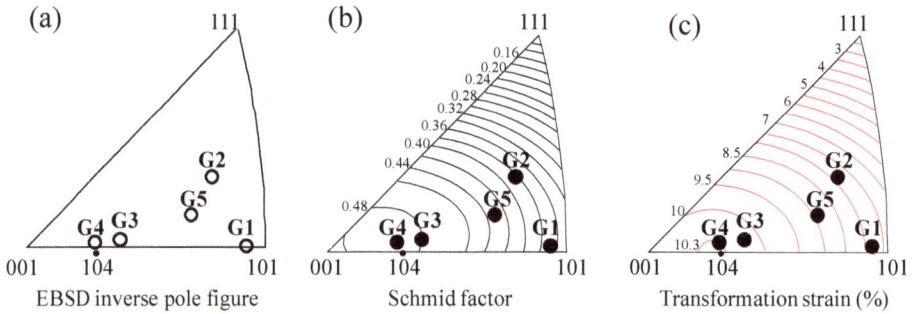

Figure 4. (**a**) EBSD inverse pole figure showing the orientation of the grains in the loading direction, which is indicated by G1, G2, G3, G4, and G5 in the EBSD image of Figure 3b; (**b**) Stereo-triangle showing the Schmid factor of the respective grains; (**c**) Stereo-triangle showing the transformation strain (%) of the respective grains.

Laue patterns diffracted from a tensile sample in the transmission (Laue) geometry were recorded *in situ* using a flat panel detector during the tensile test. Tensile strain applied on the sample was determined by the displacement of grips on a tensile stage. X-ray energy spectra for several Laue reflection regions were measured using a solid-state detector. The obtained energy spectra were used to calculate the lattice spacing of the (*hkl*) planes, d^{hkl} [8]. Laue patterns of deformed superelastic alloys would be composed of many spots that are generated from both the parent phase and SM. In this study, however, measurable Laue spots of SM were hardly obtained due to low intensity, and therefore strain/stresses were measured only for the parent phase. Parent and martensitic phases have the following orientation relationship: $[001]\beta_1//[010]\beta_1'$ and $[110]\beta_1//[001]\beta_1'$ [9].

Lattice strain in the (*hkl*) planes, ε^{hkl}, was calculated from the change in d^{hkl} with respect to the initial lattice spacing before deformation, d_0. The stress was calculated assuming a two-dimensional stress state using Equation (1) [10].

$$\varepsilon^{hkl} = A^{hkl}\sigma_x + B^{hkl}\tau_{xy} + C^{hkl}\sigma_y \tag{1}$$

where σ_x and σ_y are the normal stresses in the tensile direction (*x*) and the transverse direction (*y*), respectively, and τ_{xy} is the shear stress. *A*, *B*, and *C* are constant values determined by the elastic compliance, crystal coordinate system (*X*), diffraction plane coordinate system (or laboratory coordinate system (*L*)), and specimen coordinate system (*P*). For the elastic compliance, the elastic constants (C_{11} = 142 GPa, C_{12} = 124 GPa, and C_{44} = 95 GPa) of a Cu-Al-Ni single crystal were considered [11].

From the σ_x, σ_y, and τ_{xy} obtained, the principal stresses (σ_1 and σ_2) and the principal stress direction (θ_p) can be calculated directly. The principal stress direction defines an angle at which the shear stress becomes zero. Principal stresses were calculated using Equation (2) below.

$$\sigma_{1,2} = \frac{\sigma_x + \sigma_y}{2} \pm \sqrt{\left(\frac{\sigma_x - \sigma_y}{2}\right)^2 + \tau_{xy}^2} \tag{2}$$

3. Results and Discussion

3.1. Variation of Laue Patterns during Superelastic Behavior

An analysis of the Laue patterns provides insight into the deformation characteristics, as the shape of the Laue spots is sensitive to the orientation deviation caused by deformation. *In situ* observation of Laue patterns was performed during the tensile loading cycle to obtain information on the evolution of strain/stress and orientation deviation caused by elastic or plastic deformation. Figure 5 shows the examples of serial Laue patterns obtained from the A point in the G3 grain. Laue spots under the tensile load show pronounced streaking and shifting in the radial direction of the Laue pattern. When the applied load is removed, the Laue spots almost recover the initial shape observed before loading, although the original shape is not fully restored. The streaked Laue pattern indicates that rotation of the crystallographic orientation has occurred in the parent phase, which is attributed to stress-induced martensitic transformation, resulting in the formation of sub-domains in the deformed grains [4]. The parent phase recovers its original orientation due to the reverse martensitic transformation after removal of the load. It is thought that the reverse martensitic transformation and the resulting orientation recovery are the origin of the characteristic reversible Laue pattern change. The reversible Laue pattern change is also observable during the shape memory behavior of shape memory alloys [5]. Apparently, the reversible behavior of the Laue pattern demonstrates the excellent superelasticity of the Cu-Al-Mn alloy, which is in accordance with the stress-strain curve shown in Figure 1. However, the strain recovery is incomplete; some residual strain (0.4%) remains after unloading. The residual strain may result from internal stresses and/or the presence of residual martensite [4].

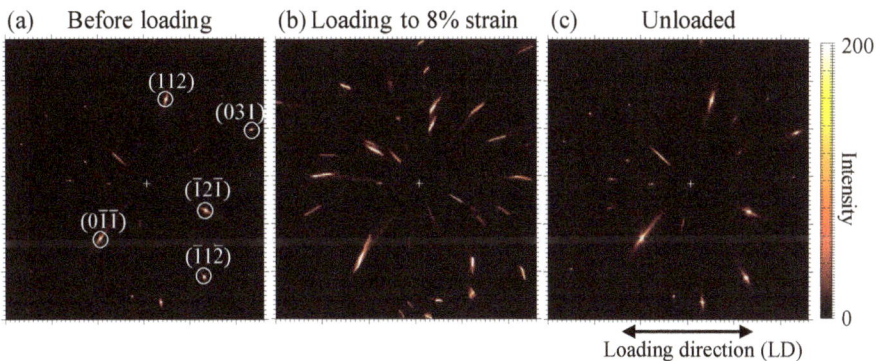

Figure 5. Laue patterns obtained from a point marked by A in Figure 3a (**a**) before loading; (**b**) under a load (8% strain); and (**c**) after unloading. Laue spots were indexed based on their lattice spacing values.

3.2. Strain and Stress Evolution during Superelastic Behavior

The evolution behavior of the lattice and residual strains in the (*hkl*) planes can be experimentally verified by observing the variation of d^{hkl} during loading and unloading, as exemplified in the data of the $(\bar{1}2\bar{1})$ plane shown in Figure 6. When the sample is tensile loaded to 8% strain, the peak position of the curve at 0% strain, d_0, shifts to a higher d value, indicating the evolution of tensile lattice strain. The d value at 8% strain then decreases upon unloading and reverts to near the initial value before loading. The imperfect recovery of d indicates the evolution of residual strain.

Strain evolution behavior was evaluated for five lattice planes (*i.e.*, (112), (031), $(\bar{1}2\bar{1})$, $(\bar{1}1\bar{2})$, and $(0\bar{1}\bar{1})$) and the results are shown in Figure 7. The strains measured at three positions (*i.e.*, A, B, and C) on a single grain are compared to examine the inhomogeneous deformation behavior. The lattice plane angle relative to the loading direction (LD) was measured by analyzing the EBSD pole figures. The lattice strain under the tensile load presents either positive (tensile strain) or negative values (compressive strain), depending on the lattice plane angles relative to the LD due to the Poisson effect. As shown in Figure 7, the magnitude of strain in the respective lattice planes varies according to the measurement position, indicating inhomogeneous strain evolution. At the region near the grain center (A), a relatively high tensile strain is observed in some lattice planes compared to those measured at the regions near the GB (B and C). The behavior of the strain release upon unloading also differs according to the position. While upon unloading, most of the lattice strain measured at the grain center region is released, a high amount of residual strain persists at most of the lattice planes measured at the GB regions.

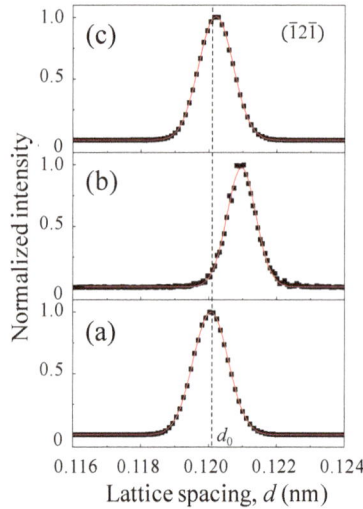

Figure 6. Lattice spacing (*d*) of the ($\bar{1}2\bar{1}$) plane as a function of applied strain (**a**) 0% (before loading); (**b**) 8% (under a load); and (**c**) 0% (unloaded). For comparison, the initial value at d_0, the state before deformation, is indicated by the vertical dotted line. The *d* value was determined from a peak position of Gaussian fit (red lines).

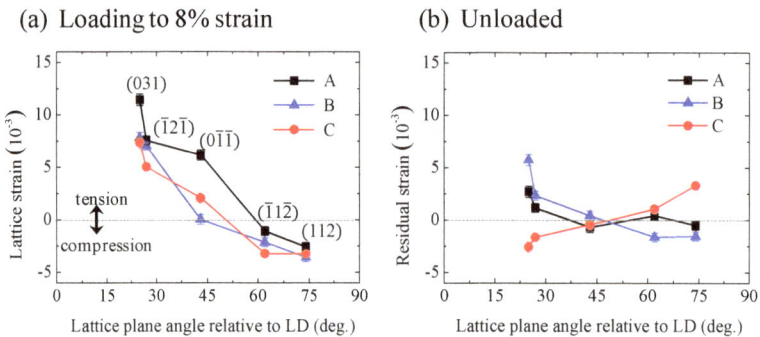

Figure 7. (**a**) Lattice strain at 8% strain and (**b**) residual strain after unloading, measured at three different positions (A, B, and C) in a grain.

The origin of the inhomogeneous strain evolution may be related to the GB constraint. In fact, the deformation of polycrystalline materials is inhomogeneous because a grain embedded within the materials deforms upon contact with neighboring grains, resulting in the geometric grain constraint effect. The GB region could exhibit large residual strain because some permanent constraint strain could be imposed by interactions with neighboring grains during deformation. On the other hand, the lattice strain applied at the grain center region could be easily released upon unloading due to a reduced grain constraint effect, thereby resulting in low residual lattice strain.

Figure 8 shows the variation of principal stresses at several points during the tensile cycle. Green and blue bars indicate tensile stress and compressive stress, respectively. The magnitude of the stresses is presented by the length of the bar. The stress data is considered to contain measurement errors of about ±50 MPa, considering experimental strain resolution of about ±0.05%. There are very low stresses before deformation, which might be induced during the preparation of the tensile sample and/or the production process of the alloy. After the sample is deformed to 8% strain, significant stresses are observed at all points. Upon unloading, the stresses are almost released. It is thought that the reversible stress evolution is a typical phenomenon that results from reversible martensitic transformation in superelastic and shape memory alloys [5].

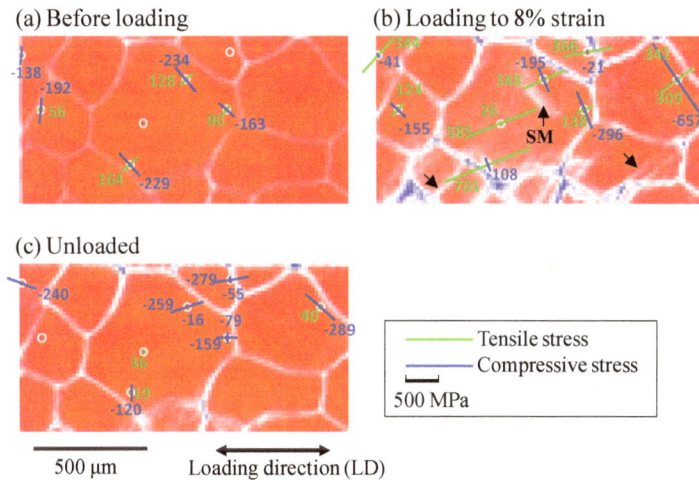

Figure 8. Principal stresses measured at several points (**a**) before loading; (**b**) during loading to 8% strain; and (**c**) after unloading. Arrows in (**b**) indicate stress-induced martensite (SM).

Figure 8b, for the condition with a tensile load, indicates the distribution of the principal stresses within the microstructure. Note that the stresses in each grain have a different magnitude and direction, indicating an inhomogeneous stress distribution at the grain scale. In Table 1, the measured stresses are listed for each grain, together with their Schmid factor and transformation strain. There is apparently no definite relationship between the magnitude of the stresses and the grain orientation (*i.e.*, Schmid factor and transformation strain). In fact, the stresses within a grain are inhomogeneous, and therefore determination of the stresses of a particular grain is reasonably difficult. Indeed, the stresses measured at four different positions within the G3 grain exhibit large variations in magnitude, ranging from a tensile stress of 701 MPa and a compressive stress of −296 MPa. This result indicates the intragranular heterogeneity of the stress distribution.

The stress incompatibility between grains may be attributed to the different deformation properties of the respective grains due to their different orientations. Different grains will be more or less compliant to the applied load, depending on their orientations, which in turn leads to geometric constraint between grains. The constraint may affect the stress evolution in a grain.

Table 1. Principal stresses ($\sigma_{1,2}$) measured at the grains of G1, G2, G3, G4, and G5 during loading to 8% strain.

Grains	Schmid Factor	Transformation Strain (%)	σ_1 (MPa)	σ_2 (MPa)
G1	0.37	7.8	344	−41
G2	0.38	7.8	124	−155
G3	0.48	10.2	348	−195
	0.48	10.2	585	26
	0.48	10.2	701	−108
	0.48	10.2	138	−296
G4	0.49	10.3	366	−21
G5	0.43	8.8	309	−657

The dependence of the martensitic transformation on the grain orientation can be verified by observing the grain images showing the formation of SM during deformation. The grain images of Figure 8 show that upon loading to 8% strain, the white contrast corresponding to SM (marked by arrows) is newly formed, which then disappears after unloading, indicating the reversible martensitic transformation. Considering the extent of the white contrast, the formation of SM appears to be favored in the G3 grain with a high Schmid factor of 0.48 compared with the G2 grain with a lower Schmid factor of 0.38. This experimental result agrees with the assumption based on the Schmid factor described in Figure 4.

(a) Loading to 8% strain   (b) Unloaded

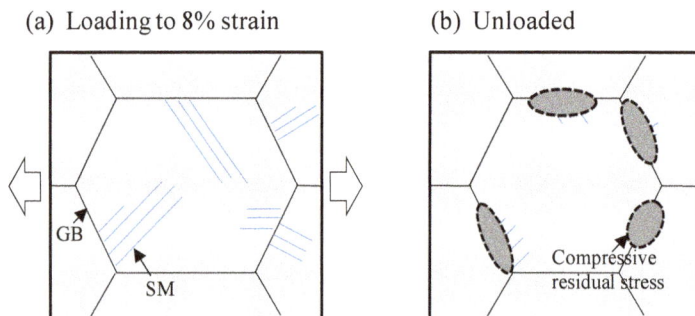

Figure 9. Schematic illustrating the microscopic deformation behavior of the Cu-Al-Mn sample (a) during loading to 8% strain and (b) after unloading.

It is also worthwhile mentioning that large compressive residual stresses form around the GB regions after unloading, as shown in Figure 8c. This may be explained by the formation of residual martensite, as illustrated schematically in Figure 9. As described in Figure 8, the SM is formed within grains after loading to 8% strain. The distribution of SM is considered inhomogeneous, as the strain/stress evolution is inhomogeneous. Upon unloading, the SM is reverse-transformed to the parent phase, but some residual martensite remains around the GB, probably due to the occurrence of geometric constraint near the GB, resulting in the formation of local compressive residual stress. As shown in the optical micrograph of Figure 10, the residual martensite, which is mostly in contact with the GB, is indeed present after unloading of the tensile sample. The residual martensite would lead to incomplete strain recovery of the tensile sample (Figure 1). According to the modeling study by Ueland *et al.*, the GB regions undergo severe grain constraint during deformation, which yields a high stress concentration, and therefore the GB area must be reduced for designing high-performing superelastic alloys [12].

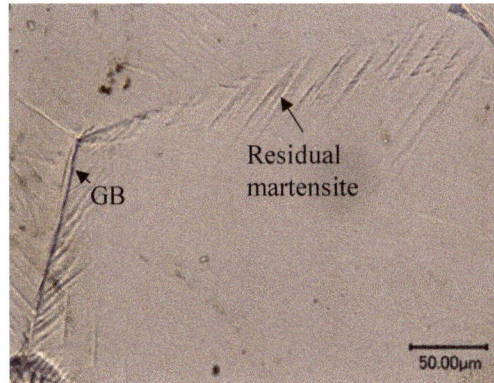

Figure 10. Optical micrograph showing residual martensite formed around grain boundaries after unloading.

4. Conclusions

In situ white X-ray microbeam diffraction experiments were conducted to investigate microscopic deformation behavior of a Cu-Al-Mn superelastic alloy during tensile testing. Evolution of Laue patterns and strain/stress showed reversible phenomena during the superelastic behavior of the tensile sample, probably because of the reversible martensitic transformation. Strain recovery of the sample was incomplete due to the residual martensite, resulting in the formation of local compressive residual stresses at GB regions. Stress data measured at different positions in several grains with different orientations verified the inhomogeneous stress distribution, both at the grain scale and the intragranular scale. The inhomogeneous deformation behavior may be attributed to the orientation dependence of the martensitic transformation as well as to the geometric GB constraint.

Acknowledgments: This study was financially supported in part by a Grant-in-Aid for Scientific Research Fund from the Japan Society for the Promotion of Science (JSPS) and research project from Korea Institute of Industrial Technology. The synchrotron radiation experiments were performed on the beamline BL28B2 at SPring-8 with the approval of the Japan Synchrotron Radiation Research Institute.

Author Contributions: E.P.K. wrote the manuscript; S.S. (Shigeo Sato), S.F., and K.S. helped with data collection and analysis; R.K. prepared the materials; K.K. and M.S. helped with the diffraction experiments. S.S. (Shigeru Suzuki) supervised the research. All authors contributed to the interpretation of the data.

Conflicts of Interest: The authors declare no conflict of interest.

References

1. Sutou, Y.; Omori, T.; Kainuma, R.; Ono, N.; Ishida, K. Enhancement of superelasticity in Cu-Al-Mn-Ni shape memory alloys by texture control. *Metall. Mater. Trans.* **2002**, *33A*, 2817–2824. [CrossRef]
2. Sutou, Y.; Omori, T.; Yamauchi, K.; Ono, N.; Kainuma, R.; Ishida, K. Effect of grain size and texture on pseudoelasticity in Cu-Al-Mn-based shape memory wire. *Acta Mater.* **2005**, *53*, 4121–4133. [CrossRef]
3. Sutou, Y.; Omori, T.; Wang, J.J.; Kainuma, R.; Ishida, K. Effect of grain size and texture on superelasticity of Cu-Al-Mn-based shape memory alloys. *J. Phys. IV* **2003**, *112*, 511–514. [CrossRef]
4. Berveiller, S.; Malard, B.; Wright, J.; Patoor, E.; Geandier, G. *In situ* synchrotron analysis of lattice rotations in individual grains during stress-induced martensitic transformations in a polycrystalline Cu-Al-Be shape memory alloy. *Acta Mater.* **2011**, *59*, 3636–3645. [CrossRef]
5. Kwon, E.P.; Sato, S.; Fujieda, S.; Shinoda, K.; Kajiwara, K.; Sato, M.; Suzuki, S. Microscopic residual stress evolution during deformation process of an Fe-Mn-Si-Cr shape memory alloy investigated using white X-ray microbeam diffraction. *Mater. Sci. Eng.* **2013**, *570*, 43–50. [CrossRef]

6.   Tomota, Y.; Harjo, S.; Lukas, P.; Neov, D.; Sittner, P. *In-situ* neutron diffraction during shape-memory behavior in Fe-Mn-Si-Cr. *JOM* **2000**, *52*, 32–34. [CrossRef]

7.   Kajiwara, K.; Sato, M.; Hashimoto, T.; Hirosawa, I.; Yamada, T.; Terachi, T.; Fukumura, T.; Arioka, K. Development of visualization method of grain boundaries in stainless steel by using white X-ray micro-beam and image detector. *Phys. Status Solidi* **2009**, *206*, 1838–1841. [CrossRef]

8.   Pyzalla, A. Methods and feasibility of residual stress analysis by high-energy synchrotron radiation in transmission geometry using a white beam. *J. Nondestruct. Eval.* **2000**, *19*, 21–31. [CrossRef]

9.   Dutkiewicz, J.; Kato, H.; Miura, S.; Messerschmidt, U.; Bartsch, M. Structure changes during pseudoelastic deformation of Cu-Al-Mn single crystals. *Acta Mater.* **1996**, *44*, 4597–4609. [CrossRef]

10.   Tanaka, K.; Suzuki, K.; Akiniwa, Y. *Evaluation of Residual Stress by X-ray Diffraction, Fundamentals and Application*; Yokendo: Tokyo, Japan, 2006; pp. 38–357.

11.   Sedlak, P.; Seiner, H.; Landa, M.; Novak, V.; Sittner, P.; Manosa, L. Elastic constants of bcc austenite and 2H orthorhombic martensite in Cu-Al-Ni shape memory alloy. *Acta Mater.* **2005**, *53*, 3643–3661. [CrossRef]

12.   Ueland, S.M.; Schuh, C.A. Grain boundary and triple junction constraints during martensitic transformation in shape memory alloys. *J. Appl. Phys.* **2013**, *114*, 053503. [CrossRef]

Review

Structural Dynamics of Materials under Shock Compression Investigated with Synchrotron Radiation

Kouhei Ichiyanagi [1] and Kazutaka G. Nakamura [2,*

[1]   Photon Factory, High Energy Accelerator Research Organization, 1-1 Oho, Tsukuba 305-8555, Japan;
     kouhei.ichiyanagi@kek.jp
[2]   Materials and Structures Laboratory, Tokyo Institute of Technology, 4259 Nagatsuta,
     Yokohama 226-8503, Japan
*   Correspondence: nakamura@msl.titech.ac.jp; Tel: +81-45-924-5397; Fax: +81-45-924-5339

Academic Editor: Klaus-Dieter Liss
Received: 30 September 2015; Accepted: 9 December 2015; Published: 15 January 2016

Abstract: Characterizing material dynamics in non-equilibrium states is a current challenge in material and physical sciences. Combining laser and X-ray pulse sources enables the material dynamics in non-equilibrium conditions to be directly monitored. In this article, we review our nanosecond time-resolved X-ray diffraction studies with 100-ps X-ray pulses from synchrotron radiation concerning the dynamics of structural phase transitions in non-equilibrium high-pressure conditions induced by laser shock compression. The time evolution of structural deformation of single crystals, polycrystals, and glass materials was investigated. In a single crystal of cadmium sulfide, the expected phase transition was not induced within 10 ns at a peak pressure of 3.92 GPa, and an over-compressed structure was formed. In a polycrystalline sample of Y_2O_3 stabilized tetragonal zirconia, reversible phase transitions between tetragonal and monoclinic phases occur within 20 ns under laser-induced compression and release processes at a peak pressure of 9.8 GPa. In polycrystalline bismuth, a sudden transition from Bi-I to Bi-V phase occurs within approximately 5 ns at 11 GPa, and sequential V–III–II–I phase transitions occur within 30 ns during the pressure release process. In fused silica shocked at 3.5 GPa, an intermediate-range structural change in the nonlinear elastic region was observed.

Keywords: structural dynamics; shock compression; time-resolved X-ray diffraction; synchrotron radiation

1. Introduction

All natural phenomena in physical, chemical, and biological systems change with time and occur away from equilibrium. Characterizing and controlling systems far from equilibrium is now recognized as a great challenge in science and engineering [1]. However, the majority of materials science is devoted to characterizing states and functions at equilibrium. Shock compression is one of the techniques to generate non-equilibrium high-pressure states. A sudden increase of pressure induces phase transitions of materials, which may be the way to study the dynamics in non-equilibrium states. Traditional techniques have limitations in investigating phase transitions that are irreversible or associated with negligible volume change from the step feature in Hugoniot curves or particle velocity profiles [2]. Then, the transient non-equilibrium structures cannot be estimated without the aid of static compression results. The duration of shock compression is short, and time-resolved structural detection is required to investigate the dynamics of phase transitions under shock compression. Combining laser and short X-ray pulses enables the transient structures of non-equilibrium states under shock compression to be monitored [3–13]. Most studies are limited to X-ray diffraction at one moment during shock compression with a single-shot measurement [6,9–11]. To elucidate the dynamics, time evolution of the structural change needs to be monitored. In the last decade, time evolution in the nanosecond time region has been studied at the Photon Factory Advanced Ring (PF-AR), which is a unique facility with a single-bunch operation. It is important to review the nanosecond dynamics under shock compression performed at the PF-AR, because investigation of the phase transition dynamics at much faster times (picoseconds or femtoseconds) using X-ray free electron lasers has recently attracted considerable attention [13–15].

Here, we review time-resolved X-ray diffraction observations of the structural dynamics in materials in non-equilibrium high-pressure states induced by shock compression at the PF-AR [16–19]. Synchrotron radiation was used as the source of the X-ray pulses. The structure of this paper is as follows. In Section 2, we describe the laser shock compression method and the time-resolved X-ray diffraction setup. In Section 3, we describe three examples of structural dynamics: a single crystal (cadmium sulfide) [16], polycrystal (zirconia ceramics and bismuth) [18,19], and glass (silica glass) samples [17].

2. Synchrotron Facility and Experimental Setup

The dynamics of structural phase transitions under high pressure were directly investigated by nanosecond time-resolved X-ray diffraction using laser induced shock compression and 100-ps pulsed X-rays from synchrotron radiation. In this section, we describe the laser shock compression technique, the synchrotron radiation facility, and the time-resolved X-ray diffraction experimental setup.

2.1. Laser Shock Compression

Since the pioneering work of Bridgman [20,21], the properties of materials under high pressure have been extensively investigated not only in materials science but also in geoscience [21,22] and planetary science. There are two ways to generate high pressure: static compression and shock compression. For static compression, a large press machine has long been used to study material properties and synthesize new functional materials. In addition, diamond anvil cells have been developed and extensively used for optical spectroscopy and structural analysis [23,24]. The main benefits of static compression are the ability to maintain high-pressure conditions and the capability of controlling the temperature. However, the pressure is limited by the fracture strength of the press machine. In addition, a reference point obtained by dynamic-compression experiments is required to determine the induced pressure. Conversely, in shock compression, the induced pressure is not limited by the fracture strength of the materials, and the pressure is determined by measuring shock and particle velocities with conservation laws [2]. The materials are compressed to a high-pressure state in a very short time (e.g., nanosecond time scale) by shock compression. The induced state is a non-equilibrium state that reverts to the equilibrium state with time. The dynamics of materials in non-equilibrium high-pressure states can be monitored using appropriate time-resolved measurements.

A conventional technique to generate a shock wave is hypervelocity impact of the target materials with a projectile, which is accelerated using a light-gas gun [25–28]. It is very difficult to synchronize firing gunpowder and triggering electric devices with high time accuracy (e.g., within nanoseconds). In recent times, a high-power laser pulse technique has evolved to induce shock compression via laser ablation, which is called laser-shock compression. Using this technique, it is very easy to synchronize the electronic measurement devices and timing of the shock compression with very high accuracy (nanoseconds) [5,29–33]. In addition, ultrahigh pressures above 1 TPa can be achieved using an intense laser pulse [34–39].

There are two main target geometries for laser-shock compression. One is a direct-irradiation target, which consists of a sample and an ablator (usually aluminum foil). When the laser pulse irradiates the ablator surface, laser ablation occurs. A shock wave induced by reaction to the laser ablation propagates into the sample. The peak pressure of this process can be estimated by the equation $P \approx 8.6 \times 10^{11}(I/10^{14})^{2/3}\lambda^{-2/3}$, where P is the pressure (Pa), I is the laser power density (W·cm^{-2}), and λ is the wavelength (μm) [40]. This direct laser-shock compression is frequently used for very high-power laser irradiation (higher than TW/cm^2). The other target is a plasma-confined target, which has a sandwich structure consisting of the sample, an ablation foil, and a cover layer, which is transparent at the laser wavelength. The laser pulse penetrates the cover layer and induces ablation at the ablator surface, and the ablation plume is confined between the sample and the cover layer. For the cover layer, glass and polymer foils are usually used. In this case, the induced pressure and the shock duration are enhanced. The peak pressure can be estimated with the equation $P \approx 3.16 \times 10^2 \sqrt{\alpha/(2\alpha + 3)}\sqrt{IZ}$, where P is the pressure (Pa), I is the laser power density (W·cm^{-2}), Z is the shock impedance (g·cm^{-2}·s^{-1}), and α is a constant [41]. Using the plasma-confined target, much higher pressures than the direct-irradiation target can be generated with the same laser power density. However, there is a limit for the applied power density because of the ablation threshold of the cover layer. The time evolution of the induced pressure can be obtained by measuring the shock and particle velocities [2].

2.2. Time-Resolved X-ray Diffraction Setup

We developed a single-shot time-resolved X-ray diffraction and scattering measurement system based on the storage-ring synchrotron X-ray source at the NW14A beamline of the Photon Factory Advanced Ring (PF-AR) in Tsukuba, Japan [42]. The PF-AR operates in a single-bunch mode at 6.5 GeV electron energy and supplies high-intensity hard X-ray pulses at a repetition rate of 794 kHz. The complete experimental setup of the single-shot time-resolved X-ray measurement system is shown in Figure 1. The pump source for shock wave generation was a Q-switched yttrium aluminum garnet (YAG) laser (Powerlite 8000, Continuum Inc., San Jose, CA, USA). The wavelength, pulse width, and energy were 1.064 µm, 8 ns (full width of half maximum and Gaussian shape), and ~1 J/pulse, respectively. The peak energy and pulse width of the probe X-ray pulse were 15.6 keV and 100 ps, respectively. The energy band width of the X-ray can be changed with the sample condition. We will describe the energy-band width of the X-ray source in more detail later. The frequency of the X-ray pulse train was divided by an X-ray pulse selector (XPS) from 794 kHz–946 Hz. Then, the single X-ray pulse was picked up by the high-speed solenoid X-ray shutter (XRS1, Uniblitz Shutter System, Rochester, NY, USA). The pump laser was synchronized with the frequency of the divided X-ray pulse at 9.46 Hz using a PF master clock of 508 MHz. The 946 Hz X-ray was reduced to 9.46 Hz by a frequency divider [16]. The single laser pulse synchronized with the X-ray pulse was also selected using a solenoid laser shutter. The delay time between the X-ray and laser pulses (Δt) was controlled by a delay generator (DG645, Stanford Research System Inc., Sunnyvale, CA, USA). The timing jitter in this measurement system is about 1 ns. The X-ray diffraction and scattering patterns were recorded on a two-dimensional (2D) charge-coupled device (CCD) detector (MarCCD 165, Rayonix, Evanston, IL, USA) with a diameter of 165 mm. The same CCD detector was used for all of the experiments, and the single-shot images were obtained without accumulation.

The pump-laser and probe X-ray were focused to 0.45 × 0.45 mm^2 or 0.45 × 0.25 mm^2 on the ablator surface. The pump-laser was irradiated at about 15°–20° normal to the sample. We carefully aligned both the X-ray and laser beams as follows. First, we checked the X-ray beam position at the sample position using the pinhole scan technique, in which we checked the beam center and beam width by measuring the X-ray intensity through the pinhole by scanning the pinhole position. The pinhole position was also monitored by using a microscope, which was fixed at a certain position. We then placed a fluorescent plate at the sample position. The laser beam was focused on the plate and the fluorescence from the focused spot was monitored by the microscope. The single laser pulse destroyed and removed the sample in the X-ray path. We changed the sample after taking a single scattering image of the shocked sample at each delay time.

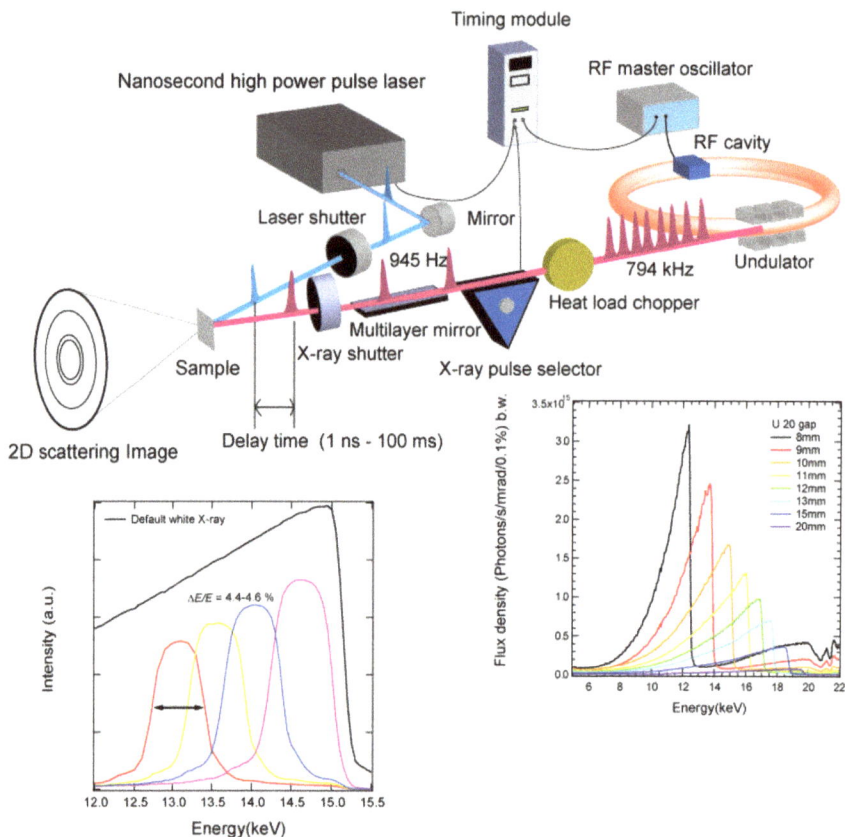

Figure 1. Schematic diagram of the single-shot time-resolved X-ray diffraction and scattering system at beamline NW14A of the PF-AR. The energy bandwidth of the X-ray is changed by the X-ray multilayer optics with the depth graded Ru/C from the default X-ray spectrum to $\Delta E/E$ = 4.4%–4.6%. The pump-laser for shock-wave generation and the X-ray pulse selector and shutter are synchronized with the RF master oscillator. The insert figures show the default X-ray spectrum from the U20 and the Gaussian-shaped and narrow energy bandwidth of the X-ray spectrum using the X-ray multilayer optics, which is modified from Figure 3 in [43].

The energy bandwidth of the probe X-ray was adjusted by the multilayer optics downstream of XPS. The default X-ray energy bandwidth that is suitable for the single-shot Laue diffraction measurement is $\Delta E/E$ = 15% with a broad asymmetric energy spectrum from an undulator with a period length of 20 mm (U20) [42]. The photon flux was 10^9 photons/pulse. However, the broad X-ray energy bandwidth is not suitable for time-resolved X-ray diffraction and scattering measurements of laser-induced shocked polycrystalline and amorphous materials. Therefore, we changed the X-ray energy bandwidth to the sample configuration. A depth-graded Ru/C layer on monocrystalline Si provided a Gaussian-shaped $\Delta E/E$ = 4.4%–4.6% X-ray energy bandwidth from the default undulator X-ray spectrum [43]. This photon flux was 3×10^8 photons/pulse. We can use the discretional energy bandwidth in the X-ray for the spectrum without reducing the photon flux per pulse.

3. Structural Dynamics

3.1. Over-Compressed State in a Single Crystal of Cadmium Sulfide

The shock-induced phase transitions of cadmium sulfide (CdS) have been studied by time-resolved spectroscopy using a light gas gun [44,45]. The wurtzite-rocksalt phase transition has been reported to occur at 2.92 and 3.25 GPa for *a*-axis and *c*-axis compression, respectively [46,47]. The dynamics of the structural phase transition of CdS under shock compression has attracted much attention because an intermediate phase with a face-centered tetragonal structure has been proposed [44]. Using nanosecond time-resolved Laue diffraction and laser shock compression, the structural dynamics of CdS under shock compression at a peak pressure of 3.92 GPa were monitored.

The target assembly had a plasma-confined geometry consisting of three layers: a PET film (25 μm thick), an Al ablator (50 nm thick), and a single crystal of CdS (50 μm thick) [16]. The CdS crystal had a (001) orientation. Laue diffraction was performed with a white X-ray pulse with a peak energy of 16 KeV and an energy width ($\Delta E/E$) of 15%. The photon flux of the X rays was 10^9 phonons/pulse. The 10 ns laser pulse with a wavelength of 1064 nm and energy of 860 mJ was focused on a 0.4×0.4 mm^2 spot on the sample. The spot size of the X-ray pulse was 0.49×0.24 mm^2, and then a small part of the non-laser-irradiated sample was also probed.

Laue diffraction images under laser-shock compression for typical time delays (0, 6, 12, and 22 ns) are shown in Figure 2. The delay between the laser and X-ray pulses was determined at the sample position with their half maximum intensities. We monitored the timing of the laser and X-ray pulses for each shot using photodiodes set in the optical path. The relative delay when the pulse reached the monitor and the sample position was calibrated. Before laser irradiation, there are diffraction spots with hexagonal symmetry corresponding to the wurtzite structure. Under shock compression, the Laue images retain this hexagonal feature and all of the peaks move to the higher angle side of 2θ and then back to their original positions, which suggest that the laser-induced shock compression is parallel to the *c*-axis direction. The changes of the positions of the 201 and 302 Bragg peaks with time are shown in Figure 3.

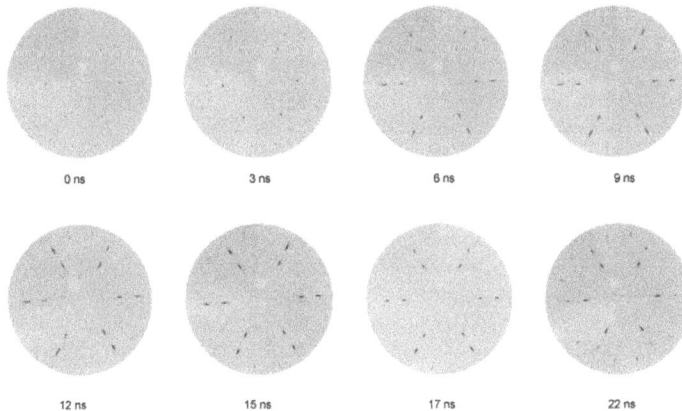

Figure 2. Laue diffraction images of CdS under laser-shock compression at Δt = 0, 6, 12, and 22 ns [16]. Reproduced with permission from [Applied Physics Letters]. Copyright [2007], AIP Publishing LLC.

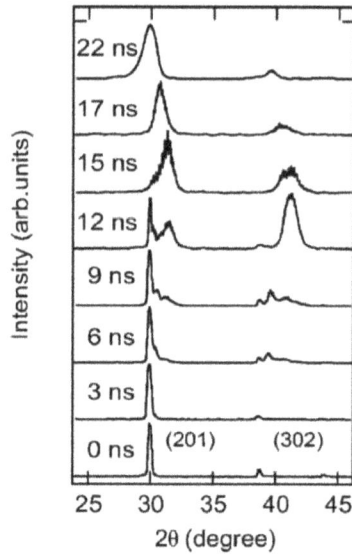

Figure 3. Positions of the 201and 302 Bragg peaks at Δt = 0, 3, 6, 9, 12, 15, 17, and 22 ns [16]. Reproduced with permission from [Applied Physics Letters]. Copyright [2007], AIP Publishing LLC.

The 201 peak intensity decreased after laser irradiation. A higher angle shifted peak appeared and its intensity increased until Δt = 15 ns. For Δt > 17 ns, the new peak shifted to a lower angle. This feature can be explained by laser ablation generating a shock wave, which propagates inside the sample with a shock speed. At a certain time delay, shock-compressed and uncompressed (pristine sample) regions exist inside the sample, and the X-ray diffraction pattern then consists of both the original peak and the higher angle shifted peak from the compressed region. As the delay increases, the shocked volume increases and the intensity of the higher angle shifted peak increases. When the shock front reaches the rear side of the sample, a release wave and shock-wave reflection account for the decreased shock pressure. The peak position then returns to the original position after 22 ns. The shock speed was estimated to be 4.2 ± 0.5 km/s from the time evolution of the higher angle shifted peak. This value is in good agreement with the previously reported elastic velocity [47]. The maximum compression was estimated to be 4.4% of the cell volume from the 201 peak shift [48]. The shock pressure was estimated to be 3.92 GPa from volume compression, which is higher than the phase transition pressure (3.25 GPa) for the wurtzite-rocksalt phase transition by *c*-axis compression.

Although the shock pressure is higher than the phase transition pressure, the phase transition did not occur within 15 ns under shock compression. This indicates that the shock-induced structural phase transition does not instantaneously occur and requires an incubation time for a single-crystal sample. From another point of view, the over-compressed structure, which is not realized in equilibrium conditions, is generated within nanoseconds in a non-equilibrium high-pressure state.

3.2. Reversible Phase Transition in Zirconia Ceramics

Y_2O_3 (3 mol %) stabilized tetragonal zirconia polycrystalline (3Y-TZP) ceramics are widely used engineering ceramics because of their high strength and toughness [49]. Although the phase diagram of pure zirconia is well established, the phase stability of 3Y-TZP under high pressure is controversial [50–52]. The tetragonal structure transforms to a disordered structure or an orthorhombic II phase via a monoclinic phase under static compression. Tetragonal zirconia directly transforms to the orthorhombic II phase during shock compression or a quenchable monoclinic phase during the shock release process. The transformation path between the tetragonal and monoclinic phases is not well established. The transient structure under shock compression can only be assessed by time-resolved X-ray diffraction.

The target assembly had a plasma-confined geometry consisting of three layers: a plastic film (25 µm thick), an Al ablation film (1 µm thick), and the 3Y-TZP sample (50 µm thick) [18]. The 3Y-TZP was a polycrystalline sample obtained from Tosoh Co. (Tokyo, Japan). The X-ray pulse used in these experiments had a peak energy of 15.6 KeV, a bandwidth of 4.4%, and a flux of 3×10^8 photons/pulse. The 10 ns laser pulse with a wavelength of 1064 nm and energy of 700 mJ was focused on a 0.4×0.4 mm^2 spot on the sample. The peak pressure in the sample was estimated to be 9.8 GPa.

Figure 4 shows a typical example of the Debye-Scherrer pattern (Figure 4a) and the rocking curves obtained by integrating the Debye-Scherrer ring of 3Y-TZP before laser irradiation (Figure 4b), which clearly shows that the pristine sample had a tetragonal structure. The error of the diffraction intensity was estimated to be approximately 5% from the fluctuation of several signals.

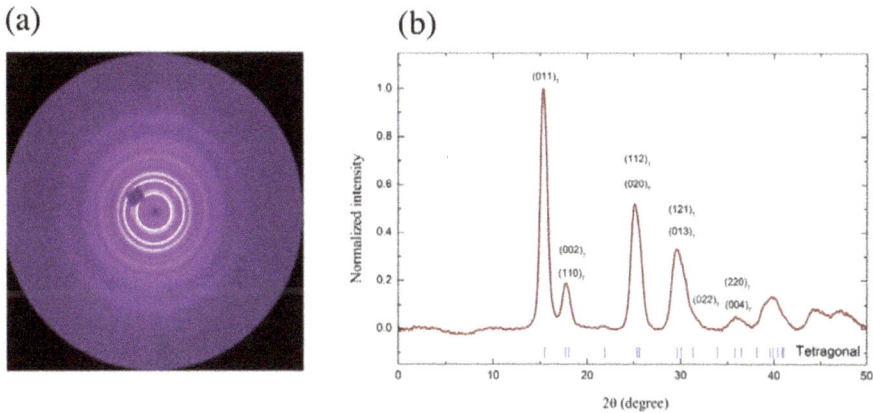

Figure 4. X-ray diffraction of the pristine sample of 3Y-TZP. (**a**) Debye-Scherrer ring detected by a CCD camera. The black square is the shadow of the laser beam block. (**b**) X-ray diffraction intensity profile obtained from the Debye-Scherrer ring. The stick diagram shows the peak positions of the tetragonal phase [18]. Reproduced with permission from [Journal of Applied Physics]. Copyright [2012], AIP Publishing LLC.

Nanosecond time-resolved X-ray diffraction experiments were performed using the pump-probe protocol. X-ray diffraction was performed before and after laser irradiation, and each rocking curve was normalized by its total intensity from $2\theta = 0$–$60°$. The change is obtained from the differential signals obtained by subtracting the rocking curve obtained after laser irradiation from that before irradiation.

Figure 5 shows the change of the differential signals of the X-ray diffraction intensity profile after laser irradiation for $\Delta t = 5$–1005 ns. For $\Delta t < 25$ ns, diffraction peaks appear at $11°$, $13°$, and $16°$, which correspond to the 110, 1$-$11, and 111 peaks of the monoclinic phase, respectively. The peak intensity increased with increasing delay for $\Delta t < 15$ ns and then decreased for $\Delta t > 15$ ns. The intensity of the 110 peak of the tetragonal phase increases with increasing delay. In this time range, both the tetragonal and monoclinic phases coexist under shock compression.

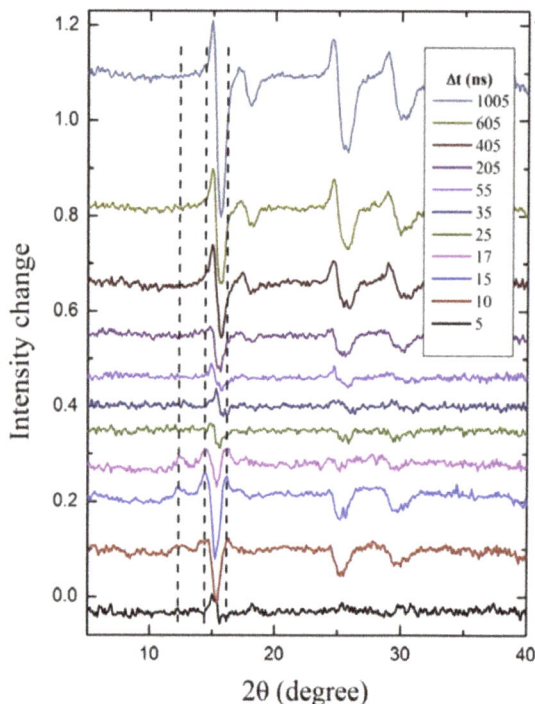

Figure 5. Change of the differential signal of the X-ray diffraction intensity profile from laser-shocked 3Y-TZP with delay time. The dashed lines represent the 110, 1–11 and 111 peak position of the monoclinic phase [18]. Reproduced with permission from [Journal of Applied Physics]. Copyright [2012], AIP Publishing LLC.

For $\Delta t > 55$ ns, the differential signal corresponding to the 110 peak of the tetragonal phase at ~15° is positive and negative at lower and higher angles, respectively. This means that the peak shifted to lower angle, indicating volume expansion induced by pressure release. The results clearly indicate that laser-shocked 3Y-TZP ceramics undergo a reversible tetragonal-monoclinic phase transition in a nanosecond time regime under shock compression and release processes, although an irreversible transition has been suggested by previous shock–recovery investigations.

3.3. Phase Transition Dynamics in Polycrystalline Bismuth

Bismuth has one of the most complicated phase diagrams, and its phase transition point is used as a pressure standard for static high-pressure experiments [53–56]. The kinetic process of the Bi-I to Bi-II transition has been extensively investigated, although it has only recently been semi-qualitatively understood through a ramp compression technique. However, information about the structural dynamics of bismuth under shock compression is quite limited. The sequence of shock-induced polymorphous transformations beyond the Bi-I to Bi-II transition has not been systematically identified. Furthermore, the dynamics during the shock release process have been proven to be almost unobtainable, owing to the complexity arising from the quasi-elastic release effect and the release-induced multiple phase transition. Here, we investigated the dynamics of the structural phase transition of bismuth under shock compression of approximately 11 GPa, which is higher than the reported phase transition pressure of 7.7 GPa for the B-V phase [55].

The plasma confined target consisted of a backup plastic film (25 μm thick), an Al foil (3 μm thick), and the sample. The sample was a 20 μm thick foil of polycrystalline bismuth (99.97% pure) obtained from Goodfellow Cambridge Limited (Huntingdon, UK) [19]. A laser pulse with energy of 1.0 J and pulse width of 8 ns was focused on the target with a spot diameter of 0.48 mm. The peak pressure was estimated to be 11 GPa using the Fabbro-Devaux model [41]. The X-ray pulse had energy of 15.6 KeV and a band width of 1.4%. X-ray diffraction showed a Debye-Scherrer ring pattern in a characteristic form of the polycrystal. The X-ray diffraction intensity profile was obtained by azimuthally averaging the Debye-Scherrer diffraction patterns. The X-ray diffraction intensity profile from the pristine sample (at a delay of 0 ns in Figure 6) is in the Bi-I (R−3m) phase. The most intense peak at 14° is the 110 line of the Bi-I phase.

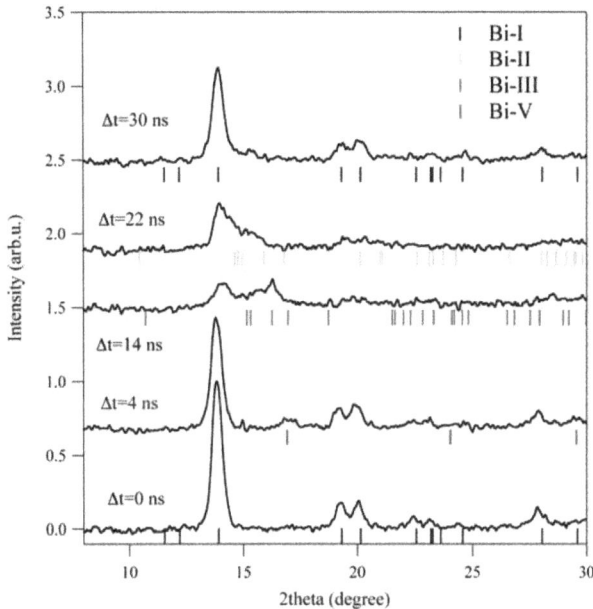

Figure 6. Time evolution of X-ray diffraction intensity profile along of laser-shocked bismuth at Δ*t* = 0, 4, 14, 22, and 30 ns [19]. Reproduced with permission from [Applied Physics Letters]. Copyright [2013], AIP Publishing LLC.

To systematically investigate the structural dynamics after laser irradiation, a series of pump-probe measurements were performed for −2 ⩽ Δ*t* ⩽ 36 ns. Selected signals at Δ*t* = 0, 4, 14, 20, and 30 ns are shown in Figure 6, and are compared with the calculated diffraction-peak positions of the Bi-I to B-V phases. The new peaks correspond to the Bi-V phase (Im−3m) at 4 ns, the Bi-III phase at Δ*t* = 14 ns, and the Bi-II phase at Δ*t* = 20 ns. At Δ*t* = 30 ns, the shocked sample transformed back to the Bi-I phase. Transient structural information at each delay was extracted by comparing the experimental diffraction profile with the calculated diffraction peaks. The time evolution of the structure is schematically summarized in Figure 7 with a pressure profile estimated using the Fabbro-Devaux model [41] and a Gaussian laser profile. It shows that the Bi-I phase transforms to the Bi-V phase within approximately 5 ns during compression and then sequentially transforms to Bi-III, Bi-II, and Bi-I within 30 ns during the release process.

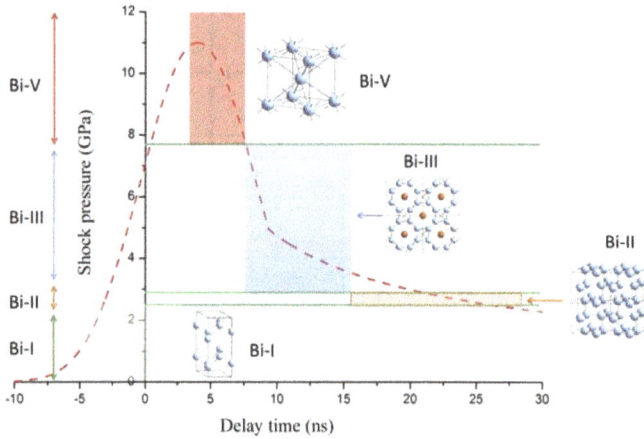

Figure 7. Time evolution of structure of bismuth under laser shock compression [19]. Reproduced with permission from [Applied Physics Letters]. Copyright [2013], AIP Publishing LLC.

The observed transformation from Bi-I to Bi-V appears to be a direct transformation. There are two possibilities. One is that the time the sample spends as the Bi-II and Bi-III phases during compression is too short for to it to be resolved with the current experimental conditions. The other arises from the reconstructive character of the Bi-I to Bi-II transition, which requires an incubation time of tens of nanoseconds. Thus, the Bi-I phase would be over-shocked to the Bi-V phase, although the displacive transformation path between the two phases is still unknown.

Recently, the shock-compression and pressure-release processes of bismuth at shock pressures up to 14 GPa have been investigated using femtoseocond X-ray diffraction with a X-ray free electron laser [13]. The Bi-V phase under compression was also observed. In the pressure-release process from the Bi-V phase, melting was observed within 3 ns. Time-resolved X-ray diffraction is thus an excellent way to determine the nature and time scale of phase transitions induced by shock compression.

3.4. Intermediate Structural Deformation in Shock-Compressed Silica Glass

The response of silica glass to shock compression has been investigated for many years. The behavior of the intermediate- and short-range structures in shock-compressed silica glass is important to understand the shock fracture process. Below 10 GPa, silica glass densification mainly occurs for changes in the intermediate-range structure, as indicated by the first sharp diffraction peak (FSDP) at 1.55 Å^{-1} with a Fourier component period of approximately 4 Å [57]. A nonlinear elastic response below 9 GPa has been observed in many types of silicate glass by free surface velocity measurement. The shock wave front in the elastic shock pressure region produces a non-discontinuous ramp wave front [58,59]. However, the shocked structure of silica glass has not been observed because the X-ray scattering signal using one X-ray pulse is very week. We used a $\Delta E / E = 4.6\%$ energy bandwidth of the probe X-ray pulse with a peak energy of 15.6 KeV. We investigated the dynamics of the intermediate-range structure in silica glass under elastic shock-wave loading of around 4 GPa by time-resolved X-ray scattering measurements.

We fabricated the sample assembly with a plasma confined geometry, as shown in Figure 8a. The sample consisted of silica glass, aluminum film, and poly(ethylene terephthalate) (PET) film to confine the plasma [17]. The sample size was $5 \times 5 \times 70 \ \mu\text{m}^3$ and the thicknesses of aluminum and the PET film were 18 and 25 μm, respectively. The laser ablation was generated at the aluminum-PET film interface, and the shock wave propagated into the aluminum and silica glass. At $\Delta t = 0$ ns, the laser intensity at the aluminum surface was 50%. We constructed a one-dimensional radial scattering curve for each delay time by integrated the 2D X-ray scattering pattern. Figure 8b shows the X-ray scattering patterns as a function of Q, where $Q = (4\pi/\lambda)\sin\theta$, before laser irradiation and at $\Delta t = 10$ ns with the X-ray scattering silica glass and aluminum film peaks labeled. We estimated the shock pressure of silica glass by the impedance matching method using the shift of the 111 aluminum diffraction peak and the compressive data of silica glass in the nonlinear elastic shock region [60,61]. The mean maximum shock pressure in silica glass was estimated to be 3.5 GPa by the impedance matching method.

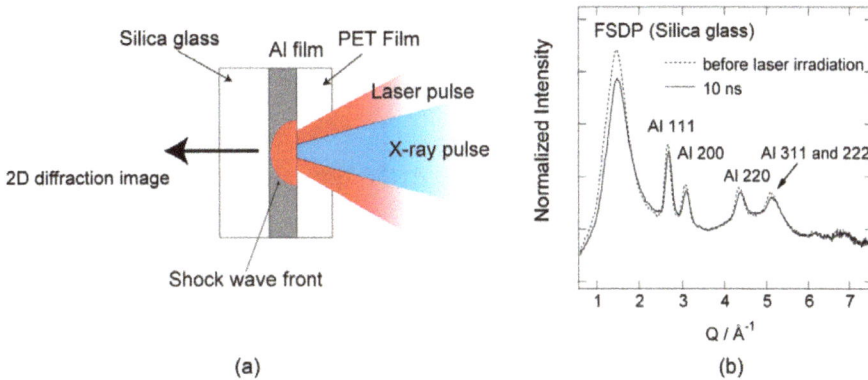

(a)   (b)

Figure 8. (**a**) Schematic drawing of the plasma confined target of silica glass and a typical X-ray scattering pattern using a single X-ray pulse. Aluminum was an ablator and pressure marker to estimate the shock pressure of silica glass. (**b**) X-ray scattering patterns before laser irradiation (dashed line) and at $\Delta t = 10$ ns (solid line). The FSDP of silica glass and X-ray diffraction peaks of aluminum 111, 200, 220, 311, and 222 of the face-centered cubic structure [17] are indicated in the figure. Reproduced with permission from [Applied Physics Letters]. Copyright [2012], AIP Publishing LLC.

Figure 9 shows differential scattering curves for each delay time, with the reference curve before laser irradiation subtracted from the curve for each delay time. After laser ablation, the shock wave generated at the aluminum-PET film interface propagated into the aluminum film. The maximum 111 and 200 peak shifts of aluminum are at $\Delta t \approx 0$ ns, and the shock wave entered the silica glass through the aluminum-silica glass interface. The FSDP shifts to the high Q side. The intermediate-range structure changed with shock wave loading and the pressure release process. At $\Delta t \approx 10$ ns, the shock wave reached and reflected at the free-surface of silica glass. For $\Delta t > 13$ ns, the shock pressure gradually released. These shifts and intensity changes were also seen in hydrostatic compressed silica glass using a synchrotron X-ray source. The intermediate range structure, such as the Si–O–Si bond angle, only changed under shock wave loading in the nonlinear elastic shock region. This time-resolved X-ray scattering method using a short X-ray pulse is able to reveal amorphous structure dynamics under laser-induced shock wave loading.

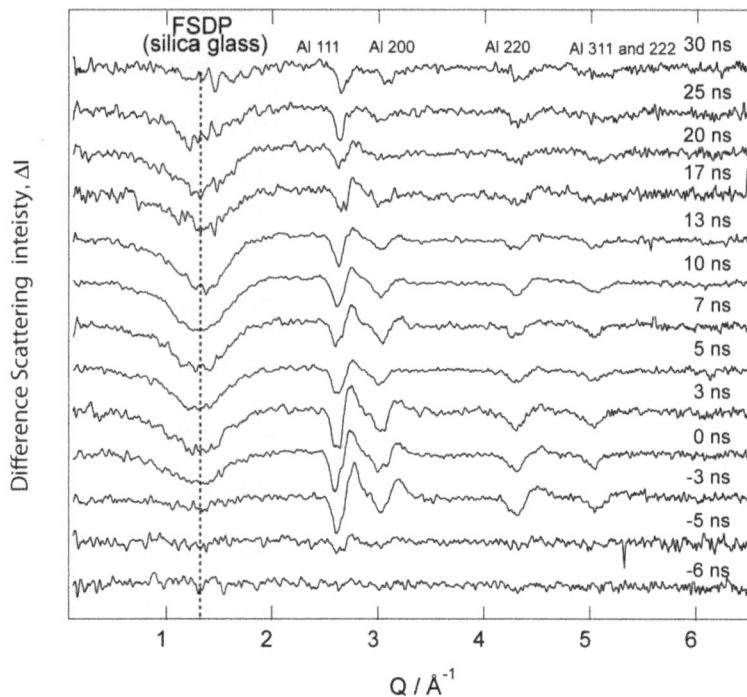

Figure 9. Difference X-ray scattering intensity ΔI as a function of Q at different delay times. The reference curve before laser irradiation was subtracted for each delay time. The gray area is the FSDP of silica glass, which is associated with the intermediate-range structure of glass. The shock wave generated on the aluminum surface occurred at $\Delta t \approx -5$ ns, and propagated into the silica glass at $\Delta t \approx 0$ ns. The shock pressure was released to the ambient pressure at $\Delta t \approx 30$ ns [17]. Reproduced with permission from [Applied Physics Letters]. Copyright [2012], AIP Publishing LLC.

4. Conclusions

Using 100-ps X-ray pulses of synchrotron radiation at the PF-AR (NW-14A beamline) with the laser-shock compression technique, the dynamics of the structural changes of solid materials under non-equilibrium high-pressure conditions were investigated with picosecond time-resolved X-ray diffraction. The present technique can monitor the dynamics of monocrystalline, polycrystalline and glass samples. In a single crystal of CdS, we found that the wurtzite-rocksalt phase transition requires an incubation time of greater than 10 ns, and the over-compressed structure forms before the transition at 3.92 GPa. In a polycrystalline sample of 3Y-TPZ ceramic, the reversible phase transition between tetragonal and monoclinic phases occurs within 20 ns under laser-induced compression and release processes at a peak pressure of 9.8 GPa. In polycrystalline bismuth, a sudden transition from the Bi-I to the Bi-V phase occurs within ~5 ns during the compression process at 11 GPa, and sequential V–III–II–I transitions occurs within 30 ns in the pressure release process. In fused silica shocked at 3.5 GPa, an intermediate-order structural change in the nonlinear elastic region was observed. The phase transitions were observed in polycrystalline samples but not in a single crystal sample within several tens of nanoseconds, which indicates that a large crystal sample requires longer shock duration for the phase transition to occur.

The sequential observation using time-resolved X-ray diffraction is a very powerful technique for monitoring the structural change in many types of materials (from glass to single crystals) under non-equilibrium high-pressure conditions. Further experiments at much higher pressures, which can be generated with the high-power laser system, are required to study the dynamics of phase transitions in metals and minerals in connection with geoscience and planetary science. The dynamics at much shorter time scale (<picoseconds) will be also studied by using time-resolved X-ray diffraction with high-power laser facility and much shorter and strong pulses from XFEL [13–15] and small-size crystals such as nanocrystals.

Acknowledgments: The authors thank Shunsuke Nozawa, Tokushi Sato, and Shin-ichi Adacchi of KEK, Nobuaki Kawai of Kumamoto University, Jianbo Hu, Katsura Norimatsu, Shinya Kosihara of TokyoTech for their experimental help and valuable discussion about the experiments presented in this review.

Author Contributions: Both co-authors contributed to writing this review article. In particular, K.I. to Sections 1, 2.2, 3.1, 3.4 and 4 and K.G.N. to Sections 1, 2.1, 3.2, 3.3 and 4.

Conflicts of Interest: The authors declare no conflict of interest.

References

1. *Directing Matter and Energy: Five Challenges for Science and the Imagination, A Report from the Basic Energy Science Advisory Committee*; U.S. Department of Energy: Washington, D.C., USA, 2007.
2. Graham, R.A. *Solids under High-Pressure Shock Compression*; Springer-Verlag: New York, NY, USA, 1993; pp. 15–138.
3. Wark, J.S.; Riley, D.; Woolsey, N.C.; Keihn, G.; Whitlock, R.R. Direct measurements of compressive and tensile strain during shock breakout by use of subnanosecond X-ray diffraction. *J. Appl. Phys.* **1990**, *68*, 4531–4534. [CrossRef]
4. Kalantar, D.H.; Chandler, E.A.; Colvin, J.D.; Lee, R.; Remington, B.A.; Weber, S.V.; Hauer, A.; Wark, J.S.; Loveridge, A.; Failor, B.H.; *et al.* Transient X-ray diffraction used to diagnose shock compresses Si crystals on the Nova laser. *Rev. Sci. Instrum.* **1999**, *70*, 629–632.
5. Hironaka, Y.; Yazaki, A.; Saito, F.; Nakamura, K.G.; Takenaka, H.; Yoshida, M. Evolving shock-wave profiles measured in a silicon crystal by picosecond time-resolved X-ray diffraction. *Appl. Phys. Lett.* **2000**, *77*, 1967–1969. [CrossRef]
6. Johnson, Q.; Mitchell, A.C. First X-ray diffraction evidence for a phase transition during shock-wave compression. *Phys. Rev. Lett.* **1972**, *29*, 1369–1371. [CrossRef]
7. Wark, J.S.; Whitelock, R.R.; Hauer, A.; Swain, J.E.; Solone, P.J. Shock launching in silicon studied with use of pulsed X-ray diffraction. *Phys. Rev. B* **1987**, *35*, 9391–9394. [CrossRef]
8. Wark, J.S.; Whitelock, R.R.; Hauer, A.; Swain, J.E.; Solone, P.J. Subnanosecond X-ray diffraction from laser-shocked crystals. *Phys. Rev. B* **1989**, *40*, 5705–5714. [CrossRef]
9. D'Almeida, T.; Gupta, Y.M. Real-time X-ray diffraction measurements of the phase transition in KCl shocked along [100]. *Phys. Rev. Lett.* **2000**, *85*, 330–333. [CrossRef] [PubMed]
10. Podurets, A.M.; Dorokhin, V.V.; Trunin, R.F. X-ray diffraction study of shock-induced phase transformations in zirconium and bismuth. *High Temp.* **2003**, *41*, 216–220. [CrossRef]
11. Kalantar, D.H.; Belak, J.F.; Collins, G.W.; Colvin, J.D.; Davies, H.M.; Eggert, J.H.; Germann, T.C.; Hawreliak, J.; Holian, B.L.; Kadau, K.; *et al.* Direct observation of the α–ε transition in shock-compressed iron via nanosecond X-ray diffraction. *Phys. Rev. Lett.* **2005**, *95*, 075502:1–075502:4.
12. Rygg, J.R.; Eggert, J.H.; Lazicki, A.E.; Coppari, F.; Hawreliak, J.A.; Hicks, D.G.; Smith, R.F.; Socrce, C.M.; Uphaus, T.M.; Yaakobi, B.; *et al.* Powder diffraction from solids in the terapascal regime. *Rev. Sci. Instrum.* **2012**, *83*, 113904:1–113904:7.
13. Gorman, M.G.; Briggs, R.; McBride, E.E.; Higginbotham, A.; Arnold, B.; Eggert, J.H.; Fratanduro, D.E.; Galtier, E.; Lazicki, A.E.; Lee, H.J.; *et al.* Direct observation of melting in shock-compressed bismuth with femtosecond X-ray diffraction. *Phys. Rev. Lett.* **2015**, *115*, 095701:1–095701:5.
14. Nagler, B.; Arnold, B.; Bouchard, G.; Boyce, R.F.; Boyce, R.M.; Callen, A.; Campell, M.; Curiel, R.; Galtier, E.; Garofoli, J.; *et al.* The matter in extreme conditions instrument at the linac coherent light source. *J. Synchrotron Radiat.* **2015**, *22*, 520–525. [PubMed]

15. Schroppe, A.; Hoppe, R.; Meier, V.; Patommel, J.; Seiboth, F.; Ping, Y.; Hicks, D.G.; Beckwith, M.A.; Collins, G.W.; Higginbotham, A.; *et al.* Imaging shock waves in diamond with both high temporal and spatial resolution at an XFEL. *Sci. Rep.* **2015**. [CrossRef]

16. Ichiyanagi, K.; Adachi, S.; Hironaka, Y.; Nakamura, K.G.; Sato, T.; Tomita, A.; Koshihara, S. Shock-induced lattice deformation of CdS single crystal by nanosecond time-resolved Laue diffraction. *Appl. Phys. Lett.* **2007**, *91*, 231918:1–231918:3. [CrossRef]

17. Ichiyanagi, K.; Kawai, N.; Nozawa, S.; Sato, T.; Tomita, A.; Hoshino, M.; Nakamura, K.G.; Adachi, S.; Sasaki, Y.C. Shock-induced intermediate-range structural change of SiO_2 glass in the nonlinear elastic region. *Appl. Phys. Lett.* **2012**, *101*, 181901:1–181901:4. [CrossRef]

18. Hu, J.; Ichiyanagi, K.; Takahashi, H.; Koguchi, H.; Akasaka, T.; Kawai, N.; Nozawa, S.; Sato, T.; Sasaki, Y.; Adachi, S.; *et al.* Reversible phase transition in laser-shocked 3Y-TZP ceramics observed via nanosecond time-resolved X-ray diffraction. *J. Appl. Phys.* **2012**, *111*, 053526:1–053526:5, Erratum, *J. Appl. Phys.* **2013**, *113*, 039901:1.

19. Hu, J.; Ichiyanagi, K.; Doki, T.; Goto, A.; Eda, T.; Norimatsu, K.; Harada, S.; Horiuchi, D.; Kabasawa, Y.; Hayashi, S.; *et al.* Complex structural dynamics of bismuth under laser-driven compression. *Appl. Phys. Lett.* **2013**, *103*, 161904:1–161904:5.

20. Bridgman, P.W. Effects of high shearing stress combined with high hydrostatic pressure. *Phys. Rev.* **1935**, *48*, 825–847. [CrossRef]

21. Bridgman, P.W. Shearing phenomena at high pressure of possible importance for geology. *J. Geol.* **1936**, *44*, 653–669. [CrossRef]

22. Ahrens, T.J. Application of shock compression science to earth and planetary physics. In *Shock Compression of Condensed Matter 1995*; AIP Conference Proceedings No. 370; American Institute of Physics: Melville, NY, USA, 1995; pp. 3–8.

23. Jayaraman, A. Diamond anvil cell and high-pressure physical investigations. *Rev. Mod. Phys.* **1983**, *55*, 65–108. [CrossRef]

24. Loubetre, P.; Occelli, F.; LeToullec, R. Optical studies of solid hydrogen to 320 GPa and evidence for black hydrogen. *Nature* **2002**, *416*, 613–617. [CrossRef] [PubMed]

25. Jones, A.H.; Isbel, W.M.; Maiden, C.J. Measurement of the very-high-pressure properties of materials using a light-gas gun. *J. Appl. Phys.* **1966**, *37*, 3493–3499. [CrossRef]

26. Holmes, N.C.; Moriarty, J.A.; Gathers, G.R.; Nellis, W.J. The equation of state of platinum to 660 GPa (6.6 Mbar). *J. Appl. Phys.* **1989**, *66*, 2962–2967. [CrossRef]

27. Moritoh, T.; Kawai, N.; Nakamura, K.G.; Kondo, K. Optimization of a compact two-stage light-gas gun aiming at a velocity of 9 km/s. *Rev. Sci. Instrum.* **2001**, *72*, 4270–4272. [CrossRef]

28. Yokoo, M.; Nkamura, K.G.; Kondo, K.; Tange, Y.; Tuchiya, T. Ultrahigh-pressure scales for gold and platinum at pressures up to 550 GPa. *Phys. Rev. B* **2009**, *80*, 104114:1–104114:9. [CrossRef]

29. Lee, I.-Y.S.; Hill, J.R.; Suzuki, H.; Dolott, D.D.; Baer, B.J.; Chronister, E.L. Molecular dynamics observed 60 ps behind a solid-state shock front. *J. Chem. Phys.* **1995**, *103*, 8313–8321. [CrossRef]

30. Wakabayashi, K.; Nakamura, K.G.; Kondo, K.; Yoshida, M. Time-resolved Raman spectroscopy of polytetrafluoroethylene under laser driven shock compression. *Appl. Phys. Lett.* **1999**, *75*, 947–949. [CrossRef]

31. Matsuda, A.; Nakamura, K.G.; Kondo, K. Time-resolved Raman spectroscopy of benzene and cyclohexane under laser-driven shock compression. *Phys. Rev. B* **2002**, *65*, 174116:1–174116:4. [CrossRef]

32. Matsuda, A.; Kondo, K.; Nakamura, K.G. Nanosecond rapid freezing of benzene under shock compression studied by coherent anti-Stokes Raman spectroscopy. *J. Chem. Phys.* **2006**, *124*, 054501:1–054501:4. [CrossRef] [PubMed]

33. Sato, A.; Oguchi, S.; Nakamura, K.G. Temperature measurement of carbon tetrachloride under laser shock compression using nanosecond time-resolved Raman spectroscopy. *Chem. Phys. Lett.* **2007**, *445*, 28–31. [CrossRef]

34. Cauble, R.; Phillion, D.W.; Hoover, T.J.; Holmes, N.C.; Kilkenny, J.D.; Lee, R.W. Demonstration of 0.75 Gbar planar shocks in X-ray driven colliding foils. *Phys. Rev. Lett.* **1992**, *70*, 2102–2105.

35. Batani, D.; Balducci, A.; Beretta, D.; Bernarinello, A.; Löwer, T.; Koenig, M.; Benuzzi, A.; Faral, B.; Hall, T. Equation of sate data for gold in the pressure range <10 TPa. *Phys. Rev. B* **2000**, *61*, 9287–9294.

36. Wakabayashi, K.; Hattori, S.; Tange, T.; Fujimoto, Y.; Yoshida, M.; Kozu, N.; Tanaka, K.A.; Ozaki, N.; Sasatani, Y.; Takenaka, H.; *et al.* Laser-induced shock compression of tantalum to 1.7 TPa. *Jpn. J. Appl. Phys.* **2000**, *39*, 1815–1816.
37. Nagao, H.; Nakamura, K.G.; Kondo, K.; Ono, T.; Takamatsu, K.; Ozaki, N.; Tanaka, K.A.; Nagai, K.; Nakai, M.; Wakabayashi, K.; *et al.* Hugonipt measurement of diamond under shock compression up to 3 TPa. *Phys. Plasma* **2006**, *13*, 052705:1–052705:5.
38. Cottet, F.; Hallouin, M.; Romain, J.P.; Fabbro, R.; Faral, B.; Pepin, H. Enhancement of a laser-driven shock wave up to 10 TPa by the impedance-match technique. *Appl. Phys. Lett.* **1985**, *47*, 678–680. [CrossRef]
39. Trainor, R.J.; Shaner, J.W.; Auerbach, J.M.; Holmes, N.C. Ultrahigh-pressure laser-driven shock-wave experiments in aluminum. *Phys. Rev. Lett.* **1979**, *42*, 1154–1157. [CrossRef]
40. Benuzzi, A.; Löwer, T.; Koenig, M.; Faral, B.; Batani, D.; Berretta, D.; Danson, C.; Pepler, D. Indirect and direct laser driven shock waves and applications to copper equation of state measurements in the 10–40 Mbar pressure range. *Phys. Rev. E* **1996**, *54*, 2162–2165. [CrossRef]
41. Devaux, D.; Fabbro, R.; Tollier, L.; Bartnicki, E. Generation of shock waves by laser-induced plasma in confined geometry. *J. Appl. Phys.* **1993**, *74*, 2268–2273. [CrossRef]
42. Nozawa, S.; Adachi, S.; Takahashi, J.; Tazaki, R.; Guérin, L.; Daimon, M.; Tomita, A.; Sato, T.; Chollet, M.; Collet, E.; *et al.* Developing 100 ps-resolved X-ray structural analysis capabilities on beamline NW14A at the Photon Factory Advanced Ring. *J. Synchrotron Radiat.* **2007**, *14*, 313–319. [PubMed]
43. Ichiyanagi, K.; Sato, T.; Nozawa, S.; Kim, K.H.; Lee, J.H.; Choi, J.; Tomita, A.; Ichikawa, H.; Adachi, S.; Koshihara, S. 100 ps time-resolve solution scattering utilizing a wide-bandwidth X-ray beam from multilayer optics. *J. Synchrotron Radiat.* **2009**, *16*, 391–394. [CrossRef] [PubMed]
44. Knudsson, M.D.; Gupta, Y.M.; Kunz, A.B. Transformation mechanism for the pressure-induced phase transition in shocked CdS. *Phys. Rev. B* **1999**, *59*, 11704–11715. [CrossRef]
45. Knudsson, M.D.; Gupta, Y.M. Transformation kinetics for the shock wave induced phase transition in cadmium sulfide crystals. *J. Appl. Phys.* **2002**, *91*, 9561–9571. [CrossRef]
46. Tang, Z.P.; Gupta, Y.M. Shock-induced phase transformation in cadmium sulfide in an elastomer. *J. Appl. Phys.* **1988**, *64*, 1827–1837. [CrossRef]
47. Tang, Z.P.; Gupta, Y.M. Phase transition in cadmium sulfide single crystals shocked along *c* axis. *J. Appl. Phys.* **1997**, *81*, 7203–7212. [CrossRef]
48. Bringa, E.M.; Rosolankova, K.; Rudd, R.E.; Remington, B.A.; Wark, J.S.; Duchaineau, M.; Kalantar, D.H.; Hawreliak, J.; Belak, J. Shock deformation of face-centered-cubic metals on subnanosecond timescales. *Nat. Mater.* **2006**, *5*, 805–809. [CrossRef] [PubMed]
49. Garvie, R.C.; Hannink, R.H.; Pascoe, R.T. Ceramic steel? *Nature* **1975**, *258*, 703–704. [CrossRef]
50. Alzyab, B.; Perry, C.H.; Ingel, R.P. High-pressure phase transitions in zirconia and yttria-doped zirconia. *J. Am. Ceram. Soc.* **1987**, *70*, 760–765. [CrossRef]
51. Ohtaka, O.; Kumi, S.; Ito, E. Synthesis and phase stability of cotunnite-type zirconia. *J. Am. Ceram. Soc.* **1988**, *71*, C448–C449. [CrossRef]
52. Igarashi, Y.; Matsuda, A.; Akiyoshi, A.; Kondo, K.; Nakamura, K.G.; Niwase, K. Laser-shock compression of an yttria-doped tetragonal zirconia studied by Raman spectroscopy. *J. Mater. Sci.* **2004**, *39*, 4371–4372. [CrossRef]
53. Yu, E.; Ponyatovsky, E.G. *Phase Transformations of Elements under High Pressure*; CRC: Boca Raton, FL, USA, 2005; p. 148.
54. Pélissier, J.L.; Wetta, N. A model-potential approach for bismuth (I). Densification and melting curve calculation. *Physica A* **2001**, *289*, 459–478.
55. McMahon, M.I.; Degtyareva, O.; Nelmes, R.J. Ba-IV-type incommensurate crystal structure in group-V metals. *Phys. Rev. Lett.* **2000**, *85*, 4896–4899. [CrossRef] [PubMed]
56. McMahon, M.I.; Degtyareva, O.; Nelmes, R.J.; van Smaalen, S.; Palatinus, L. Incommensurate modulations of Bi-III and Sb-II. *Phys. Rev. B* **2007**, *75*, 184114:1–184114:5. [CrossRef]
57. Inamura, Y.; Katayama, Y.; Utsumi, W.; Funakoshi, K. Transformation in the intermediate-range structure of SiO_2 glass under high pressure and temperature. *Phys. Rev. Lett.* **2004**, *93*, 015501:1–015501:4. [CrossRef]
58. Alexander, C.S.; Chhabildas, L.C.; Reinhart, W.D.; Templeton, D.W. Changes to the shock response of fused quartz due to glass modification. *Int. J. Impact Eng.* **2008**, *35*, 1376–1385. [CrossRef]

59. Gibbons, R.V.; Ahrens, T.H. Shock metamorphism of silicate glasses. *J. Geophys. Res.* **1971**, *76*, 5489–5498. [CrossRef]

60. Michell, A.C.; Nelis, W.J. Shock compression of aluminum, copper, and tantalum. *J. Appl. Phys.* **1970**, *52*, 3363–3374. [CrossRef]

61. Barker, L.M.; Hollenbach, R.E. Shock-wave studies of PMMA, fused silica, and sapphire. *J. Appl. Phys.* **1970**, *41*, 4208–4226. [CrossRef]

metals

MDPI

Article

Dynamic Strain Evolution around a Crack Tip under Steady- and Overloaded-Fatigue Conditions

Soo Yeol Lee [1],*, E-Wen Huang [2], Wanchuck Woo [3], Cheol Yoon [1], Hobyung Chae [1] and Soon-Gil Yoon [1]

[1] Department of Materials Science and Engineering, Chungnam National University, Daejeon 305-764, Korea; yc2013@cnu.ac.kr (C.Y.); highteen5@cnu.ac.kr (H.C.); sgyoon@cnu.ac.kr (S.-G.Y.)
[2] Department of Materials Science and Engineering, National Chiao Tung University, Hsinchu 300, Taiwan; ewenhuang@nctu.edu.tw
[3] Neutron Science Division, Korea Atomic Energy Research Institute, Daejeon 305-353, Korea; chuckwoo@kaeri.re.kr
* Author to whom correspondence should be addressed; sylee2012@cnu.ac.kr; Tel.: +82-42-821-6637; Fax: +82-42-821-5850.

Academic Editor: Klaus-Dieter Liss
Received: 2 September 2015; Accepted: 4 November 2015; Published: 12 November 2015

Abstract: We investigated the evolution of the strain fields around a fatigued crack tip between the steady- and overloaded-fatigue conditions using a nondestructive neutron diffraction technique. The two fatigued compact-tension specimens, with a different fatigue history but an identical applied stress intensity factor range, were used for the direct comparison of the crack tip stress/strain distributions during *in situ* loading. While strains behind the crack tip in the steady-fatigued specimen are irrelevant to increasing applied load, the strains behind the crack tip in the overloaded-fatigued specimen evolve significantly under loading, leading to a lower driving force of fatigue crack growth. The results reveal the overload retardation mechanism and the correlation between crack tip stress distribution and fatigue crack growth rate.

Keywords: fatigue; crack growth; overload; stress/strain; neutron diffraction

1. Introduction

A fundamental understanding of the fatigue crack growth mechanism is critical for the development of lifetime prediction methodology in structural materials. It is well recognized that variable-amplitude cyclic loading can retard or accelerate the crack propagation rate by making it difficult to predict the crack propagation rate and fatigue life [1–6]. A tensile overload, a load higher than a maximum load during constant-amplitude cyclic loading, intervened during constant-amplitude cyclic loading is one of the examples to retard the crack propagation rate and increase the fatigue lifetime significantly. Many investigations have been reported to understand the retardation mechanisms of the crack growth rate following the overload [7–17]. Among them, the plasticity-induced crack closure approach suggested by Elber [1] was well recognized, and it emphasized the importance of a crack closure in the region of a crack wake. The crack tip plasticity approach [18], based on the large overload-induced plastic zone due to large plastic deformation caused by overloading, has drawn much attention from researchers in examining the retardation phenomena.

Our previous works have shown that the combined effects of large compressive residual stresses and crack tip blunting with secondary cracks are responsible for the overload-induced crack growth retardation [15]. In addition, the strain evolution near the crack tip was systematically examined at the various crack growth stages (with a different stress intensity factor range) through the retardation period [16]. For a better understanding of the correlation between the stress/strain distributions around the crack tip and fatigue crack growth rate, the direct comparison of the evolution of the crack tip stress/strain fields between the steady- and overloaded-fatigue conditions under the same applied stress intensity factor range should be elucidated. In this regard, the two fatigued compact-tension specimens were prepared for the direct comparison of the crack tip stress/strain distributions. Importantly, they experienced the identical stress intensity factor range, but a different fatigue history. The evolution of the strain fields around the crack tip under loading was compared using *in situ* neutron diffraction. The results show the nonlinearity of the strain response as a function of distance from the crack tip under loading, and we discuss the influence of the crack tip stress distribution on the strain nonlinearity, crack opening process, crack tip driving force, and fatigue crack growth rate.

2. Experimental Section

2.1. Materials

Fatigue crack growth experiments were carried out on a compact-tension (CT) specimen of 304L stainless steel. This material has a single-phase face-centered cubic (FCC) structure (Figure 1), a yield strength of 241 MPa, an ultimate tensile strength of 586 MPa, and elongation of 55% at room temperature. The specimen, prepared according to the American Society for Testing and Materials (ASTM) Standards E647-99 [19], has a notch length of 10.16 mm, a width of 50.8 mm, and a thickness of 6.35 mm (Figure 2).

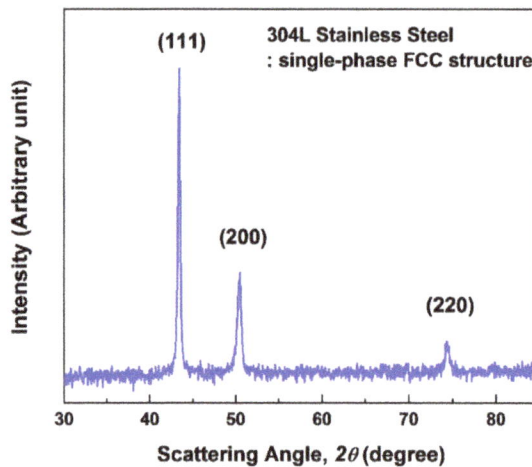

Figure 1. X-ray diffraction pattern of 304L stainless steel.

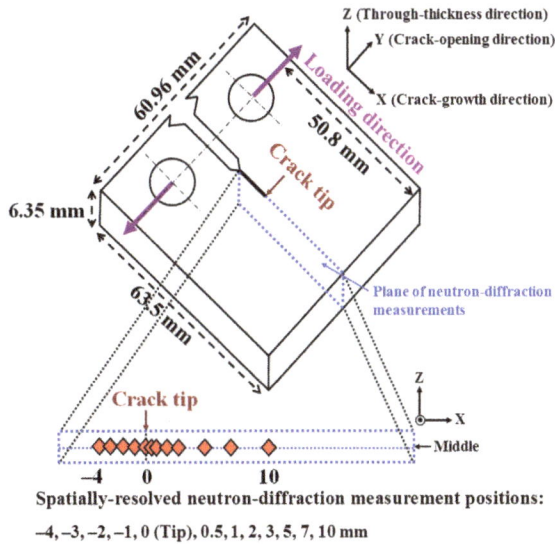

Figure 2. The geometry of compact-tension specimen. Neutron diffraction measurements were performed as a function of distance from the crack tip along the crack growth direction.

2.2. Fatigue Crack Growth Tests

Prior to the fatigue crack growth test, the CT specimens were precracked to approximately 1.27 mm, and then fatigue crack growth tests were performed in air under a constant load-range control mode (P_{max} = 7400 N, P_{min} = 740 N, a load ratio R = 0.1, frequency = 10 Hz). The crack length was measured using a compliance method, which is obtained from the load *vs.* displacement data by a crack-opening-displacement gauge [19]. The stress intensity factor, K, was obtained using the following equation [19]:

$$K = \frac{P(2+\alpha)}{B\sqrt{W}(1-\alpha)^{3/2}} \left(0.886 + 4.64\alpha - 13.32\alpha^2 + 14.72\alpha^3 - 5.6\alpha^4\right) \tag{1}$$

where P = applied load, B = thickness, $\alpha = a/W$, a = crack length, and W = width for a CT specimen. The two fatigued CT specimens are prepared: (i) steady-fatigued condition, Case 1: continuously fatigued under the same baseline condition until the crack length reaches 18 mm (ΔK = 32 MPa·m$^{1/2}$); (ii) tensile-overloaded condition, Case 2: fatigued under the same baseline condition until the crack length reaches 16 mm (ΔK = 29 MPa·m$^{1/2}$), after which a single tensile overload of 10,360 N (140% of P_{max}) was applied, and then continuously fatigued under the same baseline condition until the crack length reaches 18 mm (ΔK = 32 MPa·m$^{1/2}$). Therefore, Case 1 and Case 2 have identical crack lengths but the different fatigue histories.

2.3. In Situ Neutron Diffraction Strain Measurements

In situ neutron diffraction experiments were performed using the Residual Stress Instrument at HANARO, Korea Atomic Energy Research Institute. The neutron diffraction measurements were performed at diffraction angles (2θ) of $84°$ for the {311} diffraction peaks of the austenite phase using a wavelength of 1.46 Å on the neutron beam. The (311) plane normal is perpendicular to the crack plane, *i.e.*, crack growth direction. The change of d-spacings in the (311) orientation was measured *in situ* during loading as a function of distance from the crack tip along the crack-propagation direction. The strain mapping with a 1 mm spatial resolution was performed along the center line (mid-thickness) of the CT specimen (−4, −3, −2, −1, 0 (crack tip), 0.5, 1, 2, 3, 5, 7, 10 mm, Figure 2). The peak position was determined from the Gaussian fitting of the {311} diffraction peak in the crack-opening direction strain component. The lattice strains were then calculated by Equation 2.

$$\varepsilon = (d - d_0)/d_0 = -\cot\theta\,(\theta - \theta_0) \tag{2}$$

where d_0 is the stress-free reference d-spacing, d is the lattice spacing under the stress condition, θ_0 and θ are the diffraction angles for the stress-free and stressed conditions, respectively. The stress-free reference d-spacing was obtained 10 mm away from the corner of the annealed CT specimen. In the current study, the lattice strain evolution in the vicinity of the crack tip was examined at the 15 load levels (*i.e.*, $0.01P_{max}$, $0.1P_{max}$, $0.2P_{max}$, $0.25P_{max}$, $0.3P_{max}$, $0.35P_{max}$, $0.4P_{max}$, $0.45P_{max}$, $0.5P_{max}$, $0.55P_{max}$, $0.6P_{max}$, $0.65P_{max}$, $0.7P_{max}$, $0.85P_{max}$, and $1P_{max}$), as the sample was *in situ* deformed.

3. Results and Discussion

Figure 3 shows the crack growth rate (da/dN) as a function of the stress intensity factor range (ΔK) under the two different fatigue loading conditions. While the constant-amplitude fatigue crack growth testing shows a linear relationship of the crack growth rate *vs.* ΔK following the Paris law, the tensile overloaded testing reveals the crack growth retardation period with a ΔK range of 29 to 38 MPa·m$^{1/2}$ after the application of a single tensile overload. For a better understanding of the crack growth retardation phenomena, the two different fatigued specimens, Case 1 and Case 2, marked in Figure 3, with a ΔK range of 32 MPa·m$^{1/2}$, are used to examine the effect of fatigue history on the strain evolution around the crack tip. The crack growth rates of Case 1 and Case 2 at $\Delta K = 32$ MPa·m$^{1/2}$ were 1.72×10^{-4} and 5.13×10^{-5}, respectively.

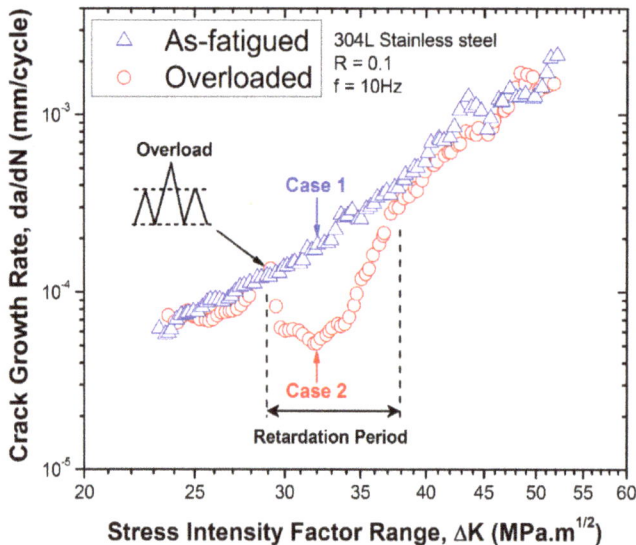

Figure 3. The crack growth rates (da/dN) *versus* stress intensity factor range (ΔK) for the as-fatigued and overloaded conditions. The compact-tension specimens of Case 1 and Case 2 marked with the same ΔK of 32 MPa·m$^{1/2}$ were used for *in situ* neutron diffraction experiments.

Figure 4 shows the crack morphology of Case 1 and Case 2 measured by the optical microscope. While Case 1 shows a relatively sharp crack tip, Case 2 shows a blunt crack at a ΔK range of 29 MPa·m$^{1/2}$ due to large plastic deformation by the overload, followed by a sharp crack tip at a ΔK range of 32 MPa·m$^{1/2}$.

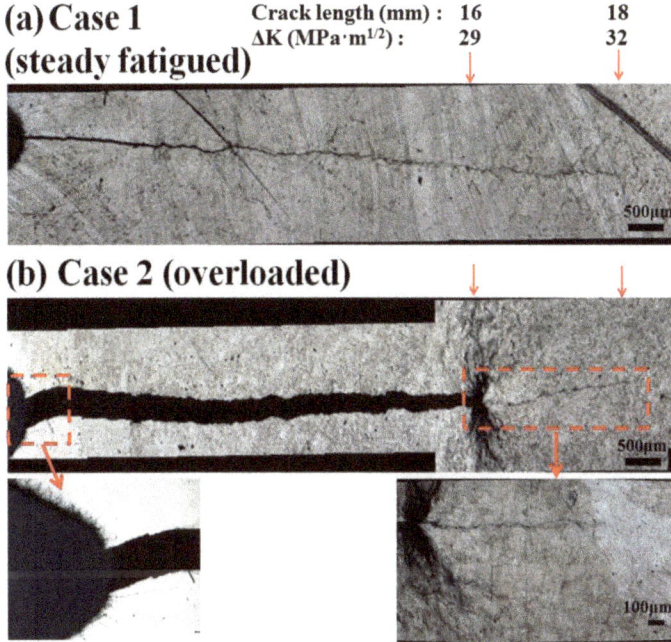

Figure 4. The crack morphology measured by optical microscope: (**a**) Case 1, (**b**) Case 2. At a tensile overload point, the crack length and ΔK are 16 mm and 29 MPa·m$^{1/2}$, respectively. The crack tip positions for both Case 1 and Case 2 are located at a crack length of 18 mm and a ΔK of 32 MPa·m$^{1/2}$.

Figure 5 shows the evolution of internal strains around the crack tip as a function of applied load. In Figure 5a, the strains were compressive from −4.5 to 2.5 mm at $0.01P_{max}$. The maximum compressive strain of ~770 μɛ (microstrain, 10^{-6}) was observed at the crack tip. As the load increases, the strains right in front of the crack tip evolve significantly, exhibiting the maximum tensile strain of ~1450 μɛ at 0.5 mm ahead of the crack tip. The change of the strains under loading becomes smaller, as the distance from the crack tip is far away. It is noted that very little change of the strains is observed in the locations behind the crack tip.

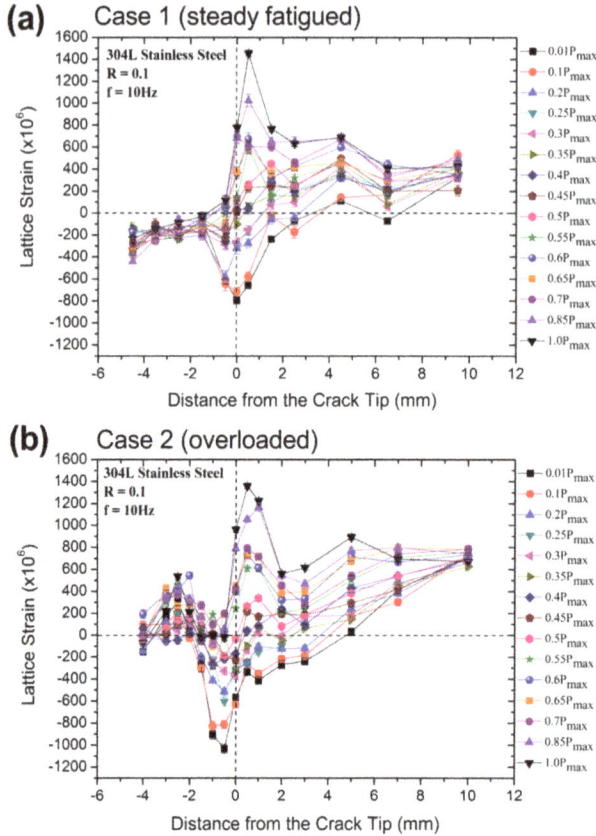

Figure 5. The lattice strain evolution as a function of distance from the crack tip under loading: (**a**) Case 1, (**b**) Case 2.

Figure 5b shows the strain profile of Case 2 upon loading. Three distinct observations are found: First, upon an applied loading of $0.01P_{max}$, much higher compressive strains with a maximum of ~1050 $\mu\varepsilon$ were observed at −2 to 0 mm behind the crack tip, which is between an overload point and the crack tip. It is suggested that the enlarged compressive residual strains in conjunction with a blunting of the crack occurring at the overload point (Figure 4b) are related to the crack growth retardation mechanism after the overloading. Secondly, unlike the strain evolution of Case 1 shown in Figure 5a, it is obvious that the strains behind the crack tip of Case 2 evolve systematically with increasing applied load. It indicates that the stresses are distributed during loading not only in the locations ahead of the crack tip, but also in the locations behind the crack tip for Case 2. Finally, the maximum tensile strain of Case 2 at 0.5 mm in front of the crack tip was slightly smaller than that of Case 1. It was also found that the change of strains right in front of the crack tip during loading from P_{min} to P_{max} was smaller in Case 2 than in Case 1. It is thought that the less stresses applied right ahead of the crack tip for Case 2 result in less driving force of the crack growth, and thus, a lower crack growth rate as shown in Figure 3.

Figure 6 shows the strain evolution as a function of applied load at various locations away from the crack tip for Case 1 (Figure 6a) and Case 2 (Figure 6b). The nonlinearity of lattice strain as a function of applied load at the various locations means a change of the stress distribution during loading by being associated with a crack opening process. Earlier works have demonstrated that the onset of the nonlinearity of the lattice strain can be used to determine the crack opening level [16].

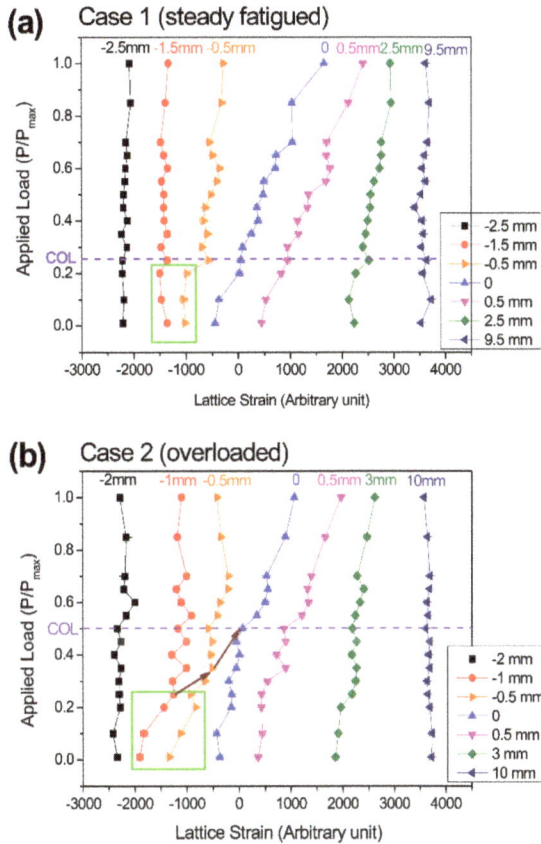

Figure 6. Lattice strain evolution as a function of applied load at the various locations away from the crack tip: (**a**) Case 1, (**b**) Case 2. "COL" indicates the crack-opening load. The square box highlights the difference of strain evolution behind the crack tip between Case 1 and Case 2. The arrow indicates the transfer of stress concentration in the locations behind the crack tip during loading.

As shown in Figure 6a, the stresses were not applied at all at -2.5 and -1.5 mm locations behind the crack tip during loading, as revealed in the invariant lattice strains during loading. From the nonlinearity of strain evolution, the crack opening load (COL) of ~$0.25P_{max}$ was determined for Case 1. On the other hand, quite different strain distributions were observed for Case 2 (Figure 6b). At a lower load, the stresses concentrate on the locations of -1 and -0.5 mm behind the crack tip, where the large compressive residual strains are observed as shown in Figure 5b. Upon loading, the stress concentration moves toward the crack tip by influencing the strain nonlinearity as indicated in Figure 6b. With the completion of a transfer of stress concentration at the crack tip, a relatively high COL of ~$0.5P_{max}$ was obtained for Case 2. The transfer of stress concentration at the crack tip should be understood in conjunction with the crack opening process, as is well described in the previous work [16]. Based on the crack opening loads determined from the strain nonlinearity shown in Figure 6, the effective stress intensity factor ranges (ΔK_{eff}) as a driving force of fatigue crack growth were calculated as 25.06 MPa·m$^{1/2}$ for Case 1 and 16.70 MPa·m$^{1/2}$ for Case 2. These correlate well with a decrease of the crack growth rate of Case 2 compared with Case 1, as shown in Figure 3. Moreover, the smaller responses of the strains right in front of the crack tip for Case 2, as compared with those of Case 1 (Figure 5), can be shown to account for a lower crack tip driving force (ΔK_{eff}) and a retardation of crack growth for Case 2.

4. Conclusions

In situ neutron diffraction was employed to compare the evolution of internal strains around the crack tip between the steady-fatigued (Case 1) and overload-fatigued (Case 2) specimens where the stress intensity factor range is identical but a different fatigue history exists. While strains behind the crack tip in Case 1 are irrelevant to increasing applied load, the strains behind the crack tip in Case 2 evolve significantly under loading, leading to smaller maximum tensile strain and strain change right in front of the crack tip. In Case 2, the transfer of stress concentration occurs toward the crack tip upon loading, resulting in a nonlinearity of the strain profile. The crack growth retardation after the overload can be attributed to a higher crack opening level measured for Case 2 by being correlated with a calculation of the effective stress intensity factor range as a driving force of fatigue crack growth.

Acknowledgments: This work was supported by the National Research Foundation of Korea (NRF) grant funded by the Korean government (MSIP) (Nos. 2014M2B2A4031983, 2013R1A1A1076023, 2013R1A4A1069528). EWH appreciates the support from Ministry of Science and Technology (MOST) Program 101-2221-E-008-039-MY3 and Atomic Energy Council (AEC) Program 10309037L.

Author Contributions: S.Y.L. wrote the initial draft of the manuscript; S.Y.L., E.-W.H., W.W., C.Y. analyzed the data; All co-authors contributed to the interpretation of the data.

Conflicts of Interest: The authors declare no conflict of interest.

References

1. Elber, W. The Significance of Fatigue Crack Closure. In *Damage Tolerance in Aircraft Structures*; ASTM STP 486; American Society for Testing Materials: West Conshohocken, PA, USA, 1971; Volume 486, pp. 230–242.
2. Gan, D.; Weertman, J. Crack closure and crack propagation rates in 7050 aluminum. *Eng. Fract. Mech.* **1981**, *15*, 87–106. [CrossRef]
3. Shin, C.S.; Hsu, S.H. On the mechanisms and behavior of overload retardation in AISI-304 stainless-steel. *Int. J. Fatigue* **1993**, *15*, 181–192. [CrossRef]
4. Sadananda, K.; Vasudevan, A.K.; Holtz, R.L.; Lee, E.U. Analysis of overload effects and related phenomena. *Int. J. Fatigue* **1999**, *21*, S233–S246. [CrossRef]
5. Makabe, C.; Purnowidodo, A.; McEvily, A.J. Effects of surface deformation and crack closure on fatigue crack propagation after overloading and underloading. *Int. J. Fatigue* **2004**, *26*, 1341–1348. [CrossRef]
6. Lee, S.Y.; Rogge, R.B.; Choo, H.; Liaw, P.K. Neutron diffraction measurements of residual stresses around a crack tip developed under variable-amplitude fatigue loadings. *Fatigue Fract. Eng. Mater. Struct.* **2010**, *33*, 822–831. [CrossRef]
7. Jones, R.E. Fatigue crack growth retardation after single-cycle peak overload in Ti/6Al/4V titanium alloy. *Eng. Fract. Mech.* **1973**, *5*, 585–604. [CrossRef]
8. Newman, J.C., Jr. *A Crack-Closure Model for Predicting Fatigue Crack Growth under Aircraft Spectrum Loading*; American Society for testing Materials: West Conshohocken, PA, USA, 1981; Volume 748, pp. 53–84.
9. Suresh, S. Micromechanisms of fatigue crack growth retardation following overloads. *Eng. Fract. Mech.* **1983**, *18*, 577–593. [CrossRef]
10. Wardclose, C.M.; Blom, A.F.; Ritchie, R.O. Mechanisms associated with transient fatigue crack-growth under variable-amplitude loading—An experimental and numerical study. *Eng. Fract. Mech.* **1989**, *32*, 613–638. [CrossRef]
11. Dougherty, J.D.; Srivatsan, T.S.; Padovan, J. Fatigue crack propagation and closure behavior of modified 1070 steel: Experimental results. *Eng. Fract. Mech.* **1997**, *56*, 167–187. [CrossRef]
12. Singh, K.D.; Khor, K.H.; Sinclair, I. Roughness- and plasticity-induced fatigue crack closure under single overloads: Finite element modelling. *Acta Mater.* **2006**, *54*, 4393–4403. [CrossRef]

13. Bichler, C.H.; Pippan, R. Effect of single overloads in ductile metals: A reconsideration. *Eng. Fract. Mech.* **2007**, *74*, 1344–1359. [CrossRef]

14. Lee, S.Y.; Choo, H.; Liaw, P.K.; Oliver, E.C.; Paradowska, A.M. *In situ* neutron diffraction study of internal strain evolution around a crack tip under variable-amplitude fatigue-loading conditions. *Scr. Mater.* **2009**, *60*, 866–869. [CrossRef]

15. Lee, S.Y.; Liaw, P.K.; Choo, H.; Rogge, R.B. A study on fatigue crack growth behavior subjected to a single tensile overload: Part I. An overload-induced transient crack growth micromechanism. *Acta Mater.* **2011**, *59*, 485–494. [CrossRef]

16. Lee, S.Y.; Choo, H.; Liaw, P.K.; An, K.; Hubbard, C.R. A study on fatigue crack growth behavior subjected to a single tensile overload: Part II. Transfer of stress concentration and its role in overload-induced transient crack growth. *Acta Mater.* **2011**, *59*, 495–502. [CrossRef]

17. Lee, S.Y.; Huang, E.W.; Wu, W.; Liaw, P.K.; Paradowska, A.M. Development of crystallographic-orientation-dependent internal strains around a fatigue-crack tip during overloading and underloading. *Mater. Charact.* **2013**, *79*, 7–14. [CrossRef]

18. Wheeler, O.E. Spectrum loading and crack growth. *J. Basic Eng.* **1972**, *94*, 181–186. [CrossRef]

19. American Society for Testing and Materials (ASTM). *Standard Test Method for Measurement of Fatigue Crack-Growth Rates*; ASTM Standard E647-99; American Society for Testing Materials: West Conshohocken, PA, USA, 2000; pp. 591–630.

metals

Article

Twinning-Detwinning Behavior during Cyclic Deformation of Magnesium Alloy

Soo Yeol Lee [1,*], Huamiao Wang [2] and Michael A. Gharghouri [3]

[1] Department of Materials Science and Engineering, Chungnam National University, Daejeon 305-764, Korea
[2] Materials Science and Technology, Los Alamos National Laboratory, Los Alamos, NM 87544, USA; wanghm@lanl.gov
[3] Canadian Neutron Beam Centre, Canadian Nuclear Laboratories, Chalk River, ON K0J 1J0, Canada; Michael.Gharghouri@cnl.ca
* Author to whom correspondence should be addressed; sylee2012@cnu.ac.kr; Tel.: +82-42-821-6637; Fax: +82-42-821-5850.

Academic Editor: Klaus-Dieter Liss
Received: 20 April 2015; Accepted: 18 May 2015; Published: 26 May 2015

Abstract: *In situ* neutron diffraction has been used to examine the deformation mechanisms of a precipitation-hardened and extruded Mg-8.5wt.%Al alloy subjected to (i) compression followed by reverse tension (texture T1) and (ii) tension followed by reverse compression (texture T2). Two starting textures are used: (1) as-extruded texture, T1, in which the basal pole of most grains is normal to the extrusion axis and a small portion of grains are oriented with the basal pole parallel to the extrusion axis; (2) a reoriented texture, T2, in which the basal pole of most grains is parallel to the extrusion axis. For texture T1, the onset of extension twinning corresponds well with the macroscopic elastic-plastic transition during the initial compression stage. The non-linear macroscopic stress/strain behavior during unloading after compression is more significant than during unloading after tension. For texture T2, little detwinning occurs after the initial tension stage, but almost all of the twinned volumes are detwinned during loading in reverse compression.

Keywords: magnesium alloy; deformation; twinning; detwinning; neutron diffraction

1. Introduction

Many investigations have been devoted to understanding the deformation behavior of magnesium and its alloys, because of their potential applications for lightweight materials in the aircraft and automotive industries, and for portable electronic devices [1,2]. Magnesium alloys have poor formability at room temperature, which can be attributed to the limited number of available slip systems [3,4]. <a> slip with a 1/3<11.0> Burgers vector on the close-packed (00.2) basal plane is the primary slip system in hexagonal close-packed (HCP) magnesium. Non-basal <a> slip on the {10.0} prismatic and {10.1} pyramidal plane has often been observed at higher stresses. While none of these slip modes can accommodate deformation along the c-axis, deformation twinning can provide limited deformation along the c-axis. {10.2}<10.1> extension twinning is commonly found during plastic deformation at room temperature in favorably oriented grains relative to the applied loading direction, leading to tension-compression yield asymmetry and strong plastic anisotropy in magnesium alloys [5–8]. When deformation twinning occurs, the lattice is reoriented approximately 86.3° relative to the parent lattice, resulting in pronounced changes in the crystallographic texture during deformation [9–12]. *In situ* neutron diffraction experiments have been employed extensively to study the plastic deformation behavior of magnesium alloys. The technique provides information on the distribution of internal stresses and strains among the various crystallographic orientations, as well as on bulk texture evolution caused by twinning [13–19].

In the current work, neutron diffraction is used to study the plastic deformation behavior of a precipitation-hardened and extruded Mg-8.5wt.%Al alloy. Lattice strains and diffraction peak intensities for several grain orientations are monitored *in situ* during deformation to examine the evolution of the stress state and the occurrence of twinning and detwinning in various grain orientations. The loading paths consist of (i) compression followed by reverse tension and (ii) tension followed by reverse compression.

2. Experimental Section

2.1. Materials

The extruded Mg-8.5wt.%Al alloy used in this study was solution treated and aged, resulting in equiaxed grains with an average size of ~60 μm. Details of the sample preparation are provided elsewhere [13]. The bulk crystallographic texture was measured using the E3 neutron diffractometer of the Canadian Neutron Beam Centre, located at the National Research Universal (NRU) reactor at Canadian Nuclear Laboratories, Canada. The orientation distribution function (ODF) for each sample was determined from the {10.0}, {00.2}, {10.1} and {10.2} pole figures. Two starting textures were used: (1) as-extruded texture, T1 (Figure 1a), in which the basal pole of most grains is normal to the extrusion axis and a very small portion of grains are oriented with the basal pole parallel to the extrusion axis; (2) a reoriented texture, T2 (Figure 1b), in which the basal pole of most grains is parallel to the extrusion axis. The reoriented texture, T2, was obtained by compressing the T1 texture along the extrusion direction to a strain of ~9%, followed by annealing at a suitable temperature [18]. Therefore, {10.2} extension twinning could be easily activated under compression along the extrusion direction for T1, and under tension along the extrusion direction for T2.

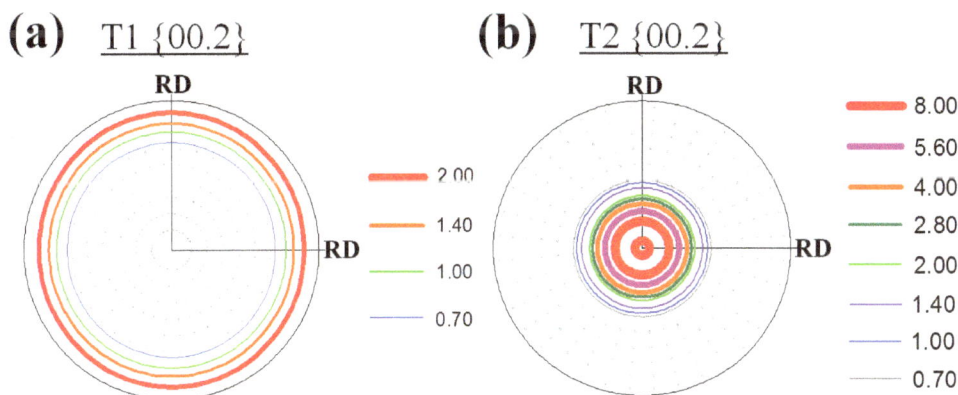

Figure 1. Initial textures of (**a**) T1 and (**b**) T2 samples determined by neutron diffraction. The center of each pole figure corresponds to the extrusion direction (ED), RD is the radial direction of the extruded bar.

2.2. Neutron Diffraction Strain Measurements

Neutron diffraction strain measurements were conducted on the L3 neutron diffractometer of the Canadian Neutron Beam Centre, Canadian Nuclear Laboratories. T1 and T2 samples were deformed in compression followed by reverse tension and in tension followed by reverse compression, respectively. Diffraction peaks for several grain families, characterized by the crystallographic plane which was normal to the loading direction, were acquired *in situ* during deformation, yielding interplanar spacings (*d*-spacings) parallel to the loading direction for each grain family as a function of applied load. The lattice strains for each family were calculated using the following equation:

$$\varepsilon_{hk.l} = \left(d_{hk.l} - d^0_{hk.l} \right) / d^0_{hk.l} \tag{1}$$

where $d^0_{hk.l}$ and $d_{hk.l}$ are the *d*-spacings of the {*hk.l*} family of planes in the unloaded and loaded conditions, respectively.

3. Results and Discussion

The macroscopic stress-strain response of the T1 sample subjected to compression followed by reverse tension is shown in Figure 2.

The material yields at ~−100 MPa, at which the elastic-plastic transition is well underway, though plasticity is not fully developed. The non-linear macroscopic stress/strain behavior during unloading after compression is more significant than during unloading after tension.

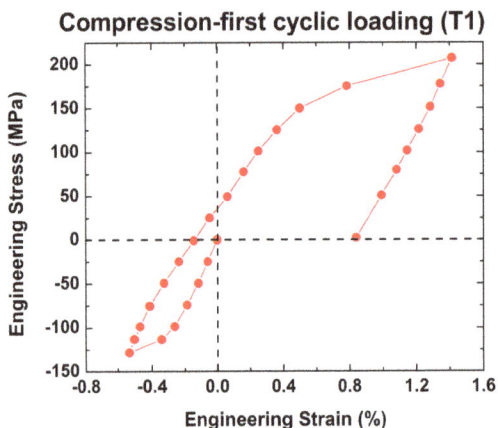

Figure 2. Macroscopic stress-strain response for the T1 sample. The symbols correspond to points in the loading history at which diffraction data were acquired.

Figure 3 shows lattice strain evolution for four different families as a function of applied stress for the T1 sample subjected to compression followed by reverse tension (Figure 2). The corresponding integrated intensity variations for all measured reflections are provided in Figure 4 [17]. A detailed discussion of the intensity variations in Figure 4 is provided in [17], based on which it is possible to identify the stress intervals over which twinning and detwinning occur. The twinning and detwinning intervals identified in Figure 3 are based on this previous analysis.

Compression-first cycling loading (T1)

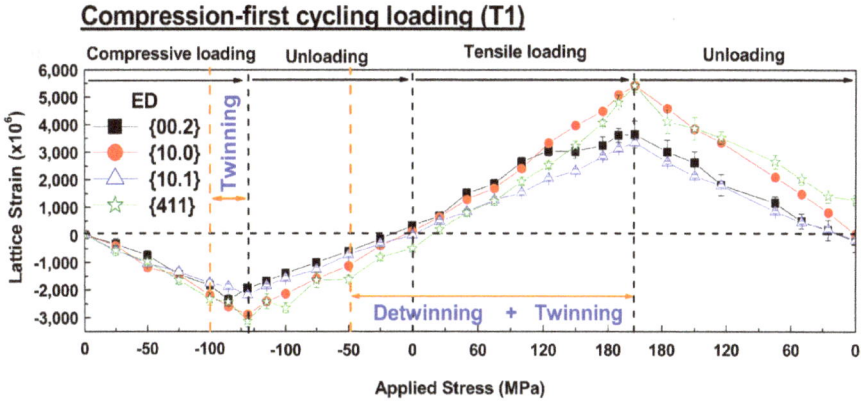

Figure 3. Lattice strain evolution for various grain families as a function of applied stress during the deformation shown in Figure 2. The {*hk.l*} plane normal is parallel to the applied loading direction, which is parallel to the extrusion direction (ED). The angles between the c-axis and the loading direction for the {00.2}, {10.0} and {10.1} families are 0°, 90° and 61.9°, respectively.

Compression-first cyclic loading (T1)

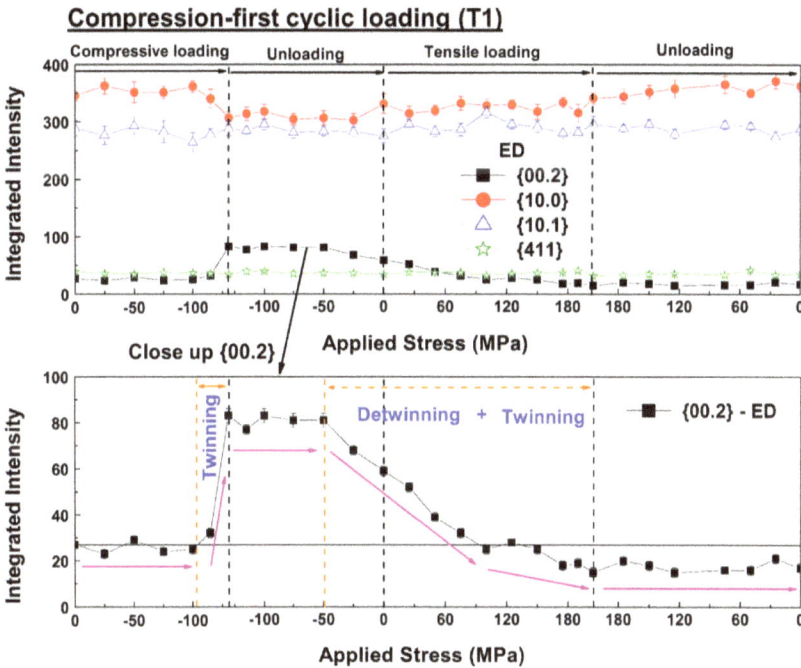

Figure 4. Intensity variations for various grain families as a function of applied stress during the deformation shown in Figure 2 [17].

In Figure 3, during the initial compressive loading step, the lattice strain increases linearly with applied stress for all four grain families up to ~−50 MPa. Beyond this stress, the lattice strain in the {10.1} family increases more slowly with applied stress compared with the other orientations, indicating that this family deforms plastically. The deformation mechanism is most likely basal <a> slip, which has the lowest critical resolved shear stress (CRSS) of all the commonly observed deformation modes, and for which the {10.1} family is favorably oriented. As the load increases beyond ~−100 MPa, the {00.2} and {10.0} lattice strains deviate from one another, corresponding to an increase in the {00.2} intensity and a concurrent decrease in the {10.0} intensity (see Figure 4 [17]). Despite being favorably oriented for {10.2}<10.1> extension twinning in compression, the strain in the {10.0} family increases linearly with the applied stress during the initial compressive loading step, resulting in almost zero residual strain upon unloading. On the other hand, the lattice strain in the {00.2} family, which is twin-related to the {10.0} grains, experiences relaxation during the final loading step, resulting in a slightly tensile residual strain at the end of unloading. The residual strain for the soft {10.1} family is slightly compressive, or zero. The strain in the β-$Mg_{17}Al_{12}$ precipitates ({411} reflection) increase linearly during loading, with no evidence of relaxation, resulting in relatively large compressive residual strains upon unloading. During unloading after compression, it was found that about 40% of the twinned volume detwins [17]. This detwinning behavior is thought to contribute significantly to the non-linear behavior observed in Figure 2.

During reverse loading in tension, the intensity data in Figure 4 [17] show that detwinning continues up to an applied stress of ~+100 MPa, after which the {00.2} minority grains undergo {10.2} extension twinning, as revealed by a decrease in the {00.2} intensity beyond ~100 MPa. The {00.2} and {10.1} families, favorably oriented for extension twinning and basal slip respectively, show stress relaxation during reverse tensile loading, resulting in slightly compressive residual strains upon unloading. Conversely, the lattice strain in the {100} family, which is not favorably oriented for basal slip or extension twinning, varies linearly throughout the test, resulting in negligible residual strain at the end of the test. The {411} precipitates accumulate high levels of lattice strain during the reverse tensile loading, but, unlike the {10.0} family, the precipitates show large tensile residual strains upon final unloading.

Figure 5 shows the macroscopic stress-strain response for the T2 sample subjected to tension followed by reverse compression. The initial stress-strain response is linear up to an applied stress of ~75 MPa, after which the slope decreases as the material starts to yield. During unloading after tension and reloading in compression, the stress-strain response is clearly non-linear. The material undergoes general yielding in compression at ~−50 MPa. The unloading behavior at the end of the test is also clearly non-linear, but the effect is less significant than the non-linear behavior after compression for T1 (Figure 2).

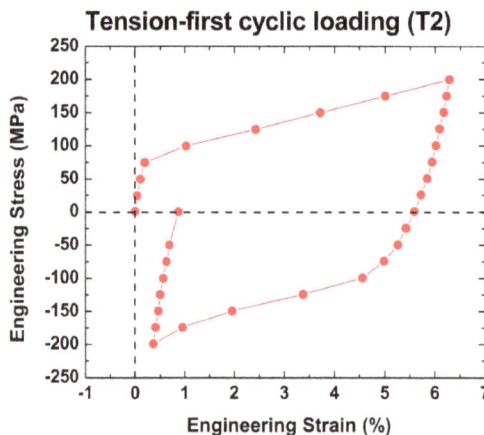

Figure 5. Macroscopic stress-strain response for the T2 sample. The symbols correspond to points in the loading history at which diffraction data were acquired.

Lattice strain and integrated intensity variations for four different grain orientations as a function of applied stress are shown in Figure 6 for the T2 sample subjected to tension followed by reverse compression.

During initial loading in tension, the lattice strain varies linearly with applied stress for all four grain orientations up to ~50 MPa showing the same slope, as expected based on the near elastic isotropy of magnesium. Beyond ~50 MPa, the {10.2} and {10.3} strains increase more slowly with stress, indicating that these grains undergo plastic deformation, most likely by basal <a> slip (Figure 6a), which is the easiest slip system to activate in Mg-Al alloys, and for which these grain families are favorably oriented. Beyond ~75 MPa, the lattice strain behavior of the {00.2} family deviates from its initial linear behavior, corresponding to a decrease in intensity for the {00.2} peak in Figure 6b, and macroscopic yielding in Figure 5. The {00.2} family, which represents a large fraction of the microstructure, has the basal pole aligned 0° relative to the applied loading direction. It is thus favorably oriented for {10.2} extension twinning in tension, which results in the observed decrease in intensity. Likewise, the changes in the {11.0} and {10.3} intensities for $75 < \sigma$ (MPa) < 200 are due to {10.2} extension twinning in the {10.3} families, as the two families are twin-related.

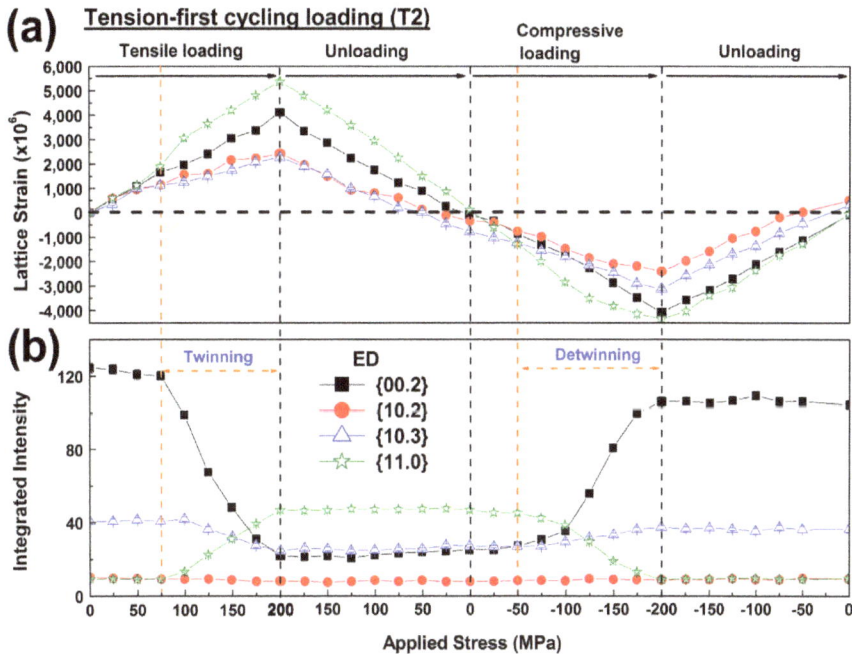

Figure 6. (a) Lattice strain evolution and (b) intensity variations for various grain families as functions of applied stress during the deformation shown in Figure 5. The {hk.l} plane normal is parallel to the applied loading direction, which is parallel to the extrusion direction (ED).

Figure 6a shows that the {10.2}, {10.3} and {00.2} families, which deform plastically by either basal slip or extension twinning during the initial tensile portion of the test, undergo stress relaxation, resulting in compressive residual strains upon unloading. On the other hand, the {11.0} family, which is unfavorably oriented for both extension twinning and basal slip, accumulates much higher lattice strains during tensile loading, resulting in slightly tensile residual strains at the end of unloading. In the compression stage of the test, the {10.2} and {10.3} families, which are favorably oriented for basal slip, accumulate relatively low levels of lattice strain during loading, resulting in small tensile residual strains at the end of the test. Conversely, the residual strain in the hard {00.2} family is slightly compressive, or zero. The {11.0} family presents an interesting case—despite being favorably oriented for extension twinning in compression, it still accumulates large lattice strains during loading, resulting in slightly negative (or zero) residual lattice strain at the end of the test. This lack of a significant relaxation effect associated with twinning is consistent with the behavior of the {10.0} family during the initial compressive loading step in T1, and is in marked contrast to the behavior of the {00.2} family in both textures, in which strong relaxation effects are associated with twinning. This difference in behavior between the {10.0}/{11.0} and {00.2} orientations may be due to the fact that extension twinning is activated by compression normal to the c-axis in the {10.0} and {11.0} family, but by tension parallel to the c-axis in the {00.2} family.

During unloading after tension, the intensities of all four reflections change little (the {00.2} and {10.3} intensities increase slightly), indicating that the twinned material generated during tension does not undergo significant detwinning. During reverse loading in compression, however, the twinned volume within the {00.2} family, which contributes to the {10.0} signal, is favorably oriented to detwin in compression. Thus, in reverse compression, detwinning starts at an applied stress of ~−50 MPa and continues up to an applied stress of −200 MPa, as revealed by the increase in the {00.2} intensity (Figure 6b). However, the {00.2} intensity at the start of the test is not fully recovered at −200 MPa. During unloading after the compressive portion of the test, the {00.2} peak intensity does not change, suggesting that the non-linear behavior observed in Figure 5 is not related to the behavior of deformation twins, but rather to dislocation phenomena. This behavior is likely due to the fact that almost all traces of twinning are gone when the highest compressive stress is reached, such that little material is favorably oriented for either twinning or detwinning when the final unloading starts.

4. Conclusions

In situ neutron diffraction has been used to investigate the plastic deformation behavior of Mg-8.5wt.%Al alloy subjected to compression followed by reverse tension (texture T1) and tension followed by reverse compression (texture T2). For texture T1, the onset of extension twinning corresponds well with the macroscopic elastic-plastic transition during the initial compression stage. The non-linear macroscopic stress/strain behavior during unloading after compression is more significant than during unloading after tension. For texture T2, little detwinning occurs after the initial tension stage, but almost all of the twinned volumes are recovered during loading in reverse compression. The development of residual strains in the various grain orientations during unloading after compression is strongly influenced by plastic anisotropy at the grain level. The residual strains observed in each grain family can be largely rationalized based on the ease of activation of basal slip and extension twinning.

Acknowledgments: This work was supported by the National Research Foundation of Korea (NRF) grant funded by the Korean government (MSIP) (No.2013R1A4A1069528, No.2013R1A1A1076023).

Author Contributions: S.Y.L. wrote the initial draft of the manuscript; S.Y.L., H.W., M.A.G. analyzed the data; All co-authors contributed to the interpretation of the data.

Conflicts of Interest: The authors declare no conflict of interest.

References

1. Avedesian, M.M.; Baker, H. *Magnesium and Magnesium Alloys*; ASM Specialty Handbook; ASM International: Materials Park, OH, USA, 1999.
2. Wang, C.; Han, P.; Zhang, L.; Zhang, C.; Yan, X.; Xu, B.S. The strengthening effect of Al atoms into Mg-Al alloy: A first-principles study. *J. Alloy. Comp.* **2009**, *482*, 540–543. [CrossRef]
3. Roberts, C.S. *Magnesium and Its Alloys*; John Wiley & Sons, Inc.: New York, NY, USA, 1960.

4. Muránsky, O.; Carr, D.G.; Barnett, M.R.; Oliver, E.C.; Šittner, P. Investigation of deformation mechanisms involved in the plasticity of AZ31 Mg alloy: *In situ* neutron diffraction and EPSC modelling. *Mater. Sci. Eng. A* **2008**, *496*, 14–24. [CrossRef]

5. Partridge, P.G. The crystallography and deformation modes of hexagonal close-packed metals. *Metall. Rev.* **1967**, *12*, 169–194. [CrossRef]

6. Agnew, S.R.; Tomé, C.N.; Brown, D.W.; Holden, T.M.; Vogel, S.C. Study of slip mechanisms in a magnesium alloy by neutron diffraction and modeling. *Scripta Mater.* **2003**, *48*, 1003–1008. [CrossRef]

7. Jain, J.; Poole, W.J.; Sinclair, C.W.; Gharghouri, M.A. Reducing the tension-compression yield asymmetry in a Mg-8Al-0.5Zn alloy via precipitation. *Scripta Mater.* **2010**, *62*, 301–304. [CrossRef]

8. Hong, S.G.; Park, S.H.; Lee, C.S. Role of {10−12} twinning characteristics in the deformation behavior of a polycrystalline magnesium alloy. *Acta Mater.* **2010**, *58*, 5873–5885. [CrossRef]

9. Clausen, B.; Tomé, C.N.; Brown, D.W.; Agnew, S.R. Reorientation and stress relaxation due to twinning: Modeling and experimental characterization for Mg. *Acta Mater.* **2008**, *56*, 2456–2468. [CrossRef]

10. Proust, G.; Tomé, C.N.; Jain, A.; Agnew, S.R. Modeling the effect of twinning and detwinning during strain-path changes of magnesium alloy AZ31. *Int. J. Plasticity* **2009**, *25*, 861–880. [CrossRef]

11. Muránsky, O.; Barnett, M.R.; Luzin, V.; Vogel, S. On the correlation between deformation twinning and Luders-like deformation in an extruded Mg alloy: *In situ* neutron diffraction and EPSC.4 modelling. *Mater. Sci. Eng. A* **2008**, *527*, 1383–1394. [CrossRef]

12. Park, S.H.; Hong, S.G.; Lee, C.S. In-plane anisotropic deformation behavior of rolled Mg-3Al-1Zn alloy by initial {10−12} twins. *Mater. Sci. Eng. A* **2013**, *570*, 149–163. [CrossRef]

13. Gharghouri, M.A.; Weatherly, G.C.; Embury, J.D.; Root, J. Study of the mechanical properties of Mg-7.7at.% Al by *in situ* neutron diffraction. *Phil. Mag. A* **1999**, *79*, 1671–1695. [CrossRef]

14. Brown, D.W.; Agnew, S.R.; Bourke, M.A.M.; Holden, T.M.; Vogel, S.C.; Tomé, C.N. Internal strain and texture evolution during deformation twinning in magnesium. *Mater. Sci. Eng. A* **2005**, *399*, 1–12. [CrossRef]

15. Wu, L.; Agnew, S.R.; Brown, D.W.; Stoica, G.M.; Clausen, B.; Jain, A.; Fielden, D.E.; Liaw, P.K. Internal stress relaxation and load redistribution during the twinning-detwinning-dominated cyclic deformation of a wrought magnesium alloy, ZK60A. *Acta Mater.* **2008**, *56*, 3699–3707. [CrossRef]

16. Muránsky, O.; Barnett, M.R.; Carr, D.G.; Vogel, S.C.; Oliver, E.C. Investigation of deformation twinning in a fine-grained and coarse-grained ZM20 Mg alloy: Combined in situ neutron diffraction and acoustic emission. *Acta Mater.* **2010**, *58*, 1503–1517. [CrossRef]

17. Lee, S.Y.; Gharghouri, M.A.; Root, J.H. Plastic deformation of magnesium alloy subjected to compression-first cyclic loading. In Proceeding of the 140th TMS Annual Meeting & Exhibition, San Diego, CA, USA, 27 February–3 March 2011; pp. 595–598.

18. Lee, S.Y.; Gharghouri, M.A. Pseudoelastic behavior of magnesium alloy during twinning-dominated cyclic deformation. *Mater. Sci. Eng. A* **2013**, *572*, 98–102. [CrossRef]

19. Lee, S.Y.; Wang, H.; Gharghouri, M.A.; Nayyeri, G.; Woo, W.; Shin, E.; Wu, P.D.; Poole, W.J.; Wu, W.; An, K. Deformation behavior of solid-solution-strengthened Mg-9 wt.% Al alloy: *In situ* neutron diffraction and elastic-viscoplastic self-consistent modeling. *Acta Mater.* **2014**, *73*, 139–148. [CrossRef]

metals

MDPI

Article

Probing Interfaces in Metals Using Neutron Reflectometry

Michael J. Demkowicz [1,2,]* and **Jaroslaw Majewski [3,4,]***

[1]   Department of Materials Science and Engineering, Massachusetts Institute of Technology, Cambridge, MA 02139, USA
[2]   Materials Science and Engineering, Texas A & M University, College Station, TX 77843, USA
[3]   MPA-CINT/Los Alamos Neutron Scattering Center, Los Alamos National Laboratory, Los Alamos, NM 87545, USA
[4]   Department of Chemical Engineering, University of California at Davis, Davis, CA 95616, USA
*   Correspondence: demkowicz@tamu.edu (M.J.D.); jarek@lanl.gov (J.M.);
    Tel.: +1-979-458-9845 (M.J.D.); +1-505-667-8840 (J.M.); Fax: +1-505-665-2676 (J.M.)

Academic Editor: Klaus-Dieter Liss
Received: 25 November 2015; Accepted: 21 December 2015; Published: 20 January 2016

Abstract: Solid-state interfaces play a major role in a variety of material properties. They are especially important in determining the behavior of nano-structured materials, such as metallic multilayers. However, interface structure and properties remain poorly understood, in part because the experimental toolbox for characterizing them is limited. Neutron reflectometry (NR) offers unique opportunities for studying interfaces in metals due to the high penetration depth of neutrons and the non-monotonic dependence of their scattering cross-sections on atomic numbers. We review the basic physics of NR and outline the advantages that this method offers for investigating interface behavior in metals, especially under extreme environments. We then present several example NR studies to illustrate these advantages and discuss avenues for expanding the use of NR within the metals community.

Keywords: neutron reactor; spallation source; metals; extreme conditions

1. Interfaces in Metals

Metals form a wide variety of interfaces, including grain and phase boundaries [1], surface-liquid interfaces [2,3], solidification fronts [4], and mechanical contacts [5]. Although they typically occupy a small fraction of the total volume, interfaces play an outsized role in determining the properties of metals [6–9]. Understanding interfaces is therefore critical to predicting and controlling the behavior of metals.

Experimental investigation of interfaces presents significant challenges. Because they are often buried within the material, accessing them frequently requires destructive characterization or sample preparation methods, such as transmission electron microscopy (TEM) [10] or atom probe tomography (APT) [11]. Interfaces in metals typically have low thickness; indeed, some are atomically sharp [12]. Thus, characterizing them requires high—sometimes Å-level—spatial resolution. Moreover, certain interfaces only exist at high temperatures and pressures [13–15] or under contact with external media, such as gases or liquids [2,16]. Investigating such interfaces requires special *in situ* characterization methods.

An expanded experimental toolbox promises to accelerate progress in understanding metal interfaces, especially in extreme environments. This paper offers a primer on neutron reflectometry (NR): a characterization method with several advantages for studying metal interfaces [17–21]. NR is a mature experimental tool. The first NR experiments were conducted by Fermi and Zinn [22] and

Fermi and Marshall [23]. The technique has experienced continuous improvement since then [17,24–26]. Nevertheless, use of this method within the metals community has been relatively limited. We illustrate the potential benefits of NR to investigations of interfaces in metals by explaining the physics of the NR experiment and by presenting several example studies.

2. The Physics of Neutron Reflectometry (NR)

Figure 1 shows a schematic of a typical NR measurement. The sample is a thin, planar film on a substrate. The experiment is usually conducted in air or vacuum, but may also be carried it out in other media (e.g., see Section 6.4) [27]. The neutron source may be a fission nuclear reactor or a spallation source. In the nuclear reactor, the sustained nuclear fission of ^{235}U- or ^{239}Pu-rich fuels immersed in H_2O, D_2O, or solid graphite produces neutrons that may be used for scattering experiments. Spallation sources usually utilize pulsed high-energy (~GeV) protons to bombard targets made of heavy elements (such as W, Hg, U) to extract neutrons [28,29].

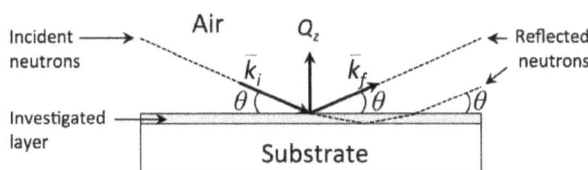

Figure 1. Schematic of a neutron reflectometry measurement.

In material property studies, the neutrons extracted from nuclear reactors or spallation sources are moderated to decrease their energies and therefore increase their wavelengths to Å ranges. Such moderation, depending on the final wavelength of neutrons required, is usually achieved by passing high-energy neutrons through H_2O, liquid H_2, or solid methane. Low energy neutrons are typically detected indirectly through absorption reactions with materials of high cross-sections for such reactions. Typically, ^{3}He, ^{6}Li, or ^{10}B is used to emit high-energy particles, whose ionization signatures may be detected by a number of means.

Upon interacting with the sample at a particular angle of incidence, θ (or a particular value of the neutron momentum transfer vector, Q_z), the incoming neutron beam can undergo absorption, reflection, transmission, or refraction. Consequently, there is a difference between the intensity of the outgoing, specularly reflected neutron beam and that of the incident beam. This difference—measured as a function of Q_z—encodes information about the distribution of the nuclear scattering length density along the direction normal to the sample surface. Moreover, the neutron is a ½-spin fermion and possesses a magnetic moment oppositely oriented to the spin. Therefore, its interaction with matter may depend on the sample's spin or magnetic field. Neutrons interact both with nuclear spins and the magnetic moments of unpaired electrons via dipole-dipole processes. Interactions with unpaired electrons may be of similar magnitude as nuclear scattering. However, they are not inherently isotropic. Rather, they depend on the orientation of the sample's magnetization *vis-a-vis* the direction of the neutron momentum wavevector transfer Q_z: only the component of sample's magnetization which is perpendicular to Q_z affects the neutron scattering. Therefore, the intensity of specularly reflected neutrons measured as a function of Q_z also encodes information about the distribution of the magnetization in the sample as a function of depth [30].

Depending on the specific NR technique, NR can take advantage of a range of different neutron-sample interactions [31]. However, this short review focuses on elastic specular NR, which is by far the most widely used NR technique. Elastic scattering conserves energy. Thus, we exclude any energy-dissipating neutron-matter interactions, except neutron absorption. In specular NR, the detector is positioned so as to measure outgoing neutrons at the same angle of incidence as the incoming neutrons, as illustrated in Figure 1. The experiment measures reflectivity, R, defined as the ratio of the number of reflected neutrons to the number of incoming neutrons.

The de Broglie expression, $\lambda = \dfrac{h}{m_n v}$, relates the neutron's wavelength λ with its momentum, $p = m_n v$ (where h is Planck's constant and $m_n = 1.6749 \times 10^{-27}$ kg is the neutron mass). Based on this formula, some simple relationships between the wavelength λ (Å), energy E (meV), and speed v (m/s) of the neutron can be developed: $E = 81.89/\lambda^2$ and $v = 3960/\lambda$. Thus, for example, a neutron with de Broglie wavelength of 1.5 Å has an energy $E = 36.4$ meV and velocity $v = 2640$ m/s. NR utilizes neutrons with wavelengths from sub-Å to tens of Å. By contrast, the size of an atomic nucleus is on the order of ~10 fm (1 fm = 10^{-15} m = 10^{-5} Å). Thus, to an excellent approximation, the incoming neutrons may be thought of as waves interacting with a uniform medium whose properties are determined by the density and type of atoms it contains. Their behavior may be described using Schrödinger's equation.

The assumption of elastic and specular conditions greatly simplifies the analysis the NR measurement. We further assume that the scattering properties of the sample vary only in one direction—namely, along the sample's normal—and therefore the components of the neutron wavevector parallel to the sample surface are not affected. Under these conditions, the component of the neutron wavevector parallel to the sample surface is conserved and the magnitude of the outgoing wavevector, $\vec{k}_f$, equals that of the incoming wavevector, $\vec{k}_i$: $\left|\vec{k}_i\right| = \left|\vec{k}_f\right| = \dfrac{2\pi}{\lambda}$. The difference between them, $\vec{Q}_z = \vec{k}_f - \vec{k}_i$, is known as the "momentum wavevector transfer" and lies perpendicular to the sample surface. From the geometry of the measurement (Figure 1), we calculate $\left|\vec{Q}_z\right| = Q_z = \dfrac{4\pi\sin(\theta)}{\lambda}$.

Quantum mechanics describes the incoming and outgoing neutron beams as a wavefunction, ψ, consisting of a superposition of plane waves:

$$\psi\left(\vec{x}\right) = e^{i\vec{k}_i \cdot \vec{x}} + re^{i\vec{k}_f \cdot \vec{x}} \tag{1}$$

Here, r is the amplitude of the outgoing (reflected) wave, normalized by the amplitude of the incoming wave (taken as unity). Knowing r, we may calculate reflectivity as $R = |r|^2$. To compute r, however, we must model the interaction of the incoming neutron beam with the sample and substrate.

Because the component of the wavevector parallel to the sample surface is conserved, we may rewrite ψ as solely a function of z—the distance perpendicular to the sample surface—and $k_i^\perp$—the component of $\vec{k}_i$ in the z-direction:

$$\psi(z) = e^{ik_i^\perp z} + re^{-ik_i^\perp z} \tag{2}$$

where we have used $k_f^\perp = -k_i^\perp$. Indeed, the entire NR measurement may be analyzed as a one-dimensional problem in the z-direction [32]. We write down wavefunctions of the type shown in Equation (2) for every distinct layer of material in the experiment, including air (or any other external medium), every layer of material in the sample, and the substrate (though in the substrate there is no reflected wave).

In free space, the neutron has kinetic energy $E_k = \dfrac{1}{2}m_n v^2$ and zero potential energy (if gravity is neglected). By contrast, within a material, it has a potential energy given by the Fermi pseudopotential [33]:

$$V_{\text{Fermi}} = \dfrac{h^2 \beta}{2\pi m_n} \tag{3}$$

where β is the nuclear scattering length density (SLD). β describes the mean neutron scattering effectiveness of the material, which depends on the number of nuclei of type i per unit volume, N_i, and the coherent neutron scattering length, b_i:

$$\beta = \sum_i N_i b_i \qquad (4)$$

b_i may be complex with its imaginary component describing absorption. Its real part may be either positive or negative, depending on the isotope [34,35]. Since neutrons have spin, their interactions with magnetic materials require an extended description that tracks changes in spin polarization [30]. The interaction of a neutron's spin with atomic magnetism (or other source of magnetic induction) can lead to magnetic scattering length density distributions that may be of the same order of magnitude as the nuclear scattering length densities. However, depending on the neutron's spin orientation (spin-up or spin-down) *vis-à-vis* the magnetic field of reference, the magnetic component is either added or subtracted from the nuclear one.

Often, samples investigated by NR may be described as a stack of discrete layers: each with its own composition, density, and thickness. In such cases, the NR experiment may be described with a 1D, time-independent Schrödinger equation on a piece-wise linear potential [19,26,32]. By matching wave functions and their derivatives at the interfaces between successive layers (as well as in the surrounding medium and in the substrate), one may solve for the amplitudes of all the waves in the setup. In particular, r—the amplitude of the outgoing wave measured at the sample surface—may be found. This amplitude comes about by coherent interference of partial waves on all interfaces in the film. From it, we determine the quantity measured by the neutron detector in Figure 1: reflectivity, R.

Thus, the SLD and thickness of the sample and substrate determine R. Since SLD in turn depends on composition and density, R is an indirect measure of these characteristics as well as the thickness of the individual material layers in the sample and substrate [17,26]. The goal of the reflectivity experiment is to measure $R(Q_z)$ and then infer $\beta(z)$ by fitting a model of the SLD distribution to the data. Q_z may be varied by changing the angle of incidence, θ (if the neutron beam is monochromatic, *i.e.*, λ = const.), or by changing the neutron wavelength, λ. The latter method of varying Q_z is typical of NR measurements at facilities where different neutron wavelengths, λ, are distinguished by the time of flight method. Figure 2 shows calculated reflectivity curves corresponding to a 500 Å Ni film on a quartz substrate in air. To illustrate the sensitivity of NR to isotopic composition, the calculation is carried out for two different isotopes: [58]Ni and [62]Ni.

Figure 2 illustrates some of the common features of reflectivity curves. Whenever the energy of the neutron is at or below the potential of the substrate (*i.e.*, whenever $k_i^2 \leqslant 4\pi\beta_{substrate}$), the neutrons are totally reflected from the surface. The onset of total reflection is called the critical edge and the value of Q_z at that point is referred to as $Q_{critical}$. The fringes in Figure 2 arise from interference between waves reflected from the top surface and the buried interface between the substrate and the layer. For this simple case, the spacing of the fringes may be calculated analytically: $d_{fringe} = \dfrac{2\pi}{t_{layer}}$, where t_{layer} is the thickness of the Ni layer. The amplitude of the fringes relates to the contrast between the layer and the substrate. The overall falloff of the curve obeys the Fresnel law: $R \sim Q_z^{-4}$. Most interfaces are not discontinuous, but rather graded due to chemical mixing or surface roughness. The surface roughness (which can be characterized by the root mean square displacement from the average interface, σ) may also be obtained from the reflectivity curve [36,37]. In general, the falloff of $R(Q_z)$ for rough or diffuse interfaces is even faster than that given by Fresnel's law.

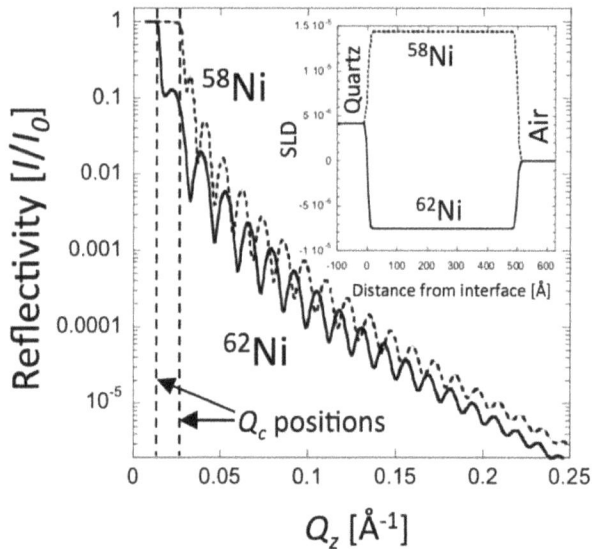

Figure 2. Calculated neutron reflectivity curves for 500 Å films of ^{58}Ni (dashed line) and ^{62}Ni (solid line) on a quartz substrate. The inset shows the SLD distributions for each isotope. The simulation was performed using MOTOFIT assuming RMS roughness parameters of 5 Å and the experimental resolution $\Delta Q_z / Q_z = 3\%$.

3. Interpreting Reflectivity Curves

$R(Q_z)$ contains information about the SLD distribution perpendicular to the sample surface, $\beta(z)$. Inferring $\beta(z)$ from $R(Q_z)$, however, is not trivial [38]. Because the NR measurement only collects the intensity of the reflected beam and not its phase, there is no unique mathematical transformation from $R(Q_z)$ to $\beta(z)$. Therefore, NR data is usually interpreted by iteratively adjusting a trial SLD distribution, $\overline{\beta}(z)$, until the reflectivity it predicts, $\overline{R}(Q_z)$, matches the measured reflectivity, $R(Q_z)$, to within a specified tolerance.

The continuous function $\beta(z)$ may often be approximated by a series of discrete layers—referred to as "boxes" or "slabs"—each with a constant SLD. Inter-layer roughness may be taken into account using an error function centered at each interface [36] or any other relevant functional form. A theoretical NR curve, $\overline{R}(Q_z)$, may be calculated from a trial SLD distribution, $\overline{\beta}(z)$, using the Parratt recursion formula [39,40], which relates the amplitudes of the reflected and transmitted waves at each interface. A number of approaches have been developed for adjusting $\overline{\beta}(z)$ to minimize the difference between $\overline{R}(Q_z)$ and $R(Q_z)$. One example is the Levenburg-Marquardt nonlinear least-squares method used in open-source reflectivity package, MOTOFIT, which runs in the IGOR Pro environment [41]. This method seeks the least-squares fit of reflectivities, corresponding to a minimum χ^2 value. SLD models with the least number of boxes are usually preferred as they involve the smallest number of fitting parameters.

Once a best-fit set of model parameters is achieved, the uncertainties of these parameters may also be quantified by measuring the increase in χ^2 that comes about from perturbing each individual fitting parameter. For example, Reference [42] defines $\tilde{\chi}^2$ as the deviation of the reflectivity calculated using the perturbed parameter values from the best-fit reflectivity:

$$\tilde{\chi}^2 = \sum_{i=1}^{N} \left(\frac{y_i^{bf} - y_i^{p}}{y_i^{bf}} \right) \tag{5}$$

Here, y_i^{bf} is the best-fit to the measured reflectivity, y_i^p is the reflectivity value obtained by perturbing one parameter of the structural model, and N is the number of data points. The uncertainties on the fitting parameters are then defined as bounds within which $\tilde{\chi}^2$ is 5% or less.

Equation (4) shows that, at any given z, $\beta(z)$ depends on the number and types of isotopes in the sample as well as their volume density. Thus, $\beta(z)$ provides information concerning the composition and density at a depth z. Inferring these quantities from $\beta(z)$ typically requires prior knowledge of some of the variables (e.g., nominal compositions or densities) or further input from other characterization methods. It should be re-emphasized that the SLD density profiles obtained from the fitting procedures described above are not unique. Due to the fact that only the intensities of scattered neutrons are measured in the NR experiment, but not their amplitudes and phases, there is no unique mathematical transformation leading from $R(Q_z)$ to SLD profile. Therefore, to resolve this problem, other data (e.g., from complementary characterization techniques) are often needed. In some cases, the phase of scattered neutrons may be resolved, as described by the work of Majkrzak and Berk [43–45] and others [46–49], enabling better inferences of SLD profiles.

4. Advantages of NR

Neutron reflectometry offers unique advantages for characterizing solid-state interfaces in metals, including in extreme environments. Some of these advantages are easily deduced by considering the dependence of coherent scattering length, b_i, on atomic number, Z. Figure 3 shows that b_i is rather weakly dependent on Z. Indeed, the scattering lengths of almost all elements (in their natural isotopic abundance) are of the same order of magnitude. Therefore, in general, no one element can dominate the scattering of a multi-component sample, drowning out the contributions of other elements. In particular, light elements—such as H/D or He—may be detected, even when embedded in a matrix of heavy elements, e.g., of actinides. Moreover, NR is often able to distinguish elements with small differences in atomic number.

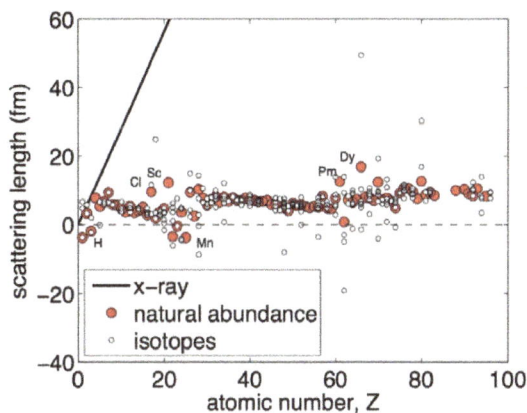

Figure 3. Neutron coherent scattering length, b_i, as a function of atomic number, Z. The X-ray scattering length is computed using Equation (6).

Figure 3 also shows that there are marked differences in b_i between different isotopes of certain elements. Thus, NR is well suited to investigations that require tracking of isotopes, such as tracer diffusion studies. Isotopic substitution may also aid in the interpretation of $\beta(z)$ profiles. For example, Ni has five different isotopes. The b_i for ^{58}Ni (abundance 68.3%) is 14.4×10^{-5} Å, whereas for ^{62}Ni (abundance 3.6%) b_i is negative and equals -8.7×10^{-5} Å. Substitution of ^{58}Ni (bulk SLD = 13.3×10^{-6} Å^{-2}) for ^{62}Ni (SLD = -7.5×10^{-6} Å^{-2}) significantly changes NR curves, as illustrated in Figure 2. Moreover, because neutron beams may be spin polarized, NR is especially well suited to investigations of magnetic properties of materials [50,51].

To appreciate the above-mentioned qualities, it is useful to compare NR with X-ray reflectometry (XRR) [52–54]. Although the mathematical description of XRR is similar to NR, the underlying physics of scattering is different: X-rays scatter from atoms' electrons while neutrons scatter off of atoms' nuclei. To interpret XRR experiments, we can replace the neutron coherent scattering lengths, b_i, in Equation (4) with the X-ray scattering lengths:

$$b_Z^{\text{X-ray}} = Z \cdot (2.81 fm) \tag{6}$$

where Z is the atomic number and 2.81 *fm* is the classical radius of the electron. Because X-rays scatter primarily from electrons, $b_Z^{\text{X-ray}}$ is directly proportional to the number of electrons per atom, *i.e.*, to the atomic number, Z, in charge-neutral materials. Consequently, XRR cannot detect differences between isotopes. Moreover, because it is a linearly increasing function of Z, $b_Z^{\text{X-ray}}$ provides very little contrast between elements with small Z differences. Finally, light elements—especially when embedded in high-Z matrices—are essentially undetectable with XRR, as the X-ray scattering length of the latter dominates reflectivity curves. However, it is often advantageous to use XRR and NR in tandem, as they may provide complementary information.

Equation (4) shows that scattering length density, β, is not only a function of b_i, but also of N_i: the number density of isotope i. Thus, depth profiles obtained by NR (and XRR) are sensitive not only to composition, but also to density. By contrast, depth profiles obtained through Rutherford backscattering are not sensitive to density [55]. NR is therefore capable of characterizing the evolution of porosity and of detecting displacive phase transformations that involve changes in density.

NR is also remarkable for its depth resolution, which is much greater than for XRR, especially for high-Z materials. Usually, NR techniques enable investigations of structures with total thickness up to ~3000 Å. For such thickness, the spacing between the scattering fringes (Figure 2) is very small, requiring very high $\Delta Q_z/Q_z$ resolution in the neutron detection. Typical $\Delta Q_z/Q_z$ values for existing neutron reflectometers are in the range from 2% to 5%. An instrumental resolution of $\Delta Q_z/Q_z = 2\%$ will result in the ability to distinguish between two thickness values which differ by 2%. Therefore differences of the film thickness on the order of Å can be readily detected. Smaller $\Delta Q_z/Q_z$ values may be achieved using detectors with higher spatial resolutions, better beam collimation, or better discrimination of the neutron wavelengths. However, increasing the resolution may result in smaller incident beam intensities, which can lead to longer measurements times and therefore higher scattering background. Thus, a proper balance between the two must be found. In general, it is advisable to adjust the resolution to match the expected thickness of the investigated films. Thick films, which give rise to dense oscillations of the interference fringes in $R(Q_z)$, require higher resolutions. By contrast, thin ones with broad interference oscillations in $R(Q_z)$ can be measured with lower resolution and therefore higher intensities, which can result in shorter measurement times and higher Q_z^{max} values.

Another advantage of NR is due to the ability of neutrons to penetrate deeply into solid matter. Several *mm* thick aluminum, quartz, silicon, or stainless steel windows absorb only a small fraction of incident neutrons with wavelengths in the Å range. Thus, NR measurements may be carried out to investigate the structure of buried interfaces as well as samples immersed in liquids or shielded from their environments by neutron-transparent containers. This quality is especially useful for investigating materials exposed to volatile media or under high pressure. Finally, NR frequently requires straightforward sample preparation and is not destructive. Therefore, samples investigated with NR may be subsequently further analyzed using other characterization methods. However, certain materials may be activated through interactions with the neutron beam, requiring some time for the radioactivity to decay before further characterization may be performed. It is also important to note that, since NR data are normalized to the incident neutron intensity, the measured SLD values are absolute.

5. Practical Considerations

Because neutron sources are inherently week (fluxes of ~10^{6-7} n/s/cm^2), the samples used for NR must be large. Samples with an area as small as 1 cm^2 may be measured, but at the expense of longer time of data acquisition and increased background noise. Neutron spallation sources usually provide some advantages enabling faster NR measurements. This is due to the polychromatic nature of the pulsed neutron beams they generate and the time of flight method used to discriminate between different neutron wavelengths, λ. NR spectra within a limited Q_z^{max} range (<0.1 Å^{-1}) can be obtained in 5–10 min. However, to obtain a full-spectrum NR data set (usually with $Q_z^{max} \approx 0.3$ Å^{-1} and $R \approx 10^{-6}$) requires from one to several hours of measurement time, regardless which neutron source is used.

Existing NR beamlines usually provide point-, line-, or 2-D neutron scattering detectors. Line- and 2-D detectors enable recording of scattering signals beyond the specular reflection: the so-called "off-specular" reflections. The "off-specular" data provides the neutron intensity distribution as function of the components of the neutron momentum transfer vector parallel to the sample's surface. This information can provide additional insight to extend the interpretation of the specular reflectivity measurements regarding in-plane correlations of the samples studied [56]. For example, these data allow correlations between the roughness of different interfaces or the growth of in-plane islands to be addressed. Reflectivities at high Q_z values are of great interest as they allow access to shorter length scales, which are important for characterizing the detailed structure of the investigated films. However, as already mentioned, the reflectivity R rapidly decays as Q_z^{-4}, making it difficult to acquire data at high Q_z. This challenge may be mitigated to some extent through the preparation of high quality samples: by minimizing the roughness of the sample surface as well as the roughness of its internal interfaces, high quality NR data at high Q_z and low R values may be collected. RMS roughness parameters up to 20 Å are usually tolerable, but detailed NR investigations typically require RMS roughness below 5 Å. Samples with such low roughness are most conveniently prepared using vapor deposition techniques. For such samples, $R \approx 10^{-6}$ and $Q_z^{max} \approx 0.2$–0.3 Å^{-1} can be routinely achieved for sample areas of several cm^2.

At the time of writing, there are several world-class NR instruments available worldwide, e.g., at the Spallation Neutron Source at Oak Ridge National Laboratory, the Lujan Center at Los Alamos National Laboratory, NCNR at NIST, the Institute Laue-Langevin in France, J-PARC in Japan, ANSTO in Australia, FRM-II in Germany, and several others. Several neutron sources are currently under construction or discussion. For example, the European Spallation Source in Sweden and the Second Target Station at SNS/ORNL will provide excellent capabilities for NR.

6. Example Applications of NR to Metals

This section provides examples of NR measurements conducted on metals. The examples are chosen to illustrate the unique advantages of the NR, namely its ability to detect density changes (Section 6.1), its sensitivity to magnetic moments and complementarity to X-ray reflectometry (XRR, Section 6.2), its sensitivity to light elements (Section 6.3), and its ability to penetrate through container walls (Section 6.4).

6.1. He in fcc/bcc Composites: Detecting Density Changes

Some nuclear transmutation reactions give rise to alpha particles, *i.e.*, nuclei of ^{4}He. When implanted into solids, these particles rapidly come to rest, pick up two electrons, and become regular He atoms. Since He is a noble gas and does not bond with surrounding atoms, it usually has negligible solubility within solids [57]. Thus, it precipitates out of solution into nanometer-scale bubbles [58]. These precipitates are usually deleterious to the properties of the solid, e.g., they lead to embrittlement in Ni-base alloys [59] and surface damage in plasma-facing materials [60]. Much effort has been invested into mitigating damage induced by implanted He, especially in materials for nuclear energy [61–64].

One way of controlling implanted He is to trap it at specially designed internal interfaces in composite materials [61]. However, investigations of this effect are limited by the difficulty of characterizing He precipitates at internal interfaces. NR (and XRR) provides a distinct advantage within this context: its sensitivity to local density changes enables detection of the onset of He precipitate formation [42,65,66].

Kashinath *et al.* investigated He precipitation at interfaces between copper (Cu) and one of three body-centered cubic (bcc) metals: niobium (Nb), vanadium (V), and molybdenum (Mo) [42]. They found that each of these interfaces has a distinct critical He dose at which precipitates begin to form. Figure 4 illustrates the findings of this study. Upper and lower bounds on best-fit SLD profiles were estimated by superimposing the upper and lower error bounds for each individual fitting parameter, as defined in Section 3. All SLD profiles with $\tilde{\chi}^2$ less than or equal to 5% are contained within these bounds, but the converse is not true: not all SLD profiles within these bounds have $\tilde{\chi}^2$ less than or equal to 5%. Therefore, these uncertainty estimates for best-fit SLD profiles are conservative.

The target is a Cu/Nb bilayer deposited on a Si substrate. Both the Cu and the Nb layer are approximately 20 nm thick. After implantation of 20 keV ^{4}He$^+$ ions to a dose of 3×10^{16}/cm^2, the reflectivity of the sample is consistent with an unaltered Cu/Nb bilayer structure, as shown in Figure 4a. However, upon implantation to a slightly higher He dose of 4×10^{16}/cm^2, there is a clear change in the reflectivity, indicated by arrows in Figure 4a,b. This change may be explained by the formation of a layer of reduced density on the Cu side of the Cu–Nb interface, as illustrated in Figure 4b.

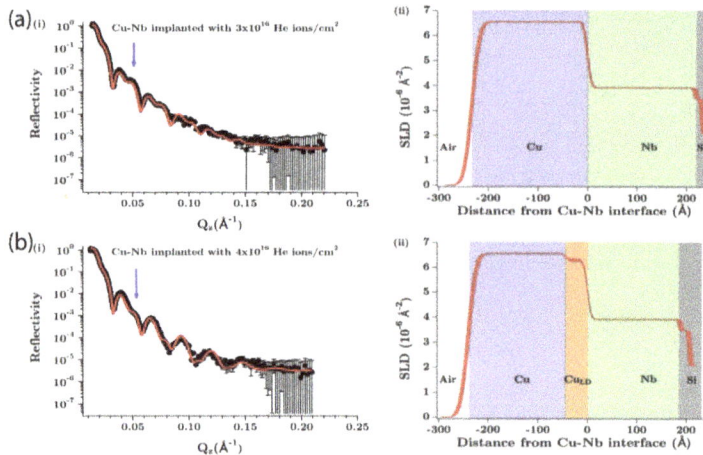

Figure 4. Reflectivity curves (**left** column) and SLD profiles (**right** column) for (**a**) 3×10^{16}/cm^2 and (**b**) 3×10^{16}/cm^2 He ions implanted in a Cu/Nb bilayer on a Si substrate. Reprinted with permission from Reference [42]. Copyright (2013), AIP publishing LLC.

At the He doses used in this study, nearly all the implanted He is believed to either escape through the Cu free surface or become trapped at the Cu–Nb or Nb–Si interfaces [42]. Precipitation within the Cu or Nb layers themselves is thought to be minimal. Thus, the low-density layer adjacent to the Cu–Nb interface in Figure 4b is thought to arise from the formation of He precipitates there. The critical He dose of 4×10^{16}/cm^2 is consistent with preceding transmission electron microscopy (TEM) studies [61,67] as well as atomistic simulations [68]. However, whereas those previous investigations merely inferred interfacial precipitation, NR is able to observe it directly.

6.2. Fe/Y₂O₃ Interface: Sensitivity to Magnetization and Complementary to XRR

The structure of interfaces between low solubility metals—such as those discussed in the previous section—is easy to describe, as these interfaces are usually atomically sharp [12]. By contrast, the structure of oxide/oxide or metal/oxide interfaces is much more difficult to assess. Such interfaces are often several nanometers wide [69], exhibit transitions in structure reminiscent of phase changes [8], and contain intrinsic defects with distinct local compositions [70]. NR provides several advantages for investigating such interfaces, including high depth resolution and sensitivity to composition.

Watkins *et al.* used NR to study the structure of an interface between α-Fe and Y_2O_3 [71]. They found that this interface is a ~64 Å-thick transitional zone containing mixtures or compounds of Fe, Y, and O. By comparing their NR data to XRR and X-ray diffraction (XRD) measurements, they further determined that the interface was likely compositionally sharp upon synthesis and only later broadened as the neighboring crystals reacted. Finally, since α-Fe is ferromagnetic while Y_2O_3 is not, Watkins *et al.* were able to track changes in magnetization across this interface. Figure 5 shows that to model the reflectivity of this interface, contributions of spin-up and spin-down states of the neutron beam must be averaged. By using comparing the SLD profiles of these two states, the exact depth at which the ferromagnetic ordering is lost may be found (marked with an "x" in the right panel in Figure 5).

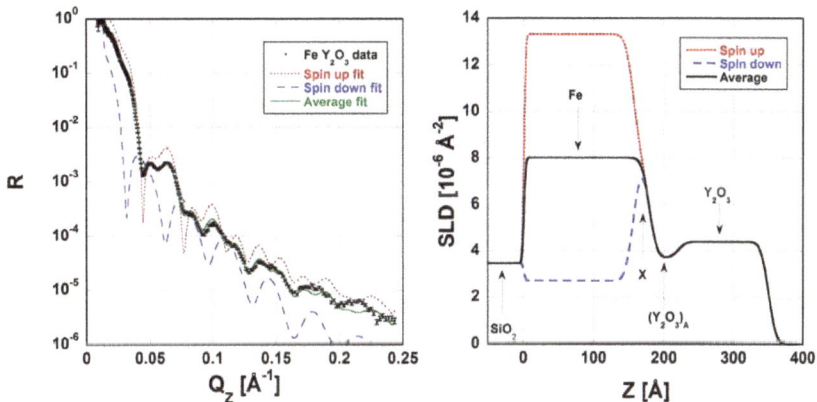

Figure 5. The neutron reflectivity (**left**) of the α-Fe and Y_2O_3 interface investigated by Watkins *et al.* is the average of SLD contributions from spin-up and spin-down states of the neutron beam (**right**). The depth at which ferromagnetic ordering is lost is marked with an "x" in the **right** panel. Also in the **right** panel, $(Y_2O_3)_A$ refers to a distinctive Y–O layer forming at this metal/oxide interface. Reprinted with permission from Reference [71]. Copyright (2014), AIP publishing LLC.

6.3. Actinides: Sensitivity to Light Elements

Actinides and their oxides exhibit some of the most intriguing and challenging chemistry known [72]. Frequently, the composition of these materials is not stoichiometrically precise. Moreover, their oxide structures can change dramatically under different environmental conditions. Neutrons provide a distinct advantage over X-rays in structural characterization of hydrides and oxides of heavy metals because they are better able to detect the lighter elements, such as H/D and O, within their actinide matrices. Figure 6 illustrates neutron scattering length densities for different uranium oxide phases, showing that NR is able to distinguish between them.

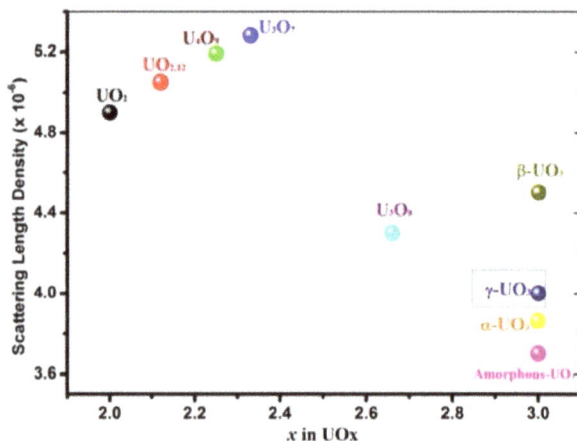

Figure 6. The calculated values of the nuclear scattering length density (SLD) for some common phases in uranium-oxide system. The calculations are based on specific densities published in literature [73,74].

The work of He *et al.* illustrates the utility of NR for investigating imperfect uranium oxide films [75]. They deposited uranium oxide on silicon substrates (with a thin native layer of oxide) using a combination of DC magnetron and reactive sputtering. In this technique, U atoms generated from a solid target by sputtering are readily oxidized by residual O_2 present in the Ar/O_2 mixture under moderate vacuum (approximately 1–3×10^{-4} torr) and then deposited on the substrate above the target. Several steps were taken to ensure high film quality: a multi-step sequential reactive deposition was used to minimize preferential film growth, the substrate was rotated to even out source distribution anomalies, and the partial pressures of Ar and O_2 were adjusted to control the composition of the uranium oxide. Nevertheless, the resulting film has a non-uniform, depth-dependent stoichiometry and structure.

The NR data for these films along with the best-fit curve according to the real-space SLD profile are shown in Figure 7a. According to these results, the total thickness of the UO_x film is about 630 Å. Figure 7b illustrates schematically the real-space structure represented by the best-fit SLD profile. The simplest model that fits the NR data has a three-layer structure. There is no heteroepitaxial growth of uranium oxide on the substrate/film interface due to the (~10 Å) native amorphous Si oxide layer on top of silicon wafer. The SLD of the layer at the film/air interface (~5.0×10^{-6} Å^{-2}) suggests the presence of hyper-stoichiometric phases. Meanwhile, the SLD of middle layer of the film (~3.8×10^{-6} Å^{-2}) together with the fact that no sharp X-ray diffraction peaks were observable (data not shown) indicates that this layer consists of amorphous α-UO_3. Overall, NR demonstrates a remarkably rich variation in structure and stoichiometry in this nominally uniform sample.

Another example of the utility of NR for studies of heavy metal (lanthanide) oxides arises from recent work on Dysprosium (Dy) oxidation [76]. They deposited Dy films on silicon substrates using the same DC magnetron sputtering technique as discussed above and characterized their structures using NR after exposure to air at two different temperatures: 25 °C (ambient temperature) and 150 °C. Figure 8 shows that, under both conditions, the film may be described three-slab model. Under ambient temperature, it consists of 20 Å silicon oxide on top of the Si substrate, 418 Å Dy, and 43 Å Dy_2O_3. After exposure to air in 150 °C for ~0.5 h, the thickness of the Dy_2O_3 increased to 114 Å while simultaneously the thickness of the Dy layer decreased to 363 Å. The total thickness of Dy and Dy_2O_3 layers increased from 461 Å to 477 Å, indicating an overall swelling of the sample. The roughness parameters of the air-Dy_2O_3 and Dy_2O_3-Dy interfaces decreased, making the top surface facing the air and the interface between the metallic Dy and its oxide smoother. For the two cases of uranium oxide and dysprosium studies described above, the approximate errors for the thickness, SLD, and roughness parameters vales were ± 5 Å, 0.1 Å^{-2}, and 2 Å, respectively.

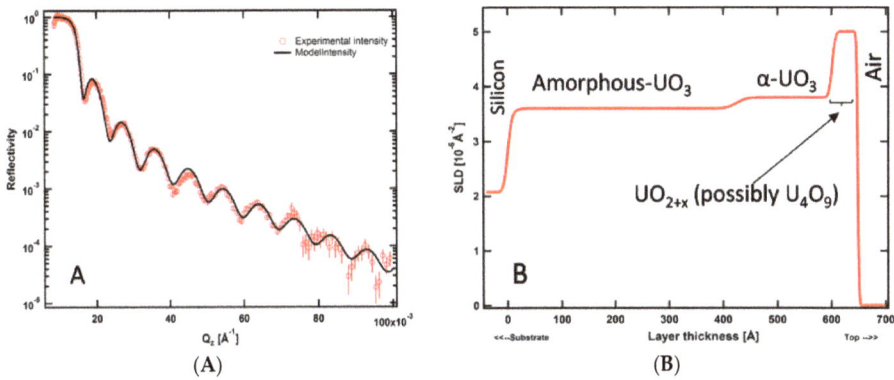

(A)   (B)

Figure 7. (**A**) NR data obtained from a UO$_x$ deposited on a silicon substrate (open circles). Error bars indicate one standard deviation. The solid line through the data points corresponds to the best-fit SLD profile shown in (**B**).

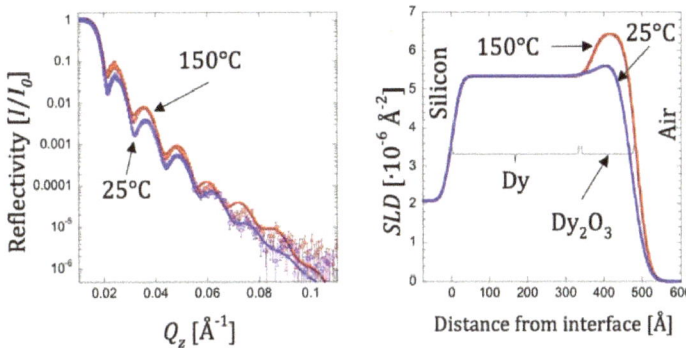

Figure 8. (**Left**) NR data (open circles) from a film of Dy covered with capping layer of Dy$_2$O$_3$ at ambient temperature (25 °C) and after expose to air at 150 °C for ~0.5 h. Error bars in the NR measurements correspond to one standard deviation. The solid line through the data points corresponds to the best-fit SLD profile shown on the (**right**).

6.4. Surfaces in Pressurized Liquids: Penetration of Neutrons through Containers

Neutrons penetrate through thick sections of solid matter with low attenuation. Thus, they are able to "see through" the walls of a high-pressure cell, enabling examination of metal surfaces in pressurized media [77]. Junghans *et al.* used this capability to study the corrosion of oxidized aluminum (Al) surfaces in pressurized seawater [27]. The corrosion of structural materials in the deep sea depends on numerous chemical and physical factors, including pH, dissolved oxygen, chloride ion activity, salinity, ocean currents, temperature, and hydrostatic pressure [78–82]. Al and its alloys find widespread use in marine environments, including in civil and defense vessels, offshore rigs, drill pipes for deep wells, and diving suits. There are several reports on corrosion of Al and its alloys at shallow depths, but investigations at high pressures are limited [78–80,82–84].

When in contact with air, Al develops a thin passivation layer of Al$_2$O$_3$. The surface oxide has a higher SLD than pure Al, providing neutron scattering contrast between the two materials (Al: 2.08×10^{-6} Å^{-2}; crystalline Al$_2$O$_3$: 5.74×10^{-6} Å^{-2}). Junghans *et al.* deposited uniform, ~900 Å-thick Al films on monocrystalline quartz wafers using DC magnetron sputtering and collected a total of 14 NR spectra over the course of 50 h (~3 h per spectrum). The film was in contact with 3.5 wt. % NaCl solution at pressures ranging from 1 to 600 atm in a specially developed solid/liquid, high pressure/temperature cell [77]. This cell provides the capabilities of solid/fluid interface investigations up to 2000 atm (~30,000 psi) and 200 °C. The cell's simple aluminum construction makes it easy to operate at high pressures and elevated temperatures, while the 13 mm thick neutron windows allow up to 74% neutron transmission. Figure 9 shows five representative NR measurements from this study [27].

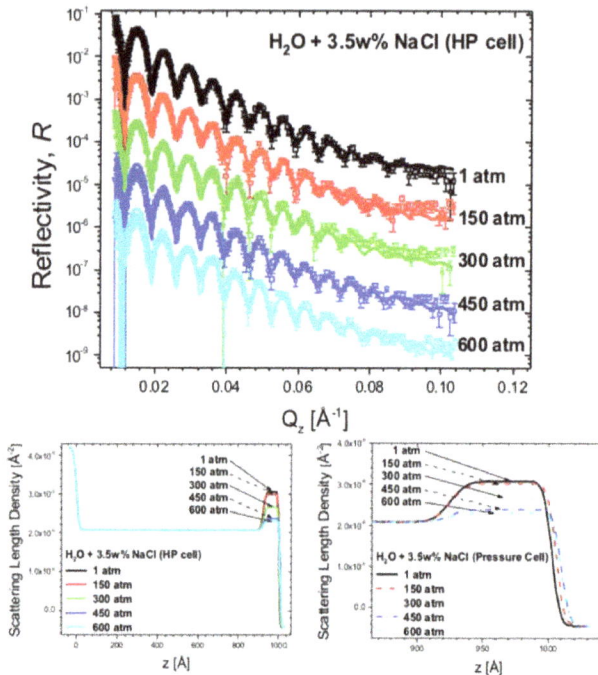

Figure 9. (**Top** panel) Five NR measurements of Al film deposited on quartz substrate and investigated at 25 °C in contact with H$_2$O + 3.5 wt. % of NaCl at different pressures. Solid curves are fits corresponding to the SLD profiles shown in the middle and right panels. Both the NR data and fits are offset by a decade along *y*-axis for clarity; (**Bottom left** panel) SLD distribution of the Al/Al$_2$O$_3$/liquid system and (**Bottom right** panel) magnified SLD distribution in the contact region. In both of the bottom panels, *z* = 0 at the quartz substrate/Al interface.

The NR results show virtually no corrosion of the Al layers. The observed decrease in the SLD of the starting Al$_2$O$_3$ passivation layer cannot be explained by the formation of new chemical compounds by the highly scattering Na, Cl, and O ions. However, this decrease is consistent with formation of stable Al–Cl–H$_2$O (or Al–O–Cl–H$_2$O) complexes or hydration of Al$_2$O$_3$ to Al(OH)$_3$. These results suggest that for the time scale of 50 h the influence of hydrostatic pressure only slightly influences interactions of the Al oxide film with Cl$^-$ ions and H$_2$O. The corrosion rate is lower than reported by Beccaria *et al.* [78–80], suggesting slower kinetics for the reactions involved.

7. Conclusions

Neutron reflectometry (NR) is a mature experimental technique that has been used extensively in condensed-matter physics. However, its potential for investigating interfaces in metals has not yet been utilized widely. The present overview is intended to raise awareness of NR in the metals community with the hope of motivating wider use of this technique. NR provides several unique advantages for investigating interfaces in metals:

- **It is non-destructive.** Thus, NR results may be combined with other, follow-on investigations, e.g., using XRR, TEM, or APT.
- **Å-level depth resolution** enables detailed investigation of interface structure: thickness, SLD, and roughness of the layers.
- **Sensitivity to composition, isotopic distribution, density, and magnetic moment** allow multiple physical characteristics to be measured simultaneously.
- **Ability to detect low-Z elements**, such as H/D, He, and other light isotopes.
- **Suitability for *in situ* studies** due to the high penetrability of neutrons through container walls and surrounding media. This capability enables investigations of a variety of buried interfaces, including solid-liquid ones, which are otherwise very difficult to access with X-rays.
- **The measured SLD values are absolute** due to the fact that the reflected beam is normalized by the incident intensity of the neutron beam.
- **Ready access** thanks to the availability of several suitable neutron sources (reactors and spallation facilities) worldwide.

Several example applications of NR to metal surfaces and buried interfaces have been discussed above. As the metals community continues to explore the structure and properties of interfaces in ever-greater detail [7–9], NR stands poised to contribute valuable new additions to this ever-growing list of examples.

Acknowledgments: J.M. thanks Heming He, Kirk Rector, Peng Wang, Ann Junghans, Erik Watkins (LANL), and David Allred (BYU). M.J.D. acknowledges Abishek Kashinath and support from the Center for Materials in Irradiation and Mechanical Extremes (CMIME), an Energy Frontier Research Center funded by the U.S. Department of Energy, Office of Science, Office of Basic Energy Sciences under Award No. 2008LANL1026. This work benefited from the use of the Lujan Neutron Scattering Center at LANSCE funded by the DOE Office of Basic Energy Sciences and Los Alamos National Laboratory under DOE Contract DE-AC52-06NA25396.

Author Contributions: Both authors contributed equally to the composition of this article.

Conflicts of Interest: The authors declare no conflict of interest.

References

1. Sutton, A.P.; Balluffi, R.W. *Interfaces in Crystalline Materials*; Oxford University Press: Oxford, UK, 1995.
2. Furukawa, S.; de Feyter, S. Two-dimensional crystal engineering at the liquid-solid interface. In *Templates in Chemistry III*; Broekmann, P., Dotz, K.H., Schalley, C.A., Eds.; Springer-Verlag Berlin: Berlin, Germany, 2009; Volume 287, pp. 87–133.
3. Kaplan, W.D.; Kauffmann, Y. Structural order in liquids induced by interfaces with crystals. *Ann. Rev. Mater. Res.* **2006**, *36*, 1–48. [CrossRef]
4. Boettinger, W.J.; Warren, J.A.; Beckermann, C.; Karma, A. Phase-field simulation of solidification. *Ann. Rev. Mater. Res.* **2002**, *32*, 163–194. [CrossRef]
5. Stachowiak, G.W.; Batchelor, A.W. *Engineering Tribology*; Butterworth-Heinemann: Boston, MA, USA, 2000.
6. Hirth, J.P.; Pond, R.C.; Hoagland, R.G.; Liu, X.Y.; Wang, J. Interface defects, reference spaces and the frank-bilby equation. *Prog. Mater. Sci.* **2013**, *58*, 749–823. [CrossRef]
7. Beyerlein, I.J.; Demkowicz, M.J.; Misra, A.; Uberuaga, B.P. Defect-interface interactions. *Prog. Mater. Sci.* **2015**, *74*, 125–210. [CrossRef]
8. Cantwell, P.R.; Tang, M.; Dillon, S.J.; Luo, J.; Rohrer, G.S.; Harmer, M.P. Grain boundary complexions. *Acta Mater.* **2014**, *62*, 1–48. [CrossRef]

9.   Mishin, Y.; Asta, M.; Li, J. Atomistic modeling of interfaces and their impact on microstructure and properties. *Acta Mater.* **2010**, *58*, 1117–1151. [CrossRef]

10.   Colliex, C.; Bocher, L.; de la Pena, F.; Gloter, A.; March, K.; Walls, M. Atomic-scale stem-eels mapping across functional interfaces. *JOM* **2010**, *62*, 53–57. [CrossRef]

11.   Schmitz, G.; Ene, C.; Galinski, H.; Schlesiger, R.; Stender, P. Nanoanalysis of interfacial chemistry. *JOM* **2010**, *62*, 58–63. [CrossRef]

12.   Mitchell, T.E.; Lu, Y.C.; Griffin, A.J.; Nastasi, M.; Kung, H. Structure and mechanical properties of copper/niobium multilayers. *J. Am. Ceram. Soc.* **1997**, *80*, 1673–1676. [CrossRef]

13.   Shaha, K.P.; Pei, Y.T.; Chen, C.Q.; Turkin, A.A.; Vainshtein, D.I.; de Hosson, J.T.M. On the dynamic roughening transition in nanocomposite film growth. *Appl. Phys. Lett.* **2009**. [CrossRef]

14.   Olmsted, D.L.; Foiles, S.M.; Holm, E.A. Grain boundary interface roughening transition and its effect on grain boundary mobility for non-faceting boundaries. *Scr. Mater.* **2007**, *57*, 1161–1164. [CrossRef]

15.   Bellon, P. Nonequilibrium roughening and faceting of interfaces in driven alloys. *Phys. Rev. Lett.* **1998**, *81*, 4176–4179. [CrossRef]

16.   Wiesler, D.G.; Majkrzak, C.F. Neutron reflectometry studies of surface oxidation. *Phys. B* **1994**, *198*, 181–186. [CrossRef]

17.   Russell, T.P. X-ray and neutron reflectivity for the investigation of polymers. *Mater. Sci. Rep.* **1990**, *5*, 171–271. [CrossRef]

18.   Dietrich, S.; Haase, A. Scattering of X-rays and neutrons at interfaces. *Phys. Rep. Rev. Sec. Phys. Lett.* **1995**, *260*, 1–138. [CrossRef]

19.   Zhou, X.L.; Chen, S.H. Theoretical foundation of X-ray and neutron reflectometry. *Phys. Rep. Rev. Sec. Phys. Lett.* **1995**, *257*, 223–348. [CrossRef]

20.   Penfold, J.; Thomas, R.K. The application of the specular reflection of neutrons to the study of surfaces and interfaces. *J. Phys. Condes. Matter* **1990**, *2*, 1369–1412. [CrossRef]

21.   Van der Lee, A. Grazing incidence specular reflectivity: Theory, experiment, and applications. *Solid State Sci.* **2000**, *2*, 257–278. [CrossRef]

22.   Fermi, E.; Zinn, W.H. Reflection of neutrons on mirrors. *Phys. Rev.* **1946**, *70*, 103.

23.   Fermi, E.; Marshall, L. Interference phenomena of slow neutrons. *Phys. Rev.* **1947**, *71*, 666–677. [CrossRef]

24.   Lekner, J. Theory of reflection of electromagnetic and particle waves. In *Developments in Electromagnetic Theory and Applications 3*; Springer Netherlands: Dordrecht, The Netherlands, 1987.

25.   Russell, T.P. The characterization of polymer interfaces. *Annu. Rev. Mater. Sci.* **1991**, *21*, 249–268. [CrossRef]

26.   Smith, G.S.; Majkrzak, C.F. Neutron reflectometry. In *International Tables for Crystallography*, 1st ed.; International Union of Crystallography, Ed.; Springer: Chester, UK; New York, NY, USA, 2006.

27.   Junghans, A.; Chellappa, R.; Wang, P.; Majewski, J.; Luciano, G.; Marcelli, R.; Proietti, E. Neutron reflectometry studies of aluminum-saline water interface under hydrostatic pressure. *Corros. Sci.* **2015**, *90*, 101–106. [CrossRef]

28.   Carpenter, J.M. Pulsed spallation neutron sources for slow-neutron scattering. *Nucl. Instrum. Methods* **1977**, *145*, 91–113. [CrossRef]

29.   Bauer, G.S. Physics and technology of spallation neutron sources. *Nucl. Instrum. Methods Phys. Res. Sect. A Accel. Spectrom. Dect. Assoc. Equip.* **2001**, *463*, 505–543. [CrossRef]

30.   Hughes, D.J.; Burgy, M.T. Reflection of neutrons from magnetized mirrors. *Phys. Rev.* **1951**, *81*, 498–506. [CrossRef]

31.   Lauter, V.; Ambaye, H.; Goyette, R.; Lee, W.T.H.; Parizzi, A. Highlights from the magnetism reflectometer at the SNS. *Phys. B* **2009**, *404*, 2543–2546. [CrossRef]

32.   Merzbacher, E. *Quantum Mechanics*; Wiley: New York, NY, USA, 1961; p. 544.

33.   Sears, V.F. Neutron scattering lengths and cross sections. *Neutron News* **1992**, *3*, 29–37. [CrossRef]

34.   Lovesey, S.W. *Theory of Neutron Scattering from Condensed Matter*; Clarendon Press: Oxfordshire, UK, 1984.

35.   Neutron SLDs. Available online: http://www.ncnr.nist.gov/resources/n-lengths/ (accessed on 23 December 2015).

36.   Nevot, L.; Croce, P. Characterization of surfaces by grazing X-ray reflection—Application to study of polishing of some silicate-glasses. *Rev. Phys. Appl.* **1980**, *15*, 761–779.

37.   Als-Nielsen, J.; McMorrow, D. *Elements of Modern X-ray Physics*, 2nd ed.; Hoboken, N.J., Ed.; Wiley: Hoboken, NJ, USA, 2011.

38. Lovell, M.R.; Richardson, R.M. Analysis methods in neutron and X-ray reflectometry. *Curr. Opin. Colloid Interface Sci.* **1999**, *4*, 197–204. [CrossRef]
39. Parratt, L.G. Surface studies of solids by total reflection of X-rays. *Phys. Rev.* **1954**, *95*, 359–369. [CrossRef]
40. Yasaka, M. X-ray thin film measurement techniques. *Rigaku J.* **2010**, *26*, 1–9.
41. Nelson, A. Co-refinement of multiple-contrast neutron/X-ray reflectivity data using motofit. *J. Appl. Crystallogr.* **2006**, *39*, 273–276. [CrossRef]
42. Kashinath, A.; Wang, P.; Majewski, J.; Baldwin, J.K.; Wang, Y.Q.; Demkowicz, M.J. Detection of helium bubble formation at fcc-bcc interfaces using neutron reflectometry. *J. Appl. Phys.* **2013**. [CrossRef]
43. Majkrzak, C.F.; Berk, N.F. Exact determination of the phase in neutron reflectometry. *Phys. Rev. B* **1995**, *52*, 10827–10830. [CrossRef]
44. Berk, N.F.; Majkrzak, C.F. Inverting specular neutron reflectivity from symmetric, compactly supported potentials. In Proceedings of the International Symposium on Advance in Neutron Optics and Related Research Facilities, Kumatori, Osaka, Japan, 19–21 March 1996; p. 107.
45. Majkrzak, C.F.; Berk, N.F. Exact determination of the phase in neutron reflectometry by variation of the surrounding media. *Phys. Rev. B* **1998**, *58*, 15416–15418. [CrossRef]
46. Dehaan, V.O.; Vanwell, A.A.; Adenwalla, S.; Felcher, G.P. Retrieval of phase information in neutron reflectometry. *Phys. Rev. B* **1995**, *52*, 10831–10833. [CrossRef]
47. Kasper, J.; Leeb, H.; Lipperheide, R. Phase determination in spin-polarized neutron specular reflection. *Phys. Rev. Lett.* **1998**, *80*, 2614–2617. [CrossRef]
48. Lipperheide, R.; Kasper, J.; Leeb, H. Surface profiles from polarization measurements in neutron reflectometry. *Phys. B* **1998**, *248*, 366–371. [CrossRef]
49. Leeb, H.; Grotz, H.; Kasper, J.; Lipperheide, R. Complete determination of the reflection coefficient in neutron specular reflection by absorptive nonmagnetic media. *Phys. Rev. B* **2001**. [CrossRef]
50. Majkrzak, C.F. Neutron scattering studies of magnetic thin films and multilayers. *Phys. B* **1996**, *221*, 342–356. [CrossRef]
51. Williams, W.G. *Polarized Neutrons*; Clarendon Press: Oxford, UK; New York, NY, USA; Oxford University Press: Oxford, UK; New York, NY, USA, 1988.
52. Renaud, G.; Lazzari, R.; Leroy, F. Probing surface and interface morphology with grazing incidence small angle X-ray scattering. *Surf. Sci. Rep.* **2009**, *64*, 255–380. [CrossRef]
53. Sinha, S.K. Reflectivity using neutrons or X-rays—A critical comparison. *Phys. B* **1991**, *173*, 25–34. [CrossRef]
54. Stoev, K.N.; Sakurai, K. Review on grazing incidence X-ray spectrometry and reflectometry. *Spectroc. Acta B Atom. Spectr.* **1999**, *54*, 41–82. [CrossRef]
55. Feldman, L.C.; Mayer, J.W. *Fundamentals of Surface and Thin Film Analysis*; North-Holland: New York, NY, USA, 1986.
56. Jablin, M.S.; Zhernenkov, M.; Toperverg, B.P.; Dubey, M.; Smith, H.L.; Vidyasagar, A.; Toomey, R.; Hurd, A.J.; Majewski, J. In-plane correlations in a polymer-supported lipid membrane measured by off-specular neutron scattering. *Phys. Rev. Lett.* **2011**. [CrossRef] [PubMed]
57. Laakmann, J.; Jung, P.; Uelhoff, W. Solubility of helium in gold. *Acta Metall.* **1987**, *35*, 2063–2069. [CrossRef]
58. Trinkaus, H.; Singh, B.N. Helium accumulation in metals during irradiation—Where do we stand? *J. Nucl. Mater.* **2003**, *323*, 229–242. [CrossRef]
59. Judge, C.D.; Gauquelin, N.; Walters, L.; Wright, M.; Cole, J.I.; Madden, J.; Botton, G.A.; Griffiths, M. Intergranular fracture in irradiated inconel X-750 containing very high concentrations of helium and hydrogen. *J. Nucl. Mater.* **2015**, *457*, 165–172. [CrossRef]
60. Baldwin, M.J.; Doerner, R.P. Helium induced nanoscopic morphology on tungsten under fusion relevant plasma conditions. *Nucl. Fusion* **2008**. [CrossRef]
61. Demkowicz, M.J.; Misra, A.; Caro, A. The role of interface structure in controlling high helium concentrations. *Curr. Opin. Solid State Mat. Sci.* **2012**, *16*, 101–108. [CrossRef]
62. Misra, A.; Demkowicz, M.J.; Zhang, X.; Hoagland, R.G. The radiation damage tolerance of ultra-high strength nanolayered composites. *JOM* **2007**, *59*, 62–65. [CrossRef]
63. Odette, G.R.; Miao, P.; Edwards, D.J.; Yamamoto, T.; Kurtz, R.J.; Tanigawa, H. Helium transport, fate and management in nanostructured ferritic alloys: *In situ* helium implanter studies. *J. Nucl. Mater.* **2011**, *417*, 1001–1004. [CrossRef]

64. Odette, G.R.; Hoelzer, D.T. Irradiation-tolerant nanostructured ferritic alloys: Transforming helium from a liability to an asset. *JOM* **2010**, *62*, 84–92. [CrossRef]

65. Zhernenkov, M.; Gill, S.; Stanic, V.; DiMasi, E.; Kisslinger, K.; Baldwin, J.K.; Misra, A.; Demkowicz, M.J.; Ecker, L. Design of radiation resistant metallic multilayers for advanced nuclear systems. *Appl. Phys. Lett.* **2014**. [CrossRef]

66. Zhernenkov, M.; Jablin, M.S.; Misra, A.; Nastasi, M.; Wang, Y.-Q.; Demkowicz, M.J.; Baldwin, J.K.; Majewski, J. Trapping of implanted he at Cu/Nb interfaces measured by neutron reflectometry. *Appl. Phys. Lett.* **2011**. [CrossRef]

67. Demkowicz, M.J.; Bhattacharyya, D.; Usov, I.; Wang, Y.Q.; Nastasi, M.; Misra, A. The effect of excess atomic volume on he bubble formation at fcc-bcc interfaces. *Appl. Phys. Lett.* **2010**. [CrossRef]

68. Kashinath, A.; Misra, A.; Demkowicz, M.J. Stable storage of helium in nanoscale platelets at semicoherent interfaces. *Phys. Rev. Lett.* **2013**. [CrossRef] [PubMed]

69. Kaplan, W.D.; Chatain, D.; Wynblatt, P.; Carter, W.C. A review of wetting *versus* adsorption, complexions, and related phenomena: The rosetta stone of wetting. *J. Mater. Sci.* **2013**, *48*, 5681–5717. [CrossRef]

70. Dholabhai, P.P.; Pilania, G.; Aguiar, J.A.; Misra, A.; Uberuaga, B.P. Termination chemistry-driven dislocation structure at SrTiO$_3$/MgO heterointerfaces. *Nat. Commun.* **2014**. [CrossRef] [PubMed]

71. Watkins, E.B.; Kashinath, A.; Wang, P.; Baldwin, J.K.; Majewski, J.; Demkowicz, M.J. Characterization of a Fe/Y$_2$O$_3$ metal/oxide interface using neutron and X-ray scattering. *Appl. Phys. Lett.* **2014**. [CrossRef]

72. Morss, L.R.; Edelstein, N.M.; Fuger, J.; Katz, J.J. *The Chemistry of the Actinide and Transactinide Elements*, 3rd ed.; Springer: Dordrecht, The Netherlands, 2006.

73. Hoekstra, H.R.; Siegel, S. The uranium-oxygen system—U$_3$O$_8$–UO$_3$. *J. Inorg. Nucl. Chem.* **1961**, *18*, 154–165. [CrossRef]

74. Loopstra, B.O.; Cordfunk, E. On structure of α-UO$_3$. *Recl. Trav. Chim. PaysBas* **1966**, *85*, 135–142. [CrossRef]

75. He, H.M.; Wang, P.; Allred, D.D.; Majewski, J.; Wilkerson, M.P.; Rector, K.D. Characterization of chemical speciation in ultrathin uranium oxide layered films. *Anal. Chem.* **2012**, *84*, 10380–10387. [CrossRef] [PubMed]

76. Watkins, E.B.; Scott, B.; Allred, D.D.; Majewski, J. Unpublished work, 2015.

77. Wang, P.; Lerner, A.H.; Taylor, M.; Baldwin, J.K.; Grubbs, R.K.; Majewski, J.; Hickmott, D.D. High-pressure and high-temperature neutron reflectometry cell for solid-fluid interface studies. *Eur. Phys. J. Plus* **2012**. [CrossRef]

78. Beccaria, A.M.; Poggi, G. Influence of hydrostatic-pressure on pitting of aluminum in sea-water. *Br. Corros. J.* **1985**, *20*, 183–186. [CrossRef]

79. Beccaria, A.M.; Poggi, G. Effect of some surface treatments on kinetics of aluminum corrosion in NaCl solutions at various hydrostatic pressures. *Br. Corros. J.* **1986**, *21*, 19–22. [CrossRef]

80. Beccaria, A.M.; Fiordiponti, P.; Mattogno, G. The effect of hydrostatic-pressure on the corrosion of nickel in slightly alkaline-solutions containing C1$^-$ ions. *Corros. Sci.* **1989**, *29*, 403–413. [CrossRef]

81. Heusler, K.E. Untersuchungen der korrosion von aluminium in wasser bei hohen temperaturen und drucken. *Mater. Corros.* **1967**, *18*, 11–15. [CrossRef]

82. Dexter, S.C. Effect of variations in sea-water upon the corrosion of aluminum. *Corrosion* **1980**, *36*, 423–432. [CrossRef]

83. Venkatesan, R.; Venkatasamy, M.A.; Bhaskaran, T.A.; Dwarakadasa, E.S.; Ravindran, M. Corrosion of ferrous alloys in deep sea environments. *Br. Corros. J.* **2002**, *37*, 257–266. [CrossRef]

84. Sawant, S.S.; Wagh, A.B. Corrosion behaviour of metals and alloys the waters of the arabian sea. *Corros. Prev. Control* **1990**, *37*, 154–157.

metals

MDPI

Article

Hydrogen Absorption in Metal Thin Films and Heterostructures Investigated *in Situ* with Neutron and X-ray Scattering

Sara J. Callori [1], Christine Rehm [2,*], Grace L. Causer [2,3], Mikhail Kostylev [4] and Frank Klose [2,5]

[1] Department of Physics, California State University, San Bernardino, CA 92407, USA; sara.callori@csusb.edu
[2] Australian Nuclear Science and Technology Organisation, Lucas Heights, NSW 2234, Australia;
     grace.causer@ansto.gov.au (G.L.C.); frank.klose@ansto.gov.au (F.K.)
[3] Institute for Superconducting and Electronic Materials, University of Wollongong,
     Wollongong, NSW 2522, Australia
[4] School of Physics, The University of Western Australia, Crawley, WA 6009, Australia;
     mikhail.kostylev@uwa.edu.au
[5] Department of Physics and Materials Science, The City University of Hong Kong, Hong Kong
*   Correspondence: christine.rehm@ansto.gov.au; Tel.: +61-2-9717-9649

Academic Editor: Hugo F. Lopez
Received: 16 April 2016; Accepted: 17 May 2016; Published: 24 May 2016

Abstract: Due to hydrogen possessing a relatively large neutron scattering length, hydrogen absorption and desorption behaviors in metal thin films can straightforwardly be investigated by neutron reflectometry. However, to further elucidate the chemical structure of the hydrogen absorbing materials, complementary techniques such as high resolution X-ray reflectometry and diffraction remain important too. Examples of work on such systems include Nb- and Pd-based multilayers, where Nb and Pd both have strong affinity to hydrogen. W/Nb and Fe/Nb multilayers were measured *in situ* with unpolarized and polarized neutron reflectometry under hydrogen gas charging conditions. The gas-pressure/hydrogen-concentration dependence, the hydrogen-induced macroscopic film swelling as well as the increase in crystal lattice plane distances of the films were determined. Ferromagnetic-Co/Pd multilayers were studied with polarized neutron reflectometry and *in situ* ferromagnetic resonance measurements to understand the effect of hydrogen absorption on the magnetic properties of the system. This electronic effect enables a novel approach for hydrogen sensing using a magnetic readout scheme.

Keywords: hydrogen absorption; neutron reflectometry; polarized neutron reflectometry; thin films; multilayers

---

1. Introduction: Hydrogen Absorption in Metals—Bulk *versus* Thin Films

Thin films and layers of metals with chemical affinity to hydrogen such as Nb, Pd, Pt, V, Mg, Ti, the rare earth metals and related alloys play an important role in many technological and engineering materials and devices. A prime area of research is chemical energy storage in the vast field of metal hydrides [1–3]. Here, the thin-film approach is of particular interest as it allows studying hydrogen-metal interactions on the nanoscale, which may show distinct novel features as well as non-equilibrium states of hydride materials that would not exist in bulk form [4–6].

Generally, the absorbed hydrogen occupies interstitial sites in the host crystal lattice. As a result, the host atoms are displaced from their regular lattice positions and the associated strain/stress fields will give rise to altered physical properties. Metal hydrides can show disordered and ordered arrangements of hydrogen sublattices, depending on the hydrogen concentration and temperature.

Formation of dislocations due to lattice coherency stresses in multi-phase regions exceeding the critical yield stress is another important reason for modified material properties [7].

As in the case of bulk materials showing the ability to absorb large quantities of hydrogen, the interaction of hydrogen with thin films can lead to significant modification of their electronic, magnetic, and structural properties [8]. Exciting results have been achieved in the past. Among these are Y and La thin films that reversibly switch their optical properties upon hydrogen absorption [9], and Fe/Nb [10] or Fe/V [11] multilayers that reversibly switch their magnetic coupling and magnetoresistivity during hydrogen charging/discharging, and fine-tuning of the spin-density-wave state in Cr/V heterostructures via hydrogen uptake [12].

In some systems, such as Fe/V [13] and Pd/Nb [14], giant lattice expansions caused by hydrogen absorption have been reported. A more recent area of thin-film research is hydrogen sensors based on gas absorption [15–17], of which there is a growing demand due to their importance in fuel-cell developments and hydrogen-powered cars. While many approaches to this challenge have been investigated, there are still obstacles in the development of higher quality sensors, for example cross-sensitivity to other gas species [18].

In some systems, hydrogen uptake can cause physical properties of the metal to deteriorate. An example is Nb-based superconducting radio-frequency cavities of particle accelerators, where metal-hydride surface layers can develop, leading to undesirable extra surface resistivity [19]. For a theoretical understanding of the remarkable changes of thin-film properties due to hydrogen charging, it is important to know how hydrogen is incorporated into the films, whether thermodynamic phase diagrams of the corresponding bulk materials are applicable, and how hydrogen absorption and its kinetics depend on various external parameters such as temperature and hydrogen pressure.

Much research has been performed on understanding the bulk metal-hydrogen systems and their temperature/concentration phase diagrams [20]. For the case of Nb, early X-ray diffraction work identified two structural phases at moderate temperatures: a gas-like phase α and a liquid-like phase α', associated with low and high hydrogen concentrations, respectively, both with disordered hydrogen distributions in which the Nb bcc lattice is retained [21]. The α-α' two-phase region terminates at a critical point at T = 444 K with 23.7% (H/Nb). The H/Nb phase diagram shows a triple point temperature of 361 K below which a wide two-phase α-β region occurs with β being an ordered Nb hydride phase with fcc orthorhombic structure [22].

The fcc-metal Pd (lattice parameter a = 0.38874 nm) can be loaded with hydrogen retaining the primary α phase up to a H/Pd ratio of 1.7% at 25 °C (corresponding to a = 0.3895 nm) [23]. Increasing the macroscopic hydrogen concentration beyond 1.7% (H/Pd) results in a two-phase α-α' region in which the fcc-α' phase has a H/Pd ratio of 60% with a distinct lattice parameter of 0.4025 nm. Beyond 60% (H/Pd), the crystal is in a single α' phase and the lattice parameter further increases with increasing concentration. Stoichiometric PdH can be approached by filling all octahedral interstices resulting in an ideal NaCl structure.

In general, charging experiments reveal that the hydrogen absorption process is altered in thin-film geometry (see Ref. [24] for a recent review on microstructural aspects of hydrogen absorption in nanoscale materials). In comparison to hydrogen charging experiments on bulk samples, two additional important effects have to be taken into account for thin films. One effect is the interaction between the hydrogen-absorbing film and the substrate. Generally, the substrate-film bond leads to a lateral clamping effect. During hydrogen uptake, the film is allowed to expand freely only in the out-of-plane direction, while in-plane expansion is strongly hindered. Figure 1 shows three-dimensional *versus* one-dimensional lattice expansion schematically.

However, in thin films, the hydrogen absorption process will build up very large lateral (in-plane) stresses, which may exceed the yield stress of the system. Should this occur, the stress will be released by non-reversible formation of dislocations or even by delamination of the film from the substrate. It is quite remarkable that in some cases thin film systems can be completely reversibly charged and discharged with hydrogen. For Nb films on sapphire substrates, for example, it has been emphasized that the adhesive forces at the metal-ceramic interface must be enormous as the in-plane stress resulting from hydrogen loading exceeded the yield stress of bulk Nb by at least one order of magnitude [25]. It turns out that the yield stress for thin film/substrate systems may significantly differ from the corresponding bulk yield stress of that particular film material. For example, it was recently demonstrated that nanometer thick ultra-thin epitaxial Nb films can absorb very high amounts of hydrogen, $c = 1$ (H/Nb), reversibly without any noticeable plastic deformation or delamination from the substrate [26]. The measured value of the hydrogen-induced stress was on the order of -10 GPa, *i.e.*, over an order of magnitude higher than the Nb bulk yield stress. The effective yield stress of a particular thin film/substrate system will depend on a number of factors, including the specific material combinations, layer thicknesses, crystalline quality and intrinsic stress levels of the films, temperature, *etc.*

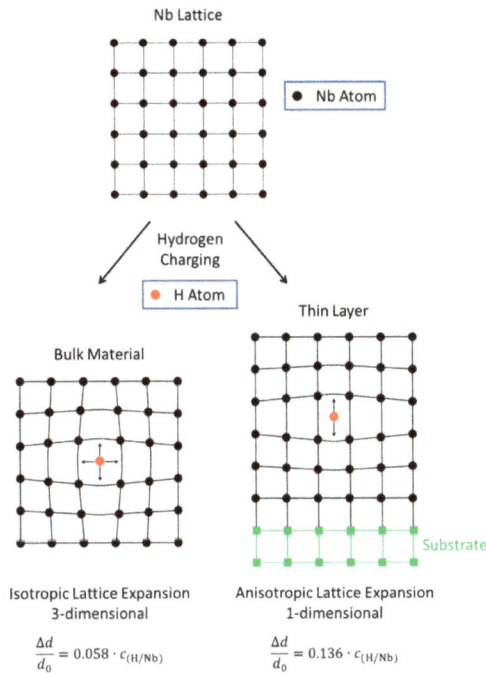

Figure 1. Hydrogen-induced three-dimensional isotropic expansion of the host lattice (bulk) *versus* one-dimensional lattice expansion (thin films).

Charging bulk Nb with hydrogen, for example, usually results in an isotropic, three-dimensional lattice expansion of the Nb. In the α and α' phases, the experimentally determined relation between relative expansion $\Delta d/d_0$ of lattice parameters and the hydrogen concentration c is $\Delta d/d_0 = 0.058 \cdot c$ (H/Nb), and it is linear up to c (H/Nb) $= 1$ [21]. For a hydrogen absorbing thin film being firmly clamped to a rigid substrate, one would find an anisotropic, one-dimensional out-of-plane lattice expansion. Due to the clamping to the substrate, no layer expansion within the layer plane is possible, and the proportionality factor will differ from the case of hydrogen charged bulk material. Theoretically, one would expect to find $\Delta d/d_0 = 0.136 \cdot c$ (H/Nb) [14].

The second important effect is the spatial proximity between a hydrogen absorbing layer and any neighboring material. Close to the interface, within the first few monolayers, drastic deviations from bulk behavior may occur due to charge-transfer processes altering the absorption potential for hydrogen [27]. Thus, one can expect a film-thickness dependence of the average hydrogen solubility. Figure 2 shows this effect schematically.

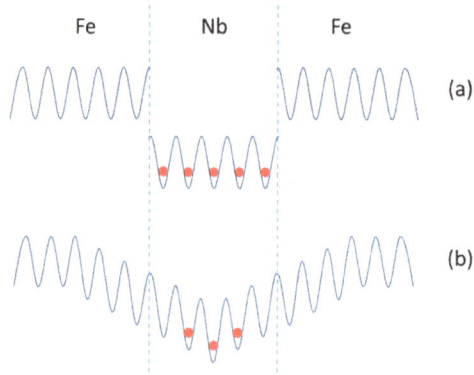

Figure 2. Schematic characteristic of the potential energy (chemical potential) of hydrogen atoms at a metal/metal-hydride/metal interface. Example: Fe (small solubility for hydrogen) and Nb (large solubility for hydrogen). The red dots indicate preferred hydrogen absorption sites. (**a**) Fe and Nb layers separated; (**b**) Fe and Nb layers in contact.

It should be noted that the absorption of hydrogen by Nb lattice vacancies is energetically strongly preferred to absorption into interstitial sites. First-principle calculations of Nb hydride formation showed that a single vacancy can accommodate six hydrogen atoms in the symmetrically equivalent lowest-energy sites and additional hydrogen in the nearby interstitial sites affected by the strain field. This indicates that a vacancy can serve as a nucleation center for hydride phase formation [19]. The trapping in vacancies was experimentally confirmed by positron annihilation experiments [28] where four hydrogen atoms were detected in one vacancy. The issue of hydrogen absorption and trapping in thin films (<100 nm) due to strain field effects is extensively discussed in the literature, see references 17–25 in [19], and in [29].

In view of the important role that the surface and surface-near layers play in the hydrogen absorption and diffusion process, it is imperative that the structural effects of hydrogen within these systems must be well understood. However, it is experimentally difficult to study hydrogen-induced structural changes in nanoscale layers. In this regard, neutrons provide a unique opportunity to directly probe hydrogen absorbed in metallic systems by scattering methods [8,10–12,30–32]. Often, X-rays are used to assess the structure of hydrogen absorbing materials and they are capable of observing effects such as lattice and layer expansion of the host crystal with increased hydrogen concentration [33–38]. However, X-rays are nearly completely insensitive to the hydrogen atoms themselves since the X-ray scattering length is determined by the atomic number. In contrast, the scattering length for neutrons varies randomly from element to element, and, as a consequence, hydrogen in a given material usually scatters as strongly, or sometimes stronger, than the absorbing medium. The fact that both hydrogen and deuterium atoms strongly scatter neutrons opens up the opportunity for isotope contrast variation [39]. These combined properties make neutron scattering a unique tool to identify the location and concentration of hydrogen in hydride materials.

Specifically for thin films, neutron reflectometry has proven to be an important technique when studying systems where the layered geometry is relevant to the behavior of the material or to their eventual applications. Neutron reflectometry is capable of determining the chemical profile of hydrogen and the host compound across thin films and multilayer samples up to roughly 200 nm in thickness. The depth sensitivity of this technique can elucidate properties such as hydrogen concentration gradients or the qualities of buried interfaces. Besides hydrogen sensitivity, neutrons have an additional advantage over X-rays in that they are strongly sensitive to magnetic moments, and, as such, polarized neutron reflectometry can be used to obtain magnetic depth profiles, which can be correlated with structural information.

This script focuses on utilizing both neutron and X-ray scattering techniques to study thin films and multilayers containing hydrogen absorbing metal components. In the first part of this review, we discuss hydrogen behavior in polycrystalline Fe/Nb multilayers. The results are compared with the case of epitaxially grown single-crystalline W/Nb [001] superlattices. The latter show distinctively different hydrogen absorption/desorption behavior due to a much stronger impact of the initial adherence to the substrate. The second portion of the work presented here reports on Pd/Co multilayers as a hybrid metallic-spintronic system for sensing hydrogen using a magnetic readout based on the ferromagnetic resonance technique. In this case, polarized neutron reflectometry proved especially useful as it allowed the observation of the magnetism of the Co layer simultaneous with hydrogen absorption studies.

2. Experimental Section

2.1. Materials and Applications

2.1.1. Hydrogen in Fe/Nb and W/Nb Films

Due to its high corrosion resistance, Nb is widely used across many types of applications, including hypoallergenic medical device and jewelry coatings, and linings in chemical plants. Despite the pure Nb surface being quite reactive, under most conditions, Nb will readily form a very thin oxide layer (Nb_2O_5 being the most stable oxide), which is responsible for its chemical stability. However, a cold-worked piece of Nb metal manufactured for a real-world application will still possess numerous vacancies, dislocations and grain boundaries, and these can very effectively transport hydrogen from an external source, e.g., from processing materials, to the Nb bulk interior. Even relatively low hydrogen concentrations at levels of 100 ppm may substantially deteriorate particular properties, e.g., the ductility of Nb alloy construction materials [40].

For the work presented here, polycrystalline thin Nb layers of variable thickness were prepared by growing Fe/Nb multilayers on oxidized Si substrates [41]. These samples were prepared by ion-beam sputtering at room temperature in an ultra-high vacuum chamber with a base pressure $<5 \times 10^{-9}$ mbar, where the Nb layers grew strongly (110) textured. Epitaxial Nb layers were prepared by growing W/Nb(001) superlattices at elevated temperatures of 200 °C on etched MgO(001) substrates by magnetron sputtering [12,13]. Both Fe/Nb and W/Nb multilayers had a 5 nm thick Pd capping layer to prevent oxidation and to facilitate hydrogen absorption. Both systems were prepared in the form of multilayers to optimize the intensity for the scattering experiments discussed below.

The effect of hydrogen on Nb thin films was initially studied by us due to observed hydrogen-induced changes of the magnetic coupling and the related giant magnetoresistivity effect in Fe/Nb multilayers [10]. The hydrogen-induced structural changes in the polycrystalline Fe/Nb system [41] were later compared with the behavior of epitaxially grown W/Nb superlattices [43].

For the Fe/Nb system, the Fe layer thickness, d_{Fe}, was kept constant at 2.6 nm, while the Nb thickness, d_{Nb}, varied from 2 to 100 nm. In the W/Nb system, the W layer thickness, d_W, was 2.6 nm, and d_{Nb} was 3 or 10 nm. The number of repeats of each Fe/Nb bilayer was chosen to provide a strong reflectivity signal allowing for the characterization of the Nb layers. Typical interface roughnesses, as determined by X-ray reflectivity, were 0.5 nm for Fe/Nb and 0.3 nm for W/Nb. A summary of the samples can be found in Table 1.

Table 1. List of samples for the Fe/Nb and W/Nb measurements. The description in the brackets is the thickness of each layer in a single bilayer. The number in subscript represents how many times each bilayer was repeated.

Fe/Nb(110)—Polycrystalline	W/Nb(110)—Single Crystal
[2.6 nm Fe/2 nm Nb]$_{20}$	[2.6 nm W/3 nm Nb]$_{120}$
[2.6 nm Fe/5 nm Nb]$_{10}$	[2.6 nm W/10 nm Nb]$_{100}$
[2.6 nm Fe/10 nm Nb]$_{10}$	-
[2.6 nm Fe/100 nm Nb]$_{10}$	-

2.1.2. Hydrogen in Co/Pd Films

Pd is one of the most utilized materials in hydrogen gas sensing due to its ability to reversibly absorb hydrogen [44]. The absorption process is accompanied by several changes to the physical properties of Pd, including the aforementioned lattice expansion of up to several percent and an increase in resistivity as high as 80% for PdH$_{0.7}$ at 25 °C [45]. Pd is also quite selective regarding hydrogen absorption and exhibits low sensitivity to other gases, such as CO, Cl$_2$, SO$_2$, H$_2$S, NO$_x$, and hydrocarbons, making it a suitable material in hydrogen sensors [46]. The challenge to reach the application stage lies in how to most effectively sense these hydrogen-induced changes to Pd. Current devices have several drawbacks: poor sensitivity, complex detection systems, slow response time, high power consumption, or potential flammability [18].

A novel approach to this problem lies in exploiting one of the lesser known aspects of Pd: magnetic properties. Although Pd itself is paramagnetic, when it neighbors ferromagnetic metals, such as Co, Fe, and NiFe, a perpendicular magnetic anisotropy (PMA) develops at the interface [47]. For ultra-thin ferromagnetic films (~0.5–1 nm), PMA aligns the magnetization of the ferromagnetic layer in the out-of-plane direction. For increased film thickness of the ferromagnet, the magnetization direction lies in the film plane, but the presence of PMA at the interface can affect a number of magnetic properties, including ferromagnetic resonance (FMR). Monitoring the magnetic response of these systems may then be a way to approach the design of novel hydrogen gas sensors.

Here, work is presented on a (10 nm) Co/(10 nm) Pd bilayer deposited via sputtering on a Si substrate [15], with the Pd layer on top allowing more Pd surface area to be exposed to external hydrogen gas.

2.2. Experimental Methods

2.2.1. Neutron Reflectometry

Reflectometry uses small-angle specular reflection ($\theta_i = \theta_f$) to probe the chemical depth profile across a surface, thin film, or multilayer as a function of depth. The geometry of this technique is shown in Figure 3 where the scattering vector, **Q**, is given by **Q** = **k**$_f$ − **k**$_i$. **Q** is related to the wavelength of the incoming neutrons, λ, and the angle of reflection, θ, by

$$Q = \frac{4\pi\sin\theta}{\lambda} \tag{1}$$

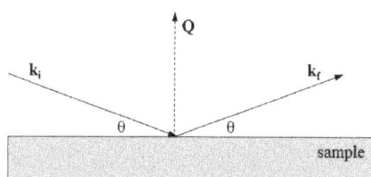

Figure 3. Schematic of reflectometry geometry.

At each surface and interface, some fraction of the incoming beam is reflected and the other fraction is transmitted and refracted. As neutrons pass through each layer, they are exposed to an effective nuclear scattering potential given as

$$V_n = \frac{2\pi\hbar^2}{m}bN \tag{2}$$

where b is the scattering length of the element, N is its atomic density and m is the neutron mass. The term bN is the scattering length density (SLD) and is related to the chemical makeup of a material as well as its physical density of the contained elements (for compounds: $b_{total} \cdot N_{total} = b_1 \cdot N_1 + b_2 \cdot N_2 + \ldots + b_i \cdot N_i$) [48]. Reflectometry allows one to model the SLD as a function of sample depth, which can be used to link features like hydrogen concentration to the change in SLD from a material's expected bulk value.

As previously mentioned, neutrons as scattering probes are particularly well suited to investigate hydrogen-sensitive materials because the magnitude of the hydrogen scattering length is of the same order as for many other metals. However, hydrogen has a negative scattering length, which can lead to its easy detection by comparing reflections from hydrogenated and non-hydrogenated materials. An example of this is demonstrated in Figure 4, where neutron reflectivity curves are simulated for a bulk Pd surface (red solid line), which would have an SLD of 402 μm^{-2}, as compared to the α' phase of PdH (blue dashed line), where the SLD changes to 126 μm^{-2}.

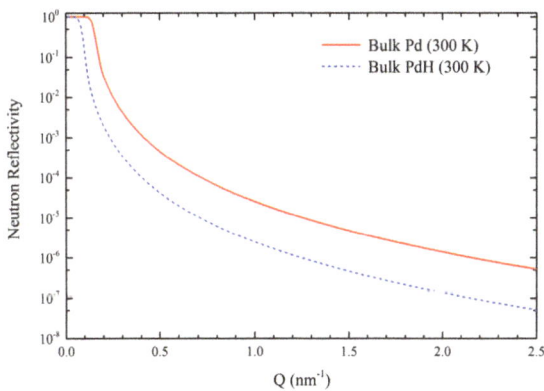

Figure 4. Simulated neutron reflectivity curves of a bulk Pd surface (red solid line) and a Pd-hydride bulk surface (blue dashed line).

Neutron reflectivity measurements on the Fe/Nb and W/Nb samples were performed at the V6 reflectometer [49,50] at the Hahn-Meitner-Institut Berlin, Germany, and the POSY2 reflectometer [51] at the Intense Pulsed Neutron Source at Argonne National Laboratory, Argonne, IL, USA.

2.2.2. Polarized Neutron Reflectometry (PNR)

Polarized neutron reflectometry (PNR) exploits the spin of the neutron as an additional tool for observing magnetic depth profiles across samples. In addition to the nuclear scattering potential in Equation (2), neutrons travelling through a magnetic layer will be exposed to a magnetic scattering potential given by

$$V_m = \frac{2\pi\hbar^2}{m}b_{mag}N = \mathbf{B} \cdot \hat{\mathbf{s}} \tag{3}$$

where $\mathbf{B}$ is the magnetic induction vector, $\hat{\mathbf{s}}$ is the neutron spin, and b_{mag} is the magnetic scattering length (1 μ_B/atom results in $b_{mag} = 2.695$ fm). Depending on the direction of the incoming neutron spin, the total scattering potential (combining Equations (2) and (3)) is then $V_{tot} = V_n \pm V_m$. By performing reflectivity measurements with two incoming, antiparallel neutron polarization states (+ and −), the difference between the R+ and R− channels essentially excludes the nuclear scattering and can be used to ascertain a profile of the magnetization across thin layers.

2.2.3. *In Situ* Ferromagnetic Resonance (FMR) Measurements and Polarized Neutron Reflectometry (PNR)

Ferromagnetic Resonance (FMR) is the effect of spin precession of the macroscopic magnetization vector in an external magnetic field and is driven experimentally by microwave power [52]. The FMR signal for a magnetic sample is observed as a sharp increase in the absorption of microwave power by the sample in a magnetic resonance condition. FMR has been applied to Pd/Co bilayers and multilayers in order to study how magnetic properties associated with perpendicular magnetic anisotropy (PMA) change with the hydrogen concentration in the Pd layer in order to investigate the feasibility of this system to act as a metallic spintronic hydrogen sensor [15]. In Ref. [15], it was reported that hydrogen absorption resulted in narrowing and shifting of the FMR absorption peak for the Co layer, indicating a coupling between these two properties.

In situ FMR and PNR measurements were carried out as a function of external hydrogen gas pressure at the PLATYPUS beamline [53–55] at the OPAL reactor at ANSTO, Lucas Heights, Australia. Figure 5 schematically shows the custom-made scattering chamber with FMR capability. The samples were placed face-up across the microwave coplanar transmission line, and aluminum windows mounted at the front and the rear of the chamber allowed the neutron beam to enter and exit. The gas atmosphere within the chamber was controlled by a mass flow controller system capable of delivering pure N_2 gas or N_2/H_2 gas mixtures of ⩽3.5% hydrogen gas at slightly above normal atmospheric pressure and at an ambient temperature of 23 °C. The external magnetic field was generated using a 1 T electromagnet.

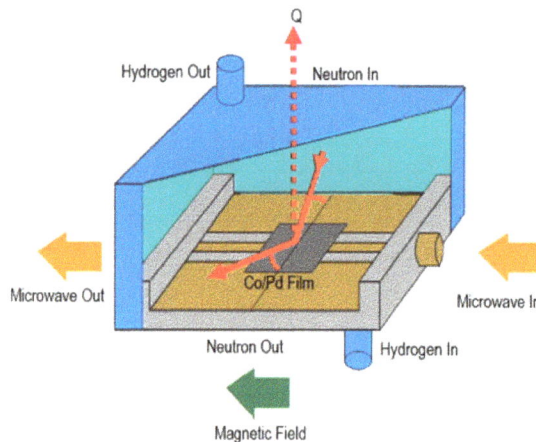

Figure 5. Schematic of the FMR/PNR experimental chamber.

3. Results and Discussion

3.1. *Nb Thin Films and Multilayers*

3.1.1. Fe/Nb Multilayers

In order to study the correlation between hydrogen absorption and resulting changes of the structural properties of Nb films, Fe/Nb heterostructures were measured with neutron reflectometry

(NR) and X-ray diffraction/reflectometry [41–43]. The multilayer geometry is advantageous as the scattering signals are strongly amplified. Using Fe as a partnering element is an apt choice, as it does not absorb hydrogen but is sufficiently transparent for hydrogen diffusion, which was experimentally confirmed. Multilayer samples were charged *in situ* with hydrogen at various pressures from 10^{-4} to 900 mbar. Measurements were made at a sample temperature of 185 °C to allow for relatively fast hydrogen absorption and to avoid phase separation that can occur at lower temperatures. For the Fe/Nb system (and for W/Nb discussed below), we present experimental data collected in the first cycle of hydrogen absorption.

Sample NR curves for polycrystalline samples are shown in Figure 6, comparing the systems in the virgin state and at the largest hydrogen pressure. The strong peaks in the top three sections are Bragg peaks resulting from the superlattice geometry. There is a noticeable shift in the peak position to a lower Q value with increased hydrogen content, indicating an expansion of the Nb layers within the heterostructures. In addition, we observe a significant increase in the Bragg peak intensity with increasing hydrogen absorption due to the increase in scattering contrast between Fe and Nb + H layers.

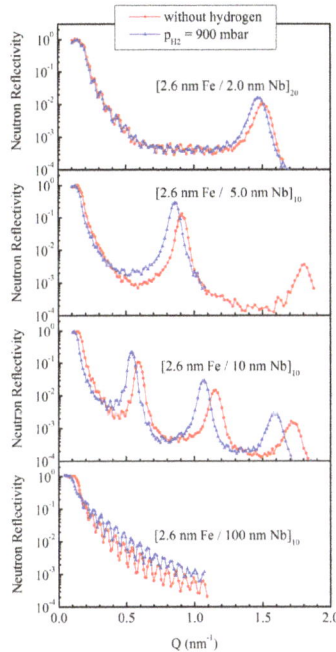

Figure 6. Unpolarized neutron reflectivity data for several polycrystalline Fe/Nb heterostructures. The red curves represent data taken in the virgin (pre-hydrogen) state, while the blue curves represent data taken at the highest hydrogen pressure of 900 mbar.

PNR was also performed due to the ferromagnetism of the Fe layers. While the Fe magnetism was found to be almost entirely unaffected by the Nb hydrogen concentration, the ability to measure two data channels for each pressure allowed for more certainty in the determination of the SLD of each material.

The data were fitted to ascertain the SLDs and thicknesses of the Nb and Fe layers. From the SLDs, the hydrogen concentrations can be calculated from [41]:

$$c_H = (\frac{SLD_{Nb+H}}{SLD_{Nb}} \times \frac{d_{Nb+H}}{d_{Nb}} - 1)\frac{b_{Nb}}{b_H} \qquad (4)$$

where the subscripts represent Nb before and after the introduction of hydrogen, d is the thickness of the Nb layer, and c_H is the number of hydrogen atoms per Nb metal atom. Data sets could be fitted well assuming a homogenous distribution of hydrogen throughout the Nb thickness. A sample of the fitted PNR data is shown in Figure 7. The resulting SLDs and layer thicknesses are shown in Table 2.

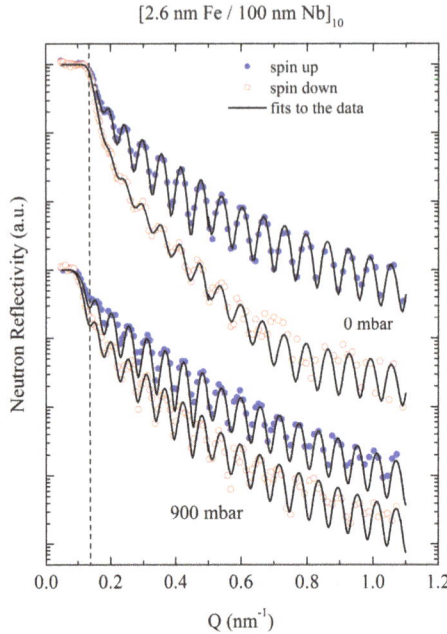

Figure 7. PNR data (circles) and fits (black lines) for a polycrystalline Fe/Nb heterostructure. Figure adapted with permission from Ref. [41]. Copyrighted by the American Physical Society.

Table 2. SLD and layer thicknesses for the sample data shown in Figure 7.

Layer Material	SLD at 0 mbar (μm^{-2})	SLD at 900 mbar (μm^{-2})	Bulk SLD (μm^{-2})	d at 0 mbar (nm)	d at 900 mbar (nm)
Pd	402	402	402	5	5
Nb	380	170	392	93.6	103.5
Fe (R+)	1010	1030	1300	2.6	2.6
Fe (R−)	490	570	303	2.6	2.6
Cr	303	303	303	5	5
Si	207	207	207	-	-

The most striking differences between the high and low hydrogen pressure data occur in the Nb layer. First, there is a decrease in SLD with hydrogen, from 380 to 170 μm^{-2}, which is accounted for by the negative hydrogen scattering length. There is also an increase in the Nb thickness by nearly 10 nm or 10%. This data can be used with Equation (4) to calculate the hydrogen concentration, and the example data indicated a hydrogen concentration of 95% (H/Nb) at 900 mbar external hydrogen pressure, *i.e.*, nearly one hydrogen atom for each Nb atom. This analysis was performed for several samples over a range of hydrogen pressures, leading to the development of a Nb/hydrogen concentration/pressure phase diagram shown in Figure 8 [56]. Here, it is demonstrated that in thin-film form, the phase diagram is drastically different than in bulk, particularly for samples with Nb thicknesses below 5 nm, where the peak hydrogen concentration reached is much less than the bulk value, and the pressures for obtaining a particular hydrogen concentration are orders of magnitude higher. As the W/Nb system shows similar behavior, we will discuss the causes for thickness dependence of the hydrogen absorption of both systems together in the Section 3.1.2.

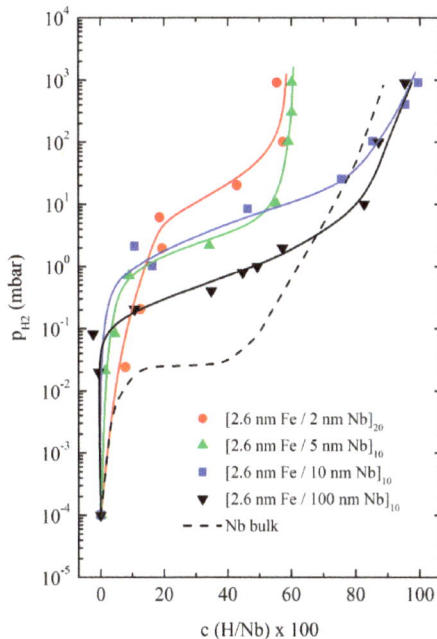

Figure 8. Phase diagram for hydrogen absorbing Nb thin films. Reprinted from [56] with permission from Elsevier.

3.1.2. W/Nb Multilayers

This section describes hydrogen absorption of Nb layers embedded in single-crystal (epitaxial) W/Nb multilayers and demonstrates how the degree of structural perfection impacts the hydrogen solubility in comparison to the previously presented polycrystalline Fe/Nb system. Similar to Fe, W does not accumulate hydrogen in any significant way due to its positive solution enthalpy for hydrogen. However, at $T = 185\,°C$, W as well as Fe is sufficiently transparent to allow a homogeneous charging of all the Nb layers in the layer stack.

In contrast to the Fe/Nb system, the W/Nb samples were charged with deuterium gas. The reason is as follows: In neutron reflectivity measurements of the Fe/Nb system, the scattering contrast between Fe (high SLD) and Nb + H (lower SLD) continuously increases with increasing hydrogen absorption due to the negative scattering length of hydrogen (the averaged b_H of hydrogen gas with naturally occurring isotope fractions is -3.739 fm). In contrast, hydrogen absorption in Nb layers of W/Nb would decrease the scattering contrast between W and Nb which would have made analysis of the data more difficult. Using deuterium with its positive scattering length of $b_D = 6.671$ fm avoids this problem. It has previously been shown on bulk [57,58] and on thin layers [59] that the solubility of deuterium in Nb deviates insignificantly from that of hydrogen. Note that, in order to facilitate the readability of the text, we continue to use the term hydrogen.

Results of *in situ* neutron reflectivity measurements on W/Nb multilayers are shown in Figure 9. The measurement on the uncharged sample is represented by red circles while the blue triangles represent the spectrum measured after hydrogen charging at $P_{H_2} = 900$ mbar.

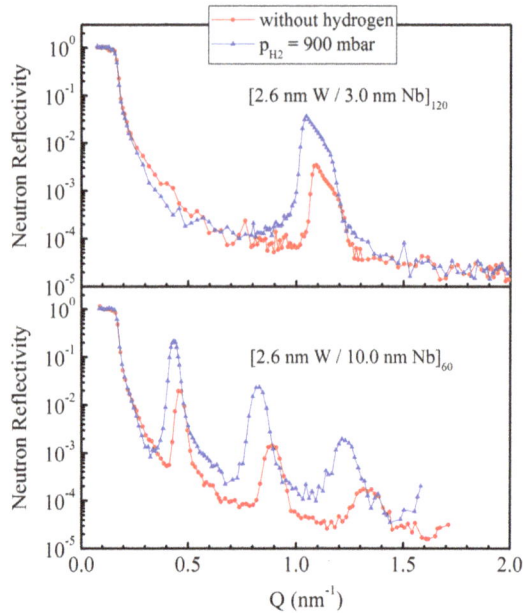

Figure 9. Neutron reflectivity curves of W/Nb multilayers with different Nb layer thicknesses as indicated, measured in the as-prepared state (red circles), and at a hydrogen pressure of 900 mbar (blue triangles).

The influence of the hydrogen is clearly reflected in both the shift of the Bragg reflections which originate from the W/Nb bilayer period and their apparent increase in intensity due to the change in scattering contrast. Due to the lower double-layer thickness of the sample with 3 nm Nb, however, only one Bragg reflection can be found in the experimentally accessible Q region.

The neutron reflectivity measurement of the sample with [2.6 nm W/3.0 nm Nb]$_{60}$ features a specific characteristic: in the Q region of about 1.1 nm^{-1}, the structural Bragg reflection is significantly asymmetrical. Such an effect can be observed if the sample possesses a slight layer thickness gradient in the lateral direction, which is present here and needed to be taken into account in the data analysis process.

Figure 10 shows the experimentally determined hydrogen solubility curves of the two W/Nb samples with 3 nm and 10 nm Nb, respectively, together with the theoretical curve of Nb bulk material.

For the sample with 3 nm Nb, a maximal hydrogen concentration of about c (H/Nb) = 40 at. % can be found, whereas this value amounts to about 80 at. % for the sample with 10 nm Nb (both at P_{H_2} = 900 mbar). The latter approaches the solubility of bulk samples at high pressures. As for the Fe/Nb multilayer system, the plateau-like regions—where most of the hydrogen is absorbed—depend on the Nb layer thickness. For 10 nm Nb, such a region is one order of magnitude, and, for 3 nm Nb, such a region is two orders of magnitude higher in pressure than for the Nb bulk material. The concentration/pressure gradient within this region increases with thinner Nb layer thickness. This indicates that the critical temperature for phase separation is significantly reduced for thin Nb layers compared to thicker Nb layers.

In order to elucidate the impact of hydrogen loading on the Nb crystal structure, *in situ* X-ray diffraction measurements were performed on the samples [2.6 nm W/10 nm Nb]$_{60}$ and [2.6 nm W/3 nm Nb]$_{120}$, see Figure 11, during the hydrogen charging at increasing hydrogen pressures. The data were taken with the scattering vector directing out-of-plane.

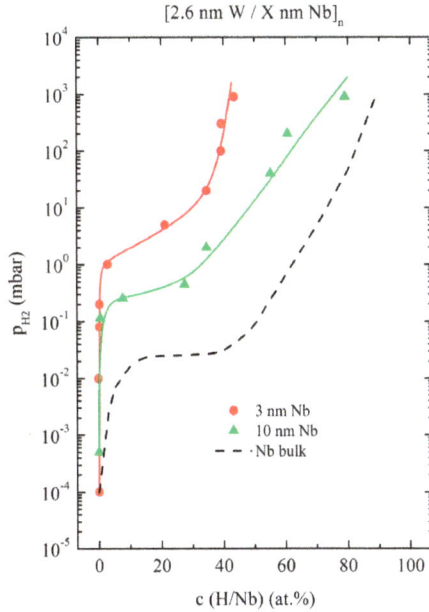

Figure 10. Hydrogen solubility curves of epitaxial [2.6 nm W/dNb]$_n$ multilayers and Nb bulk material as indicated.

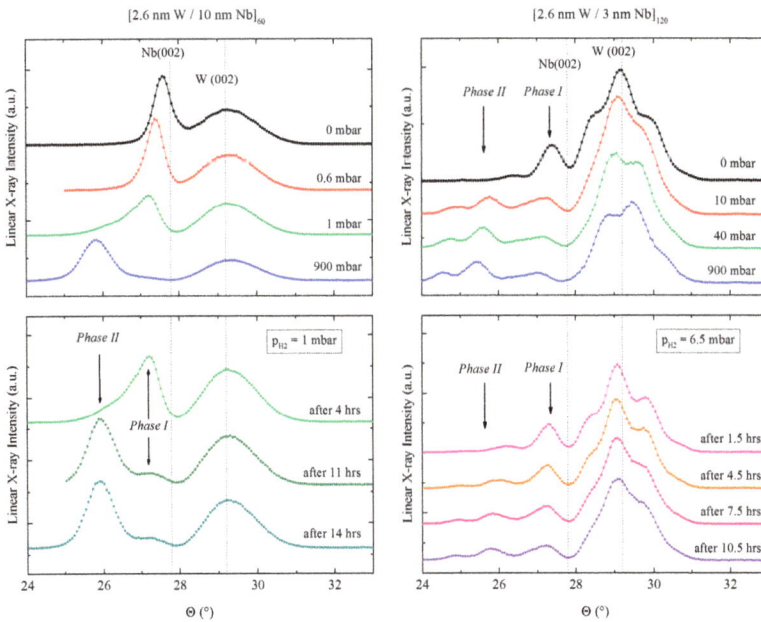

Figure 11. Cu K$_\alpha$ X-ray diffraction on the multilayers [2.6 nm W/10 nm Nb]$_{60}$ (**left**); and [2.6 nm W/3 nm Nb]$_{120}$ (**right**) as a function of increasing external hydrogen pressure, and (**bottom**) time-resolved measurement at 1 mbar and 6.5 mbar, respectively. The dashed lines indicate the positions of the Nb(002) and W(002) bulk reflections.

A reflection characteristic of the thin-film Nb crystal structure is located at a Bragg angle of $\Theta \approx 27.3°$ for both samples, a value that is shifted to smaller reflection angles when compared to bulk samples indicating larger Nb/Nb lattice plane distances in the out-of-plane direction. This is most likely due to a close matching of the W and Nb in-plane lattice parameters due to the epitaxy. We call the peak at this scattering angle *Phase I* reflection. Charging the sample with hydrogen changes the spectrum as expected. Most significantly the intensity of the *Phase I* reflection is highly reduced at higher pressures, its full width at half maximum increased, and its position shifted to smaller Θ values. At higher gas pressures, a new reflection can be identified at $\Theta \approx 25.9°$. We assume that a transition occurs between these two phases: first a slight one-dimensional lattice expansion occurs (the *Phase I* peak shifts towards lower angles), followed by a transition of Nb volume into a vertically strongly expanded phase with significantly higher hydrogen content (*Phase II*). Note that *Phase II* grows at the expense of *Phase I* (see bottom panels of Figure 11) which shows the time dependence of the Nb crystal transformation upon hydrogenation. *Phase II* is most pronounced at $P_{H_2} = 900$ mbar for both samples.

Our interpretation of this transformation is that above a Nb thickness dependent critical hydrogen pressure, the Nb lattice cannot keep its structural coherence to the W and the substrate lattice. It seems that the initially only slightly out-of-plane expanded Nb (marked as *Phase I*) undergoes a non-coherent transition to a largely out-of-plane expanded *Phase II* at hydrogen pressures above 1 mbar (see below for details). We note that only a very small fraction of the Nb is still in *Phase I* at 900 mbar, but the sum of the integrated intensities of the *Phases I* and *II* peaks stays constant for all hydrogen pressures.

Figure 12 compares relative changes of the Nb interplanar spacing and the Nb layer thickness for the two *in situ* hydrogen-charged W/Nb samples. As evidenced by the *in situ* X-ray diffraction measurements, the data sets reveal a second, unexpected hydrogen absorption mechanism which seems to be of equal importance to the hydrogen absorption on interstitial sites.

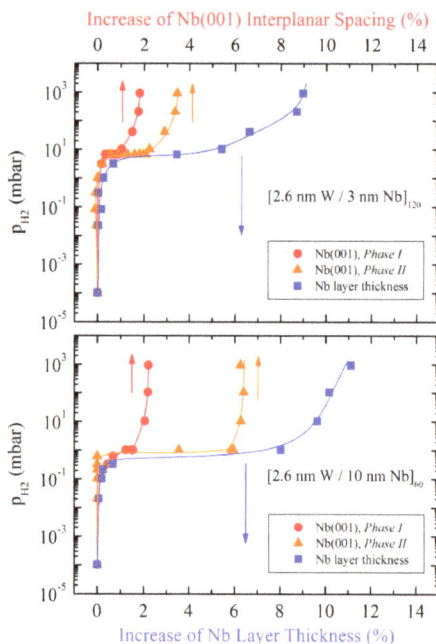

Figure 12. Comparison between hydrogen-induced relative increase of the out-of-plane Nb(110) interplanar spacing (**top**); and increase of the Nb layer thickness (**bottom**) in W/Nb multilayers for various hydrogen pressures.

The data show that the relative increase in the Nb layer thickness of both samples is significantly larger than the expansion of the Nb interplanar spacing. This result is consistent with what was found in the Fe/Nb system except that during the hydrogen absorption at no time a distinct transformation from a low out-of-plane state to a high out-of-plane state could be observed in the Fe/Nb system. Overall, the effects of changes in Nb layer thickness and Nb interplanar spacing through hydrogen charging during a first charging cycle in Fe/Nb and W/Nb (*Phase II*) are comparable. Also numerically, the same values can approximately be found. In both systems, the layer expansion in the direction of growth is considerably larger than the corresponding lattice expansion. This clearly demonstrates that, besides the well-known process of absorption at interstitial sites that entails expansion of the atomic lattice planes, there is an additional hydrogen absorption mechanism at work, which leads to a large macroscopic swelling of the Nb film.

An explanation for the large macroscopic swelling effect and for the distinct transformation between *Phase I* and *Phase II* in W/Nb was found by performing extended X-ray absorption fine structure (EXAFS) studies on the same series of samples [43]. It turned out that *Phase II* is related to a state in which, due to the large hydrogen concentrations, the in-plane Nb lattice parameter is significantly expanded indicating loss of coherency with the W lattice. In contrast, the Nb in-plane lattice parameter related to the low hydrogen concentration *Phase I* is smaller, still epitaxially matching the W in-plane lattice parameter. For the (001) orientated epitaxial [2.6 nm W/10 nm Nb]$_{60}$ multilayer, EXAFS revealed a small 2.5% in-plane expansion of the Nb lattice under hydrogen absorption. The in-plane expansion of 2.5% and the corresponding out-of-plane lattice expansion of 6.2% result in a total increase in volume of 11.5%. This amount coincides with the relative increase of the Nb layer thickness expansion of 11.1% as determined from reflectivity experiments. Hence, one only has to take into account the boundary condition of a fixed in-plane area (*i.e.*, a one-dimensional degree of freedom to expand macroscopically in the out-of-plane direction) in order to explain quantitatively the results in Figure 12. This model implies a three-dimensional rearrangement of Nb atoms caused by the massive mechanical lateral strain and its relaxation with increasing hydrogen charging within the Nb layers, especially near the interfaces to the substrate and the unloaded Fe or W layers. As a result, individual Nb atoms are squeezed out of existing lattice planes and start to form additional planes, which finally causes the anomalous large expansion of the layer thickness. Note that the newly created partial lattice planes imply a large amount of additional dislocations. The latter are evidenced by the fact that the out-of-plane X-ray diffraction peaks broaden significantly upon hydrogen absorption. For [2.6 nm W/10 nm Nb]$_{60}$, for example, we find a decrease in coherence length in growth direction from 9.0 nm after preparation, to 5.5 nm after hydrogen absorption at 900 mbar. Since lattice imperfections effectively trap hydrogen atoms, the new effect likely also explains the extraordinarily high hydrogen concentrations found in our Nb films and the partial non-reversibility of the hydrogen absorption process itself.

3.1.3. Co/Pd Bilayers

A combination of *in situ* PNR and FMR measurements was performed on Co/Pd bilayer films in order to study the magnetic effects of hydrogen absorption in this ferromagnet/metallic system and to correlate FMR changes with structural and/or magnetic changes in each individual layer.

Figure 13 shows a comparison of the experimental R+ reflectivity data and the theoretical model fits of a Co/Pd bilayer before (solid red line) and after (solid blue line) absorption of hydrogen from an ambient pressure N_2/H_2 gas mixture containing 3.5% hydrogen partial pressure (this corresponds to an absolute hydrogen partial pressure of $\approx$35.5 mbar within the loading chamber). The experimental data set prior to hydrogen absorption was fitted using a model consisting of d_{Pd} = 13.50 nm and d_{Co} = 8.36 nm. However, a best fit to the data was obtained by considering an interfacial region of thickness $d_{interface}$ = 1.43 nm and $SLD_{interface}$ = 162 μm^{-2} in the virgin state. This interface layer had a rather large roughness of 1.0 nm at its transition to the Pd layer. After exposure to a 3.5% hydrogen gas partial pressure the interfacial layer was found to undergo a 25% reduction in SLD accompanied by an increase in the in-plane magnetic moment by 66%, and a sharpening of its interfacial roughness to the Pd layer from 1.0 nm to 0.2 nm. Due to the negative scattering length of hydrogen, these observed changes to structure and magnetism highlight the presence of hydrogen at the Co/Pd interface, which have additionally resulted in modifications to the PMA strength within the Pd lattice.

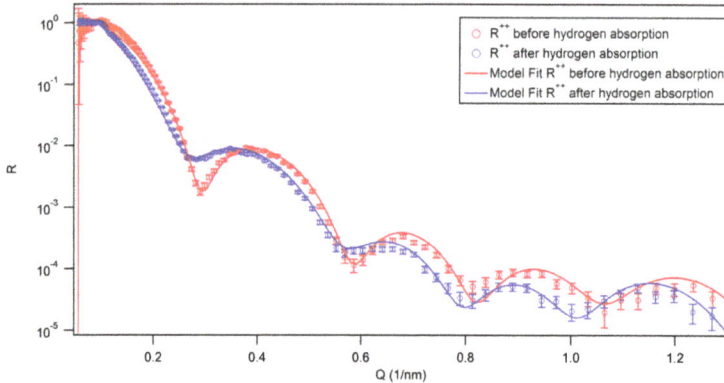

Figure 13. Comparison of R+ polarized neutron reflectivity data (circles) and fits (solid lines) for a Co/Pd bilayer before (red) and after (blue) exposure to 3.5% hydrogen gas partial pressure.

Determining the Pd SLDs and exact layer thicknesses before and after hydrogen exposure and using Equation (4) established the number density ratio of hydrogen atoms to Pd atoms to be 30% in the 3.5% hydrogen partial pressure state. Similar to the Fe/Nb (see Figure 2 in Ref. [41]) and W/Nb (see Figure 5 in Ref. [43]) systems, we found that, after the initial hydrogen absorption and desorption cycle was performed on the Co/Pd sample, the Pd layer did not return to its original thickness or density, demonstrating that the presence of hydrogen within the Pd led to the creation of irreversible deformations. It is significant, however, to note that a reproducible and reversible Pd structure was established between the second and third hydrogen absorption and desorption cycling experiments. This "training" effect needs to be taken into account for technical applications. Reversibility and training effects of the PdH_x system have been extensively studied in References [60,61].

The *in situ* FMR signal responses of the 8.36 nm Co/13.50 nm Pd bilayer obtained on PLATYPUS during two hydrogen absorption and desorption cycles are shown in Figure 14. During these measurements, the microwave frequency was kept constant and equal to 3.4 GHz, and the applied field was swept across the resonance to produce the plots shown in the figure. Ensuring efficient neutron reflection from the film requires taking FMR measurements in the "film facing up" geometry (see Section 2.2.3). For this film orientation, the FMR signal is significantly reduced because of film separation from the microwave transmission line by the 0.5-mm-thick silicon substrate of the film. In order to compensate for this signal reduction, the applied-field modulation technique combined with lock-in signal detection was utilized [62]. For this reason, the registered FMR traces have the shape of the first derivative of a Lorentzian instead of the Lorentzian itself.

The characteristic resonance changes its location with respect to the applied magnetic field depending on the hydrogen concentration within the Pd layer, shifting to lower applied fields upon hydrogen absorption, and shifting upwards upon hydrogen desorption. The downshift of the resonance peak implies that the PMA strength is reduced when hydrogen atoms are incorporated into the Pd crystal lattice. Reproducible responses analogous to those observed in the PNR data between the second and third hydrogen absorption and desorption cycling stages were also evident in the FMR data.

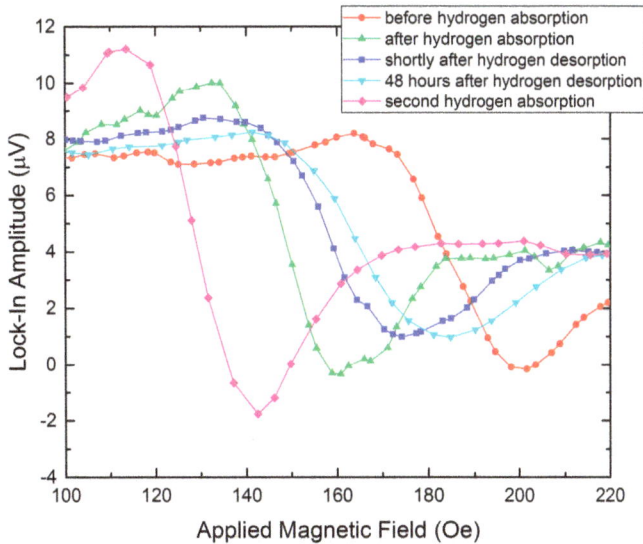

Figure 14. *In situ* FMR response of a 8.36 nm Co/13.50 nm Pd bilayer at various stages during the hydrogen absorption and desorption cycling PNR experiment.

4. Conclusions

In summary, the combination of the scattering methods neutron reflectometry, X-ray reflectometry, high-angle X-ray diffraction and EXAFS, all performed *in situ* during various hydrogen charging and discharging cycles, revealed substantial new insights in understanding the detailed physics related to hydrogen absorption and desorption in thin metal films.

Our case studies of the individual metal systems Fe/Nb, W/Nb and Co/Pd revealed that such structures absorb hydrogen in large quantities. Focusing on structural property changes upon hydrogenation, for Fe/Nb and W/Nb, we found that the relative out-of-lane expansion of the Nb layers is considerably larger than the relative increase of the Nb interplanar spacing, which indicates two distinctly different mechanisms of hydrogen absorption. In polycrystalline Fe/Nb multilayers, hydrogen expands the Nb interplanar spacing in a continuous way as a function of the external hydrogen pressure. In contrast, in epitaxial W/Nb multilayers—due to a much stronger impact of the initial adherence to the substrate—the Nb lattice expansion is discontinuous and can be regarded as a structural phase transition from exclusively out-of-plane to a three-dimensionally expanded state at low and high hydrogen pressures, respectively.

The hydrogen-induced structural changes, such as layer and lattice plane expansions, that we present in this manuscript are most pronounced during the first hydrogen absorption process. The initial absorption process results in large plastic deformations as the original state of the layer thickness and lattice constants cannot be recovered during the following hydrogen desorption process. The second process and following hydrogen absorption/desorption cycles, however, show, to a large extent, reversible behaviors of layer thickness and lattice spacings.

Our studies of the effect of hydrogen absorption on the magnetic properties of our systems revealed that the magnetic exchange coupling in Fe/Nb multilayers can reversibly be switched by hydrogen charging and discharging. In the Co/Pd system, the magnetic anisotropy at the interfaces can reversibly be changed by absorption and desorption of hydrogen. These two observed electronic effects enable a novel approach for hydrogen sensing using a magnetic readout scheme.

Acknowledgments: The authors would like to acknowledge technical support provided during their research work carried out at: Hahn-Meitner-Institut (now Helmholtz-Zentrum) Berlin, Germany; Hamburg Synchrotron Radiation Laboratory (HASYLAB) at Deutsches Elektronen-Synchrotron (DESY), Hamburg, Germany; Intense Pulsed Neutron Source (IPNS) at Argonne National Laboratory, Argonne, IL, USA; Australian Nuclear Science and Technology Organisation (ANSTO), Lucas Heights, Australia.

Author Contributions: S.J.C., C.R. and F.K. prepared the manuscript. C.R. and F.K. performed data collection and analysis of the Fe/Nb and W/Nb systems. S.J.C., G.C., M.K. and F.K. performed data collection and analysis of the Co/Pd systems.

Conflicts of Interest: The authors declare no conflict of interest.

References

1.   Sakintuna, B.; Lamari-Darkrim, F.; Hirscher, M. Metal hydride materials for solid hydrogen storage: A review. *Int. J. Hydrog. Energy* **2007**, *32*, 1121–1140. [CrossRef]

2.   Van den Berg, A.W.C.; Areán, C.O. Materials for hydrogen storage: Current research trends and perspectives. *Chem. Commun.* **2008**, 668–681. [CrossRef]

3.   Liu, C.; Li, F.; Ma, L.P.; Cheng, H.M. Advanced materials for energy storage. *Adv. Mater.* **2010**, *22*, E28–E62. [CrossRef] [PubMed]

4.   Chen, X.; Li, C.; Grätzel, M.; Kostecki, R.; Mao, S.S. Nanomaterials for renewable energy production and storage. *Chem. Soc. Rev.* **2012**, *41*, 7909–7937. [CrossRef] [PubMed]

5.   Pukazhselvan, D.; Kumar, V.; Singh, S.K. High capacity hydrogen storage: Basic aspects, new developments and milestones. *Nano Energy* **2012**, *1*, 566–589. [CrossRef]

6.   Baldi, A.; Gonzalez-Silveira, M.; Palmisano, V.; Dam, B.; Griessen, R. Destabilization of the Mg–H System through Elastic Constraints. *Phys. Rev. Lett.* **2009**, *102*, 226102. [CrossRef] [PubMed]

7.   Peisl, H. Lattice strains due to hydrogen in metals. In *Hydrogen in Metals I*; Springer Berlin Heidelberg: Berlin, Germany, 1978; pp. 53–74.

8.   Zabel, H. Thin films: Hydrogen. In *Encyclopedia of Materials: Science and Technology*, 2nd ed.; Buschow, K.H.J., Cahn, R.W., Flemings, M.C., Ilschner, B., Kramer, E.J., Mahajan, S., Veyssière, P., Eds.; Elsevier: Oxford, UK, 2001; pp. 9246–9250.

9.   Huiberts, J.N.; Griessen, R.; Rector, J.H.; Wijngaarden, R.J.; Dekker, J.P.; de Groot, D.G.; Koeman, N.J. Yttrium and lanthanum hydride films with switchable optical properties. *Nature* **1996**, *380*, 231–234. [CrossRef]

10. Klose, F.; Rehm, C.; Nagengast, D.; Maletta, H.; Weidinger, A. Continuous and reversible change of the magnetic coupling in an Fe/Nb multilayer induced by hydrogen charging. *Phys. Rev. Lett.* **1997**, *78*, 1150–1153. [CrossRef]

11. Hjörvarsson, B.; Dura, J.A.; Isberg, P.; Watanabe, T.; Udovic, T.J.; Andersson, G.; Majkrzak, C.F. Reversible tuning of the magnetic exchange coupling in Fe/V(001) superlattices using hydrogen. *Phys. Rev. Lett.* **1997**, *79*, 901–904. [CrossRef]

12. Kravtsov, E.; Nefedov, A.; Nowak, G.; Zhernenkov, K.; Zabel, H.; Hjörvarsson, B.; Liebig, A.; Hoser, A.; McIntyre, G.J.; Paolasini, L.; *et al.* Fine-tuning of the spin-density-wave state in Cr/V heterostructures via hydrogen uptake. *J. Phys. Condens. Matter* **2009**, *21*, 336004. [CrossRef] [PubMed]

13. Andersson, G.; Hjörvarsson, B.; Zabel, H. Hydrogen-induced lattice expansion of vanadium in a Fe/V(001) single-crystal superlattice. *Phys. Rev. B* **1997**, *55*, 15905. [CrossRef]

14. Yang, Q.M.; Schmitz, G.; Fähler, S.; Krebs, H.U.; Kirchheim, R. Hydrogen in Pd/Nb multilayers. *Phys. Rev. B* **1996**, *54*, 9131–9140. [CrossRef]

15. Chang, C.S.; Kostylev, M.; Ivanov, E. Metallic spintronic thin film as a hydrogen sensor. *Appl. Phys. Lett.* **2013**, *102*, 142405. [CrossRef]

16. Westerwaal, R.J.; Rooijmans, J.S.A.; Leclercq, L.; Gheorghe, D.G.; Radeva, T.; Mooij, L.; Mak, T.; Polak, L.; Slaman, M.; Dama, B.; *et al.* Nanostructured Pd-Au based fiber optic sensors for probing hydrogen concentrations in gas mixtures. *Int. J. Hydrog. Energy* **2013**, *38*, 4201–4212. [CrossRef]

17. Ngene, P.; Radeva, T.; Slaman, M.; Westerwaal, R.J.; Schreuders, H.; Dam, B. Seeing hydrogen in colors: Low-cost and highly sensitive eye readable hydrogen detectors. *Adv. Funct. Mater.* **2014**, *24*, 2374–2382. [CrossRef]

18. Hübert, T.; Boon-Brett, L.; Black, G.; Banach, U. Hydrogen sensors—A review. *Sens. Actuators B* **2011**, *157*, 329–352. [CrossRef]
19. Ford, D.C.; Cooley, L.D.; Seidman, D.N. First principles calculations of niobium hydride formation in superconducting radio-frequency cavities. *Supercond. Sci. Technol.* **2013**. [CrossRef]
20. Fukai, Y. *The Metal-Hydrogen System: Basic Bulk Properties*; Springer: Berlin, Germany, 2005.
21. Zabel, H.; Peisl, H. X-ray study of the phase diagram of hydrogen in niobium. *Phys. Stat. Sol. A* **1976**, *37*, K67–K70. [CrossRef]
22. Walter, R.J.; Chandler, W.T. The columbium-hydrogen constitution diagram. *Trans. Met. Soc. AIME* **1965**, *233*, 762–765.
23. Manchester, F.D.; San-Martin, A.; Pitre, J.M. The H-Pd (hydrogen-palladium) system. *J. Phase Equilib.* **1995**, *15*, 62–83. [CrossRef]
24. Pundt, A.; Kirchheim, R. Hydrogen in Metals: Microstructural Aspects. *Annu. Rev. Mater. Res.* **2006**, *36*, 555–608. [CrossRef]
25. Song, G.; Remhof, A.; Theis-Bröhl, K.; Zabel, H. Extraordinary adhesion of niobium on sapphire substrates. *Phys. Rev. Lett.* **1997**, *79*, 5062–5065. [CrossRef]
26. Hamm, M.; Burlaka, V.; Wagner, S.; Pundt, A. Achieving reversibility of ultra-high mechanical stress by hydrogen loading of thin films. *Appl. Phys. Lett.* **2015**. [CrossRef]
27. Hjörvarsson, B.; Andersson, G.; Karlsson, E. Metallic superlattices: Quasi two-dimensional playground for hydrogen. *J. Alloy. Compd.* **1997**, *51*, 253–254. [CrossRef]
28. Čížek, J.; Procházka, I.; Bečvář, F.; Kužel, R.; Cieslar, M.; Brauer, G.; Anwand, W.; Kirchheim, R.; Pundt, A. Hydrogen-induced defects in bulk niobium. *Phys. Rev. B* **2004**. [CrossRef]
29. Romanenko, A.; Edwardson, C.J.; Coleman, P.G.; Simpson, P.J. The effect of vacancies on the microwave surface resistance of niobium revealed by positron annihilation spectroscopy. *Appl. Phys. Lett.* **2013**. [CrossRef]
30. Fritzsche, H.; Saoudi, M.; Haagsma, J.; Ophus, C.; Luber, E.; Harrower, C.T.; Mitlin, D. Neutron reflectometry study of hydrogen desorption in destabilized MgAl alloy thin films. *Appl. Phys. Lett. B* **2008**. [CrossRef]
31. Maxelon, M.; Pundt, A.; Pyckhout-Hintzen, W.; Barker, J.; Kirchheim, R. Interaction of hydrogen and deuterium with dislocations in palladium as observed by small angle neutron scattering. *Acta Mater.* **2001**, *49*, 2625–2634. [CrossRef]
32. Munter, A.E.; Heuser, B.J. Deuterium phase behavior in thin-film Pd. *Phys. Rev. B* **1998**, *58*, 678–684. [CrossRef]
33. Song, G.; Geitz, M.; Abromeit, A.; Zabel, H. Solubility isotherms of hydrogen in epitaxial Nb(110) films. *Phys. Rev. B* **1996**, *54*, 14093–14101. [CrossRef]
34. Reimer, P.M.; Zabel, H.; Flynn, C.P.; Matheny, A.; Ritley, K. Elastic properties of hydrogen-loaded epitaxial films. *Z. Phys. Chem.* **1993**, *181*, 367–373. [CrossRef]
35. Reimer, P.M.; Zabel, H.; Flynn, C.P.; Dura, J.A. Extraordinary alignment of Nb films with sapphire and the effects of added hydrogen. *Phys. Rev. B* **1992**, *45*, 11426–11429. [CrossRef]
36. Laudahn, U.; Pundt, A.; Bicker, M.; Hülsen, U.; Geyer, U.; Wagner, T.; Kirchheim, R. Hydrogen-induced stress in Nb single layers. *J. Alloy. Compd.* **1999**. [CrossRef]
37. Hjörvarsson, B.; Rydén, J.; Karlsson, E.; Birch, J.; Sundgren, J.E. Interface effects of hydrogen uptake in Mo/V single-crystal superlattices. *Phys. Rev. B* **1991**, *43*, 6440–6445. [CrossRef]
38. Stillesjö, F.; Hjörvarsson, B.; Zabel, H. Hydrogen-induced lattice expansion in a (001)-oriented Mo/V superlattice. *Phys. Rev. B* **1996**, *54*, 3079–3083. [CrossRef]
39. Munter, A.E.; Heuser, B.J.; Ruckman, M.W. *In situ* neutron reflectometry measurements of hydrogen and deuterium absorption in a Pd/Nb/Pd layered film. *Phys. Rev. B* **1997**, *55*, 14035–14038. [CrossRef]
40. Gahr, S.; Grossbeck, M.L.; Birnbaum, H.K. Hydrogen embrittlement of Nb I-macroscopic behavior at low temperatures. *Acta Met.* **1977**, *25*, 125–134. [CrossRef]
41. Rehm, C.; Fritzsche, H.; Maletta, H.; Klose, F. Hydrogen concentration and its relation to interplanar spacing and layer thickness of 1000-Å Nb(110) films during *in situ* hydrogen charging experiments. *Phys. Rev. B* **1999**, *59*, 3142–3152. [CrossRef]

42. Klose, F.; Rehm, C.; Fieber-Erdmann, M.; Holub-Krappe, E.; Bleif, H.J.; Sowers, H.; Goyette, R.; Tröger, L.; Maletta, H. Hydrogen absorption in epitaxial W/Nb(001) and polycrystalline Fe/Nb(110) multilayers studied *in situ* by X-ray/neutron scattering techniques and X-ray absorption spectroscopy. *Phys. B* **2000**, *283*, 184–188. [CrossRef]

43. Rehm, C.; Maletta, H.; Fieber-Erdmann, M.; Holub-Krappe, E.; Klose, F. Anomalous layer expansion in thin niobium films during hydrogen absorption. *Phys. Rev. B* **2002**. [CrossRef]

44. Buttner, W.J.; Post, M.B.; Burgess, R.; Rivkin, C. An overview of hydrogen safety sensors and requirements. *Int. J. Hydrog. Energy* **2011**, *36*, 2462–2470. [CrossRef]

45. Lewis, F.A. *The Palladium Hydrogen System*; Academic Press: New York, NY, USA, 1967.

46. Gupta, R.; Sagade, A.A.; Kulkarni, G.U. A low cost optical hydrogen sensing device using nanocrystalline Pd grating. *Int. J. Hydrog. Energy* **2012**, *37*, 9443–9449. [CrossRef]

47. Engel, B.N.; England, C.D.; van Leeuwen, R.A.; Wiedmann, M.H.; Falco, C.M. Interface magnetic anisotropy in epitaxial superlattices. *Phys. Rev. Lett.* **1991**, *67*, 1910–1913. [CrossRef] [PubMed]

48. Neutron Activation Calculator. Available online: https://www.ncnr.nist.gov/resources/activation/ (accessed on 12 April 2016).

49. Mezei, F.; Golub, R.; Klose, F.; Toews, H. Focussed beam reflectometer for solid and liquid surfaces. *Phys. B* **1995**. [CrossRef]

50. Paul, A.; Krist, T.; Teichert, A.; Steitz, R. Specular and off-specular scattering with polarization and polarization analysis on reflectometer V6 at BER II, HZB. *Physica B* **2011**, *406*, 1598–1606. [CrossRef]

51. Crawford, R.K.; Felcher, G.P.; Kleb, R.; Epperson, J.E.; Thiyagarajan, P. *New Instruments at IPNS: POSY II and SAD II*; Hyer, D.K., Ed.; IOP Publishing Ltd.: Bristol, UK, 1989.

52. Gurevich, A.G.; Melkov, G.A. *Magnetization Oscillation and Waves*; C.R.C. Press: New York, NY, USA, 1996.

53. James, M.; Nelson, A.; Brule, A.; Schulz, J.C. Platypus: A time-of-flight neutron reflectometer at Australia's new research reactor. *J. Neutron Res.* **2006**, *14*, 91–108. [CrossRef]

54. James, M.; Nelson, A.; Holt, S.A.; Saerbeck, T.; Hamilton, W.A.; Klose, F. The multipurpose time-of-flight neutron reflectometer "Platypus" at Australia's OPAL reactor. *Nucl. Instrum. Methods Phys. Res. Sec. A* **2011**, *632*, 112–123. [CrossRef]

55. Saerbeck, T.; Klose, F.; LeBrun, A.P.; Füzi, J.; Brule, A.; Nelson, A.; Holt, S.A.; James, M. Polarization "Down Under": The polarized time-of-flight neutron reflectometer PLATYPUS. *Rev. Sci. Instrum.* **2012**. [CrossRef] [PubMed]

56. Maletta, H.; Rehm, C.; Klose, F.; Fieber-Erdmann, M.; Holub-Krappe, E. Anomalous effects of hydrogen absorption in Nb films. *J. Magn. Magn. Mater.* **2002**, *240*, 475–477. [CrossRef]

57. *Hydrogen in Metals I*; Alefeld, G., Völkl, J., Eds.; Springer: Berlin, Germany, 1978; Volume 28.

58. *Hydrogen in Metals III*; Wipf, H., Ed.; Springer: Berlin, Germany, 1997; Volume 73.

59. Remhof, A.; Song, G.; Sutter, C.; Schreyer, A.; Siebrecht, R.; Zabel, H.; Güthoff, F.; Windgasse, J. Hydrogen and deuterium in epitaxial Y(0001) films: Structural properties and isotope exchange. *Phys. Rev. B* **1999**, *59*, 6689–6699. [CrossRef]

60. Pivak, Y.; Gremaud, R.; Gross, K.; Gonzalez-Silveira, M.; Walton, A.; Book, D.; Schreuders, H.; Dam, B.; Griessen, R. Effect of the substrate on the thermodynamic properties of PdHx films studied by hydrogenography. *Scr. Mater.* **2009**, *60*, 348–351. [CrossRef]

61. Gremaud, R.; Gonzalez-Silveira, M.; Pivak, Y.; de Mana, S.; Slaman, M.; Schreuders, H.; Dam, B.; Griessen, R. Hydrogenography of PdHx thin films: Influence of H-induced stress relaxation processes. *Acta Mater.* **2009**, *57*, 1209–1219. [CrossRef]

62. Maksymov, I.S.; Kostylev, M. Broadband stripline ferromagnetic resonance spectroscopy of ferromagnetic films, multilayers and nanostructures. *Phys. E* **2015**, *69*, 253–293. [CrossRef]

![metals logo] *metals*

MDPI

Article

Resistance of Hydrogenated Titanium-Doped Diamond-Like Carbon Film to Hyperthermal Atomic Oxygen

Kengo Kidena [1], Minami Endo [1], Hiroki Takamatsu [1], Masahito Niibe [1], Masahito Tagawa [2], Kumiko Yokota [2], Yuichi Furuyama [3], Keiji Komatsu [4], Hidetoshi Saitoh [4] and Kazuhiro Kanda [1,*]

[1] Laboratory of Advanced Science and Technology for Industry, University of Hyogo, 3-1-2 Koto Kamigori 678-1205, Japan; k.kidena@gmail.com (K.K.); hadoholi1130@gmail.com (M.E.); renren@lasti.u-hyogo.ac.jp (H.T); niibe@lasti.u-hyogo.ac.jp (M.N.)
[2] Faculty of Engineering, Kobe University, Nada Kobe 657-8501, Japan; tagawa@mech.kobe-u.ac.jp (M.T.); yokota@mech.kobe-u.ac.jp (K.Y.)
[3] Faculty of Maritime Sciences, Kobe University, Higashi-Nada Kobe 658-0022, Japan; furuyama@maritime.kobe-u.ac.jp
[4] Department of Materials Science and Technology, Nagaoka University of Technology, Kamitomioka Nagaoka 940-2188, Japan; keiji_komatsu@mst.nagaokaut.ac.jp (K.K.); hts@nagaokaut.ac.jp (H.S.)
* Author to whom correspondence should be addressed; kanda@lasti.u-hyogo.ac.jp; Tel.: +81-79-158-0476; Fax: +81-79-158-0242.

Academic Editor: Klaus-Dieter Liss
Received: 2 September 2015; Accepted: 14 October 2015; Published: 23 October 2015

Abstract: The effect of irradiation by a hyperthermal-atomic-oxygen beam on hydrogenated titanium-doped diamond-like carbon (hydrogenated Ti-DLC) films, applied as a solid lubricant for equipment used in low-earth orbit was investigated. Unlike the film thickness of hydrogenated non-doped DLC films, that of hydrogenated Ti-DLC films was found to be constant after the films were exposed to atomic oxygen. In addition, bulk composition of the hydrogenated Ti-DLC film stayed constant, and in particular, hydrogen content in the film did not decrease. These results indicate that a hydrogenated Ti-DLC film can keep its low friction properties under vacuum. Surface chemical analysis showed that a titanium-oxide layer is form on the film by exposure to atomic oxygen. The thickness of the titanium oxide layer was estimated to be about 5 nm from the element distribution in the depth direction of the hydrogenated Ti-DLC films. The titanium-oxide layer was interpreted to protect the bulk film from erosion by hyperthermal atomic oxygen.

Keywords: titanium-doped diamond-like carbon film; solid lubricant; hyperthermal atomic-oxygen beam; Rutherford backscattering spectrometry; elastic-recoil detection analysis; X-ray photoelectron spectroscopy; near-edge X-ray-absorption fine structure; glow-discharge optical-emission spectroscopy

1. Introduction

In space, solid lubricants are required to replace oil, because oil evaporates and freezes in such an environment. To use lubricant in space, their resistances to the extreme environment of space, namely, ultra-high vacuum, high temperature, ultraviolet light and X-rays, atomic oxygen, and their synergistic effects, must be improved. Most artificial satellites are positioned in the region called the "low-earth orbit" (LEO), which is less than 2000 km above the ground. In the LEO, hyperthermal atomic oxygen is the dominant species and the main cause of deterioration in the properties of astronautical materials [1–4]. Solid lubricant used in artificial satellites must therefore have sufficient resistance against irradiation by atomic oxygen.

Diamond-like carbon (DLC) films are amorphous carbon films and have been used in the tribological field as a coating material on edged-tools, computer hard disks, and automobile components because they have a wide range of excellent properties including low friction coefficient, high hardness, and good corrosion resistance [5–10]. Hydrogenated DLC films were expected to be used as lubricants in LEO, because it was known that DLC films with a hydrogen content greater than 40 at.% provide ultra-low friction (friction coefficient less than 0.001) even under air and vacuum conditions [11]. However, we previously reported that hydrogenated DLC films are etched by the collision with hyperthermal atomic oxygen [12]. As a result, hydrogenated DLC films cannot be used as a solid lubricant in space as is.

In the last decade, many working groups reported that several properties of DLC films are improved by doping with a "hetero element" [13–27]. Especially, titanium (Ti)-doped DLC films have been widely investigated because they are expected to improve sliding properties, burning resistance, and oxidation resistance [22–27]. In a previous study, we investigated the effect of irradiation by a hyperthermal atomic-oxygen beam on hydrogenated a Ti-doped diamond-like carbon (Ti-DLC) film [28]. Under the irradiation of atomic oxygen, carbon atoms were desorbed from the hydrogenated Ti-DLC films and formed gas species (CO and/or CO_2), and oxygen atoms and titanium atoms were found to generate titanium oxide on the surface of the films. However, the surface-modification mechanism cannot be observed, because the fluence of atomic oxygen was too high. As for a more important point regarding use as a solid lubricant, the variation of bulk properties of a hydrogenated Ti-DLC film has not be confirmed. In this study, the purpose to clarify the hydrogenated Ti-DLC film used as a solid lubricant in space, dependence of surface and bulk properties of a hydrogenated Ti-DLC film irradiated by a hyperthernal atomic-oxygen beam on fluence was investigated. In particular, hydrogenated Ti-DLC films were exposed to atomic oxygen with a collisional energy of 5 eV by using a laser-detonation beam apparatus. Film thickness and bulk composition of the hydrogenated Ti-DLC film were estimated by a combination of Rutherford backscattering spectrometry (RBS) and elastic-recoil detection analysis (ERDA). The chemical analysis on the DLC film surface was performed by X-ray photoelectron spectroscopy (XPS) and near-edge X-ray-absorption fine structure (NEXAFS). Element distribution in depth direction was determined by the glow-discharge optical-emission spectroscopy (GD-OES).

2. Experimental Section

The hydrogenated Ti-DLC films used in this study, commercially available form Nippon ITF, were deposited to a thickness of 400 nm on silicon wafers using amplitude modulated RF plasma-enhanced chemical-vapor deposition (PECVD) [29], which hydrogen content was greater than 40 at.%. The hydrogenated Ti-DLC films were exposed to hyperthermal atomic oxygen by a laser-detonation beam apparatus [12,30,31]. Pure oxygen gas was introduced into the nozzle throat through a pulsed supersonic valve. A pulse from a CO_2 laser (wavelength: 10.6 μm, laser power: >5 J/pulse) was focused onto the oxygen gas at the nozzle throat. By absorbing the laser energy, high-density and high-temperature oxygen plasma was formed at the nozzle throat. Once the plasma was formed, it propagated and absorbed the energy in the tail of the laser pulse. The plasma propagated along the incident laser axis, and oxygen molecules were decomposed and accelerated at the shock front of the propagating plasma. The atomic-oxygen beam was irradiated perpendicularly onto the sample surface at room temperature. The dependence of the surface reaction of the hyperthermal atomic oxygen beam sample on the sample temperature is little known, because the irradiated oxygen atoms have translation energy of approximately 50,000 degrees [31]. The typical atomic oxygen flux at the sample position (46 cm from the nozzle) was estimated to be 3.51×10^{15} atoms·cm^{-2}s^{-1} by using a silver-coated quartz-crystal microbalance (QCM). The average translational energy of hyperthermal atomic oxygen was estimated to be 5.46 eV by using a time-of-flight (TOF) measurement system consisting of a quadrupole mass spectrometer with a scintillation detector and a multichannel scalar. This energy is the same as the atomic oxygen energy in the LEO [1–4]. In the present study, six sheets of hydrogenated Ti-DLC film were exposed to a hyperthermal-atomic-oxygen beam corresponding to fluences of 0, 1.2×10^{17}, 1.2×10^{18}, 3.5×10^{18}, 1.2×10^{19}, and 1.2×10^{20} atoms·cm^{-2}. The fluence step of the atomic oxygen beam was made smaller than that of a previous work [28] so that the reaction mechanism on the surface of hydrogenated Ti-DLC films could be investigated.

Film thickness and bulk composition of the hydrogenated Ti-DLC film were estimated by a combination of ERDA and RBS. These measurements were performed using a tandem electrostatic accelerator (5SDH, National Electrostatic Corporation, Middleton, WI, USA) located in the Faculty of Maritime Sciences, Kobe University [12,32]. The sample was irradiated by a 4.2 MeV He^{2+} beam to accelerate the negative ions of helium (^{4}He) generated by the RF-discharge negative-charge-exchange ion source. The incident angle of the beam was 15° with respect to the surface of the sample. As for the RBS, high-energy He^{2+} ions, scattered elastically by the sample, were captured with a solid-state detector (SSD) positioned at 160° with respect to the beam direction. As for the ERDA, He^{2+} ions collided elastically with hydrogen atoms in the sample. The hydrogen atoms ejected from the sample were detected with an SSD positioned at 30° with respect to the beam direction.

Elementary composition of the hydrogenated Ti-DLC film surface was estimated by a conventional X-ray photoelectron spectroscopy (XPS) apparatus (Shimadzu, ESCA-1000, Kyoto, Japan) mounted with a hemispherical electron energy analyzer. The Mg $K\alpha$ line (1253.6 eV), used as the X-ray source, was incident at 45° with respect to the surface normal. The emission angle was 45° with respect to the incident angle.

NEXAFS measurements were carried out on beamline 09A (BL09A) at the NewSUBARU synchrotron-radiation facility in University of Hyogo, which has a 1.5-GeV electron-storage ring [33–36]. Synchrotron radiation emitted by an 11-m undulator was extracted using a varied-line-spacing plane grating and irradiated onto the sample film at the "magic angle" of 54.7° with respect to the surface. The energy resolution was estimated to be less than 0.5 eV (FWHM). The electrons coming from the sample were detected in total-electron-yield mode. The intensity of the incident X-rays was measured by detecting the photocurrent from a gold mesh. The signal strength was derived from the ratio of the photocurrent from the sample to that from the gold mesh. The NEXAFS spectra of the C K edge absorption, Ti L absorption, and O K absorption were measured in the range 275–300 eV, 450–480 eV, and 520–560 eV, respectively.

Element distribution in depth direction of the hydrogenated Ti-DLC films was estimated by glow-discharge optical-emission spectroscopy (GD-OES) performed using a GD-Profiler2 (HORIBA, Kyoto, Japan). The sample was sputtered by argon plasma. Quantitative-chemical-depth profiles were estimated by detecting atomic emission from sputtering atoms.

3. Results and Discussion

Film thickness and bulk composition containing hydrogen of the hydrogenated Ti-DLC films were analyzed by a combination of RBS and ERDA. Dependence of film thickness (determined by measuring the hydrogenated Ti-DLC films by RBS) on atomic-oxygen beam fluence is shown in Figure 1. The red circles represent thickness of the hydrogenated Ti-DLC film. Film thicknesses of a hydrogenated non-doped DLC film, which were estimated in our previous study [12], are shown for reference. According to the figure, thickness of the hydrogenated non-doped DLC film decreased from 800 to 350 nm by irradiation of atomic oxygen with fluence of 5.0×10^{19} atoms cm^{-2}. On the other hand, thickness of the hydrogenated Ti-DLC film was found to be constant, $\cong$400 nm, after exposure to atomic oxygen with fluence of more than 5.0×10^{19} atoms cm^{-2}. These results demonstrate that the doping with titanium atoms gives the hydrogenated DLC films high etching tolerance against exposure to atomic oxygen. The dependence of relative hydrogen content in the film on atomic oxygen beam fluence was estimated by comparing the signal intensity of hydrogen in the ERDA spectra with that of carbon in the RBS spectra. Dependence of intensity ratio of hydrogen in the hydrogenated Ti-DLC films on atomic-oxygen beam fluence is shown in Figure 2. It is clear from the figure that the hydrogen ratio did not decrease after the film was exposed to atomic oxygen. That is to say, hydrogen is not desorbed from the hydrogenated Ti-DLC films by the irradiation with atomic oxygen. This result indicates that the hydrogen content in the hydrogenated Ti-DLC film exceeded 40% after the irradiation with atomic oxygen. A hydrogenated Ti-DLC film is therefore expected to exhibit low-friction properties in a vacuum.

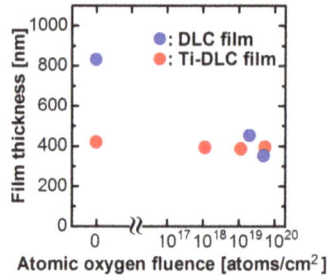

Figure 1. Dependence of film thickness determined from Rutherford backscattering spectrometry (RBS) of titanium-doped diamond-like carbon (Ti-DLC) films (red circles) and non-doped DLC films (blue circles) on atomic-oxygen fluence.

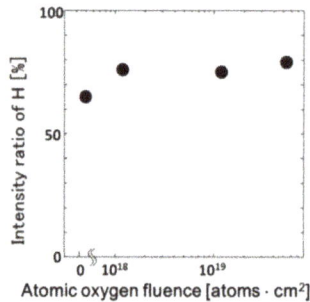

Figure 2. Dependence of relative hydrogen content determined from combination of RBS and elastic-recoil detection analysis (ERDA) in Ti-DLC films on atomic-oxygen fluence.

Surface chemical analysis on hydrogenated Ti-DLC films was performed by XPS and NEXAFS study. In our previous work, change of surface chemical states could not be traced, because the step in atomic-oxygen fluence was too large. Only XPS spectra over a wide range were reported; the detailed structure of each peak in the XPS spectra was not discussed [28]. In the present study, the fluence of an atomic-oxygen beam irradiated onto a hydrogenated Ti-DLC film was varied in smaller steps. In Figures 3–8, atomic-oxygen-beam fluences are (a) 0, (b) 1.2×10^{17}, (c) 1.2×10^{18}, (d) 3.5×10^{18}, (e) 1.2×10^{19}, and (f) 1.2×10^{20} atoms·cm^{-2}.

Change in elementary composition of the surface of hydrogenated Ti-DLC films due to the exposure to an atomic oxygen beam was investigated by XPS. Dependence of the C1s spectra of hydrogenated Ti-DLC films on fluence is shown in Figure 3. It is clear from figure that intensity of the peak at 285 eV decreased with increasing fluence of atomic oxygen. This result means that the amount of carbon atoms on the surface of hydrogenated Ti-DLC film is decreased by irradiation with atomic oxygen. Dependence of the O1s spectra of the hydrogenated Ti-DLC films on fluence is shown in Figure 4. The O1s peak at 532 eV is still observable in spectrum (a) [0 atoms·cm^{-2}]. It is ascribed to natural oxidation at the surface of the hydrogenated Ti-DLC film. The peak at 530 eV increases with increasing atomic-oxygen fluence, and it is assigned to the Ti–O bond [37]. This result indicates that the amount of oxygen atoms at the surface of the hydrogenated Ti-DLC film is increased by the irradiation by atomic oxygen. Dependence of the Ti 2p spectra of the hydrogenated Ti-DLC films on fluence is shown in Figure 5. The Ti 2p peak is not observed in spectra (a) [0 atoms·cm^{-2}] and (b) [1.2×10^{17} atoms·cm^{-2}]. However, the peaks at 460 eV and 465 eV, which were derived from TiO$_2$ [37], increase with increasing atomic-oxygen fluence. The results of estimations by elementary analysis, which took into account a relative-sensibility coefficient, are listed in Table 1. From the above-described XPS results, the amount of titanium and oxygen atoms in the hydrogenated Ti-DLC films increases with increasing atomic oxygen fluence. However, the amount of carbon atoms decreases monotonically. These changes in peak intensity and peak profile are observed for the peaks of spectra (a) [0 atoms·cm^{-2}] to (e) [1.2×10^{19} atoms·cm^{-2}]. On the other hand, the difference is hardly observed for every peak of spectra (e) [1.2×10^{19} atoms·cm^{-2}] and (f) [1.2×10^{20} atoms·cm^{-2}]. Accordingly, change in surface composition due to the irradiation by atomic oxygen is regarded to be completed by the irradiation with fluence of 1.2×10^{19} atoms·cm^{-2}.

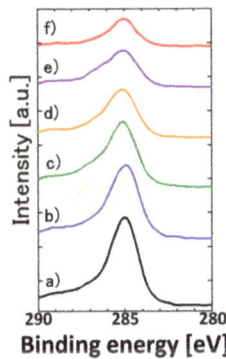

Figure 3. C 1s XPS spectra of Ti-DLC films before and after exposure to atomic oxygen. Fluences of atomic oxygen beam are (**a**) 0 atoms cm^{-2}, (**b**) 1.2×10^{17} atoms cm^{-2}, (**c**) 1.2×10^{18} atoms cm^{-2}, (**d**) 3.5×10^{18} atoms cm^{-2}, (**e**) 1.2×10^{19} atoms cm^{-2}, and (**f**) 1.9×10^{20} atoms cm^{-2}.

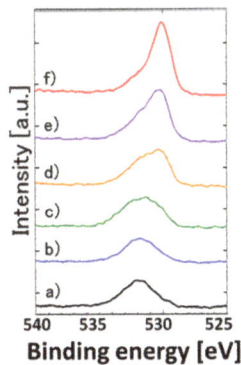

Figure 4. O 1s XPS spectra of Ti-DLC films before and after exposure to atomic oxygen. Fluences of atomic oxygen beam are (**a**) 0 atoms cm^{-2}, (**b**) 1.2×10^{17} atoms cm^{-2}, (**c**) 1.2×10^{18} atoms cm^{-2}, (**d**) 3.5×10^{18} atoms cm^{-2}, (**e**) 1.2×10^{19} atoms cm^{-2}, and (**f**) 1.9×10^{20} atoms cm^{-2}.

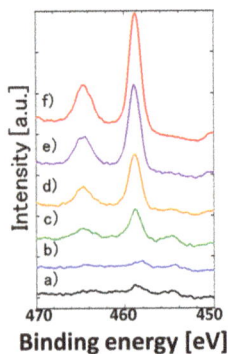

Figure 5. Ti 2p XPS spectra of Ti-DLC films before and after exposure to atomic oxygen. Fluences of atomic oxygen beam are (**a**) 0 atoms cm^{-2}, (**b**) 1.2 × 10^{17} atoms cm^{-2}, (**c**) 1.2 × 10^{18} atoms cm^{-2}, (**d**) 3.5 × 10^{18} atoms cm^{-2}, (**e**) 1.2 × 10^{19} atoms cm^{-2}, and (**f**) 1.9 × 10^{20} atoms cm^{-2}.

Table 1. Elementary composition of Ti-DLC films before and after exposure to atomic oxygen.

Fluence	C	O	Ti
atoms·cm^{-2}		at.%	
0	79	18	3
1.16 × 10^{17}	77	19	4
1.16 × 10^{18}	68	25	7
3.52 × 10^{18}	59	29	13
1.16 × 10^{19}	44	38	18
2.74 × 10^{19}	37	36	27
5.47 × 10^{19}	35	35	30
8.3 × 10^{19}	22	40	38
1.85 × 10^{20}	24	41	35

Change of the local structure of a hydrogenated Ti-DLC film due to exposure to atomic oxygen was previously discussed on the basis of the NEXAFS measurements [38]. NEXAFS spectroscopy using synchrotron radiation is known to be sensitive to the local structure around the absorber atoms. Dependence of the C *K*-edge NEXAFS spectra of hydrogenated Ti-DLC films on fluence is shown in Figure 6. The spectrum of a TiC powder is also shown for reference. The pre-edge resonance at 285.4 eV originates from sp^2 (C=C) sites [12,36]. The transitions from C 1s level to unoccupied σ* states are observed in the photon-energy region of 290 to 320 eV. Intensity of the sharp peak at 285.4 eV decreases with increasing atomic-oxygen fluence. On the other hand, intensity of the peak at 289 eV, which was assigned to C–O bonds [39], increases. In other words, the C–C bonds are changed to C–O bonds by irradiation with atomic oxygen. Dependence of the O *K*-edge NEXAFS spectra of hydrogenated Ti-DLC films on fluence is shown in Figure 7. The spectrum of a TiO$_2$ amorphous film is also shown for reference [40]. A peak is observable in spectrum (a) [0 atoms·cm^{-2}]. It is due to natural oxidation at the surface of the hydrogenated Ti-DLC film as described in XPS results. With increasing atomic-oxygen fluence, intensity of the peak derived from TiO$_2$ increases. Ti *L*-edge NEXAFS spectra of the hydrogenated Ti-DLC films, before and after exposure to atomic oxygen, are shown in Figure 8. The spectra of TiC powder and amorphous TiO$_2$ are also shown for reference [40]. The spectrum of the non-irradiated hydrogenated Ti-DLC film almost accords with that of TiC powder. Intensity of the peak derived from TiO$_2$ increases with increasing fluence of atomic oxygen. In other words, the Ti–C bonds are changed to Ti–O bonds by atomic oxygen irradiation. These results indicate that the C–C bonds in the surface neighborhood are changed to C–O bonds and that the Ti–C bonds in the surface neighborhood are changed to Ti–O bonds by atomic-oxygen irradiation. In other words, a titanium-oxide layer is formed on the surface of the hydrogenated Ti-DLC films by atomic-oxygen irradiation. This oxide layer is considered to protect the bulk film from erosion by hyperthermal atomic oxygen. In accord with the XPS results, various changes in NEXAFS spectra (a) [0 atoms·cm^{-2}] to (e) [1.2×10^{19} atoms·cm^{-2}] are observed, but little difference in spectra (e) [1.2×10^{19} atoms·cm^{-2}] and (f) [1.2×10^{20} atoms·cm^{-2}].

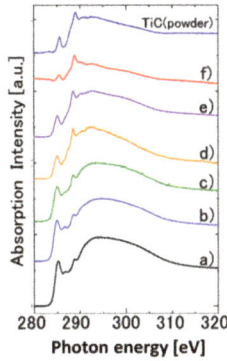

Figure 6. NEXAFS spectra of the C K-edge, before and after exposure to atomic oxygen. Fluences of atomic oxygen beam are (**a**) 0 atoms cm^{-2}, (**b**) 1.2×10^{17} atoms cm^{-2}, (**c**) 1.2×10^{18} atoms cm^{-2}, (**d**) 3.5×10^{18} atoms cm^{-2}, (**e**) 1.2×10^{19} atoms cm^{-2}, and (**f**) 1.9×10^{20} atoms cm^{-2}. The spectrum of TiC powder is also shown for reference.

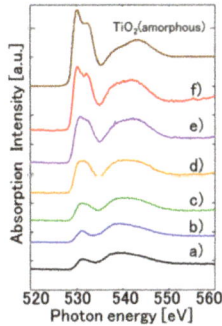

Figure 7. NEXAFS spectra of the O K-edge, before and after exposure to atomic oxygen. Fluences of atomic oxygen beam are (**a**) 0 atoms cm^{-2}, (**b**) 1.2×10^{17} atoms cm^{-2}, (**c**) 1.2×10^{18} atoms cm^{-2}, (**d**) 3.5×10^{18} atoms cm^{-2}, (**e**) 1.2×10^{19} atoms cm^{-2}, and (**f**) 1.9×10^{20} atoms cm^{-2}. The spectrum of a TiO$_2$ amorphous is also shown for reference.

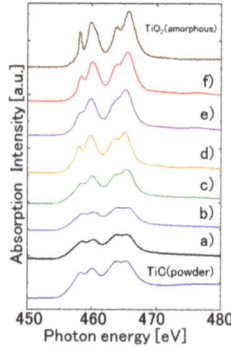

Figure 8. NEXAFS spectra of the Ti *L*-edge, before and after exposure to atomic oxygen. Fluences of atomic oxygen beam are (**a**) 0 atoms cm^{-2}, (**b**) 1.2×10^{17} atoms cm^{-2}, (**c**) 1.2×10^{18} atoms cm^{-2}, (**d**) 3.5×10^{18} atoms cm^{-2}, (**e**) 1.2×10^{19} atoms cm^{-2}, and (**f**) 1.9×10^{20} atoms cm^{-2}. The spectra of TiC powder and TiO$_2$ amorphous are also shown for reference.

To estimate the thickness of titanium-oxide layer formed by the irradiation of atomic oxygen, element distribution in the depth direction of the hydrogenated Ti-DLC films was measured by GD-OES. Element analysis results (in the depth direction) for titanium and oxygen in the hydrogenated Ti-DLC films are shown in Figures 9 and 10 respectively. In both figures, the black line and red line show the analysis results for elements in the depth direction of the hydrogenated Ti-DLC films exposed to atomic oxygen with fluences of (a) 0 atoms cm^{-2} and (f) 1.2×10^{20} atoms cm^{-2}, respectively. As for Figure 9, the amount of titanium in the surface neighborhood of the hydrogenated Ti-DLC film (*i.e.*, depth of less than 5 nm) before irradiation is small. However, it was increased in this region by irradiation of atomic oxygen. As shown in Figure 10, the amount of oxygen in the 5-nm-deep region was increased by irradiation of atomic oxygen. It is concluded from these results that the thickness of the titanium-oxide layer formed on the film surface by irradiation of atomic oxygen is about 5 nm.

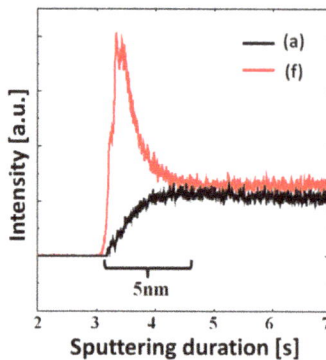

Figure 9. Element analysis (in depth direction) of titanium in Ti-DLC films. Fluence of atomic oxygen is (a) 0 atoms cm^{-2} and (f) 1.9×10^{20} atoms cm^{-2}.

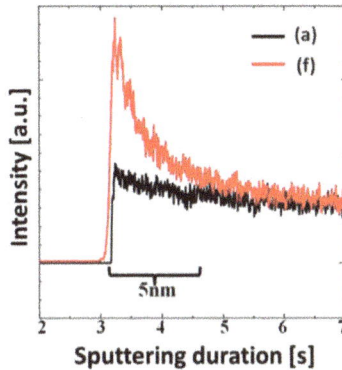

Figure 10. Element analysis (in depth direction) of oxygen in Ti-DLC films.

According to the above-described experimental results, the resistance of hydrogenated Ti-DLC films to atomic oxygen can be explained as follows. Film thickness and hydrogen content of the hydrogenated Ti-DLC films become constant after sufficient irradiation with atomic oxygen. This result differs from that reported in the case of non-doped hydrogenated DLC films [12]. Hydrogenated Ti-DLC films have resistance to etching by hyperthermal-atomic-oxygen irradiation, so they are expected to keep low-friction properties in a vacuum, because the hydrogen content in the film exceeded 40% after the irradiation. This result demonstrates that hydrogenated Ti-doped DLC films are useful as a lubricant for use in the LEO. The XPS study showed that the amounts of titanium and oxygen atoms in the surface of the hydrogenated Ti-DLC films increases, while the amount of carbon atoms decreases, with increasing fluence of atomic oxygen. The NEXAFS study showed that C–C bonds and Ti–C bonds in the surface neighborhood are changed to C–O bonds and Ti–O bonds, respectively, by atomic-oxygen irradiation. These XPS and NEXAFS results indicate that titanium-oxide layer forms on the surface of hydrogenated Ti-DLC films by irradiation with atomic oxygen. Namely, carbon atoms on the hydrogenated Ti-DLC film surface desorb from the films to form gas species (CO and/or CO_2), and the titanium atoms remain on the surface of the film and generate nonvolatile titanium oxide. The resistance of hydrogenated Ti-DLC films to atomic oxygen can be ascribed to this titanium-oxide layer formed on the surface by irradiation with atomic oxygen. This titanium-oxide layer is formed by irradiation of atomic oxygen with fluence of 1.2×10^{19} atoms·cm^{-2}. The fluence of atomic oxygen in the LEO region was estimated to be 2.3×10^{20} atoms·cm^{-2} to 6.6×10^{21} atoms·cm^{-2} per year [41]. In the LEO region, it therefore takes 16 h to 19 days to form titanium-oxide layer as a protective coating against irradiation by atomic oxygen. The thickness of the titanium-oxide layer, formed on the surface by irradiation with atomic oxygen, was found to be about 5 nm by GD-OES. Generally, solid lubricants function by surface exfoliating. It is therefore concluded that a titanium-oxide layer with thickness of 5 nm will not degrade the friction properties of solid lubricants.

4. Conclusions

Application of hydrogenated Ti-DLC films as a usable solid lubricant of components in artificial satellites positioned in the LEO region was investigated, and the resistance of hydrogenated Ti-DLC films to irradiation by hyperthermal atomic oxygen was evaluated. This investigation demonstrated that hydrogenated Ti-DLC films have etching tolerance to irradiation by hyperthermal-atomic-oxygen. Accordingly, hydrogenated Ti-DLC films can be expected to exhibit low-friction properties under vacuum, because the hydrogen content in the film does not decrease after the film is exposed to atomic oxygen. These resistances were ascribable to a titanium-oxide layer formed on the surface of the hydrogenated Ti-DLC films by irradiation with atomic oxygen. Forming the titanium-oxide layer in the LEO region was estimated to take 16 h to 19 days. The thickness of the titanium-oxide layer formed was found to 5 nm, which can be regarded as not having any influence on the lubricity of the lubricant. It is concluded from these results, that a hydrogenated Ti-DLC film can function as a solid lubricant in the LEO region. In future work, it is planned to examine the friction coefficient of the hydrogenated Ti-DLC film after it is exposed to atomic oxygen.

Acknowledgments: We thank the staff for use of 5SDH-2 at the Accelerator and Particle Beam Experimental Facility, Faculty of Maritime Sciences, Kobe University during the ERDA/RBS measurements. This work was supported in part by the Hyogo Science and Technology Association.

Author Contributions: K. Kidena conceived and designed the experiments under the instruction of K. Kanda. M. Endo and H. Takamatsu helped K. Kidena to perform the experiments. Especially, H. Takamatsu took over this work after the graduation of K. Kidena. M. Niibe helped with acquiring the NEXAFS spectra at NewSUBARU BL09A and discussed on the results. M. Tagawa, K. Yokota and Y. Furuyama obtained the ERDA and RBS spectra by a tandem electrostatic accelerator (5SDH), analyzed these data, and gave their general arguments concerning the study. K. Komatsu and H. Saitoh acquired the GD-OES spectra andinterpreted the results.

Conflicts of Interest: The authors declare no conflict of interest.

References

1. Crutcher, E.R.; Nishimura, L.S.; Warner, K.J.; Wascher, W.W. Silver Teflon Blanket: LDEF Tray C-08. In Proceedings of LDEF–69 Months in Space: First Post-Retrieval Symposium, Kissimmee, FL, USA, 2–8 June 1991; pp. 861–874.

2. Banks, B.A.; de Groh, K.K.; Miller, S.K. Low Earth Orbital Atomic Oxygen Interactions with Spacecraft Materials. In Proceedings of the 2004 Fall Meeting, Boston, MA, USA, November 29–December 3, 2014; pp. 1–12.

3. Han, J.H.; Kim, C.G. Low earth orbit space environment simulation and its effects on graphite/epoxy composites. *Compos. Struct.* **2006**, *72*, 218–226. [CrossRef]

4. Samwel, S.W. Low Earth Orbit Atomic Oxygen Erosion Effect on Spacecraft Materials. *Space Res. J.* **2014**, *7*, 1–13.

5. Aisenberg, S.; Chabot, S. Deposition of carbon films with diamond properties. *Carbon* **1972**, *10*, 356. [CrossRef]

6. Robertson, J. Properties of diamond-like carbon. *Surf. Coat. Technol.* **1992**, *50*, 185–203. [CrossRef]

7. Robertson, J. Diamond-like amorphous carbon. *Mater. Eng. R* **2002**, *37*, 129–281. [CrossRef]

8. Podgornik, B.; Vizintin, J. Tribological reaction between oil additives and DLC coatings for automotive applications. *Surf. Coat. Technol.* **2005**, *200*, 1982–1989. [CrossRef]

9. Bewilogua, K.; Hofmann, D. History of diamond-like carbon films-from first experiments to worldwide applications. *Surf. Coat. Technol.* **2014**, *242*, 214–225. [CrossRef]

10. Luo, J.K.; Fu, Y.Q.; Le, H.R.; Williams, J.A.; Spearing, S.M.; Milne, W.I. Diamond and daiamond-like carbon MEMS. *J. Micromech. Microeng.* **2007**, *17*, S147–S163. [CrossRef]

11. Donnet, C.; Belin, M.; Auge, J.C.; Martin, J.M.; Grill, A.; Patel, V. Tribochemistry of diamond-like carbon coatings in various environments. *Surf. Coat. Technol.* **1994**, *68–69*, 626–631. [CrossRef]

12. Tagawa, M.; Kumiko, K.; Kitamura, A.; Matsumoto, K.; Yoshigoe, A.; Teraoka, Y.; Kanda, K.; Niibe, M. Synchrotron radiation photoelectron spectroscopy and near-edge X-ray absorption fine structure study on oxidative etching of diamond-like carbon films by hyperthermal atomic oxygen. *Appl. Surf. Sci.* **2010**, *256*, 7678–7683. [CrossRef]

13. Oguri, K.; Arai, T. Tribological properties and characterization of diamond-like carbon coatings with silicon prepared by plasma-assisted chemical vapor deposition. *Surf. Coat. Technol.* **1991**, *47*, 710–721. [CrossRef]

14. Voevodin, A.A.; Rebholz, C.; Matthews, A. Comparative tribology studies of hard ceramic and composite metal-DLC coatings in sliding fraction conitions. *Tribol. Trans.* **1995**, *38*, 829–836. [CrossRef]

15. Silva, S.R.P.; Robertson, J.; Amaratunga, G.A.J. Nitrogen modification of hydrogenated amorphous carbon films. *J. Appl. Phys.* **1997**, *81*, 2626–2634. [CrossRef]

16. Kim, M.; Lee, K.; Eun, K. Tribological behavior of silicon-incorporated diamond-like carbon films. *Surf. Coat. Technol.* **1999**, *112*, 204–209. [CrossRef]

17. Varma, A.; Palshin, V.; Meletis, E.I. Structure-property relationship of Si-DLC films. *Surf. Coat. Technol.* **2001**, *148*, 305–314. [CrossRef]

18. Baba, K.; Hatada, R. Deposition and characterization of Ti- and W-containing diamonod-like carbon films by plasma source ion implantation. *Surf. Coat. Technol.* **2003**, *169*, 287–290. [CrossRef]
19. Liu, C.; Li, G.; Chen, W.; Mu, Z.; Zhang, C.; Wang, L. The study of doped DLC films by Ti ion implantation. *Thin Solid Films.* **2005**, *475*, 279–282.
20. Ouyang, J.H.; Sasaki, S. Friction and wear characteristics of a Ti-containing diamond-like carbon coating with an SRV tester high contact load and elevated temperature. *Surf. Coat. Technol.* **2005**, *195*, 234–244. [CrossRef]
21. Iseki, T.; Mori, H.; Hasegawa, H.; Tachikawa, H.; Nakanishi, K. Structural analysis of Si-containing diamond-like carbon. *Diam. Relat. Mater.* **2006**, *15*, 1004–1010. [CrossRef]
22. Zhao, F.; Li, H.; Jo, L.; Wang, Y.; Zhou, H.; Chen, J. Ti-DLC films with superior friction performance. *Diam. Relat. Mater.* **2010**, *19*, 342–349. [CrossRef]
23. Caschera, D.; Federici, F.; Pandolfi, L.; Kaciulis, S.; Sebastiani, M.; Bemporad, E.; Padeletti, G. Effect of composition on mechanical behaviour of diamond-like carbon coatings modified with titanium. *Thin Solid Films* **2011**, *519*, 3061–3067. [CrossRef]
24. Hofmann, D.; Kunkel, S.; Bewilogua, K.; Wittorf, R. From DLC to Si-DLC layer systems with properties for tribological applications. *Surf. Coat. Technol.* **2013**, *215*, 357–363. [CrossRef]
25. Amanov, A.; Watave, T.; Tsuboi, R.; Sakaki, S. Improvement in the tribological characteristics of Si-DLC coating by laser surface texturing under oil-lubricated point contacts at various temperatures. *Suf. Coat. Technol.* **2013**, *232*, 549–560. [CrossRef]
26. Lubwama, M.; Corcoran, B.; McDonnell, K.A.; Dowling, D.; Kirabira, J.B.; Sebbit, A.; Sayers, K. Flexibility and frictional behaviour of DLC and Si-DLC films deposited on nitrile rubber. *Surf. Coat. Technol.* **2014**, *239*, 84–94. [CrossRef]
27. Wang, S.; Wang, F.; Liao, Z.; Wang, Q.; Tyagi, R.; Liu, W. Tribological behaviour of titanium alloy modified by carbon–DLC composite film. *Surf. Eng.* **2015**. [CrossRef]
28. Kanda, K.; Fukuda, K.; Kidena, K.; Imai, R.; Niibe, M.; Fujimoto, S.; Yokota, K.; Tagawa, M. Hyperthermal atomic oxygen beam irradiation effect on the Ti-containing DLC film. *Diam. Relat. Mater.* **2014**, *41*, 49–52. [CrossRef]
29. Nakahigashi, T.; Tanaka, Y.; Miyake, K.; Oohara, H. Properties of flexible DLC film deposited by amplitude-modulated RE P-CVD. *Triol. Int.* **2004**, *37*, 907–912. [CrossRef]
30. Tagawa, M.; Yokota, K.; Ohmae, N.; Kinoshita, H.; Umeno, M. Oxidation Properties of Hydrogen-Terminated Si(001) Surfaces Following Use of a Hyperthermal Broad Atomic Oxygen Beam at Law Temperatures. *Jpn. J. Appl. Phys.* **2001**, *40*, 6152–6156. [CrossRef]
31. Yokota, K.; Tagawa, M.; Ohmae, N. Temperature Dependence in Erosion Rates of Polyimide under Hyerthermal Atomic Oxygen Exposures. *J. Spacecr. Rockets* **2003**, *40*, 143. [CrossRef]
32. Kitamura, A.; Tamai, T.; Taniike, A.; Furuyama, Y.; Maeda, T.; Ogiwara, N.; Saidoh, M. Simulation of ERD spectra for a surface with a periodic roughness. *Nucl. Instrum. Methods Phys. Res. Sect. B* **1998**, *134*, 98–106.
33. Niibe, M.; Mukai, M.; Miyamoto, S.; Shoji, Y.; Hashimoto, S.; Ando, A.; Tanaka, T.; Miyai, M.; Kitamura, H. Characterization of light radiated from 11 m long undulator. *AIP Conf. Proc.* **2004**, *705*, 576–579.
34. Niibe, M.; Mukai, M.; Kimura, H.; Shoji, Y. Polarization property measurement of the long undulator radiation using Cr/C multilayer polarization elements. *AIP Conf. Proc.* **2004**, *705*, 243–246.
35. Saikubo, A.; Kanda, K.; Niibe, M.; Matsui, S. Near-edge X-ray absorption fine-structure characterization of diamond-like-carbon thin films formed by various method. *New Diam. Front. Carbon Technol.* **2006**, *16*, 235–244.
36. Saikubo, A.; Yamada, N.; Kanda, K.; Matsui, S.; Suzuki, T.; Niihara, K.; Saitoh, H. Comprehensive classification of DLC films formed by various methods using NEXAFS measurement. *Diam. Relat. Mater.* **2008**, *17*, 1743–1745. [CrossRef]
37. Nam, S.H.; Cho, S.J.; Jung, C.K.; Boo, J.H.; Šícha, J.; Heřman, D.; Musil, J.; Vlček, J. Comparison of hydrophilic properties of TiO$_2$ thin films prepared by sol-gel method and reactive magnetron sputtering system. *Thin Solid Films* **2011**, *519*, 6944–6950. [CrossRef]
38. Kanda, K.; Niibe, M.; Wada, A.; Ito, H.; Suzuki, T.; Ohana, T.; Ohtake, N.; Saitoh, H. Comprehensive Classification of Near-Edge X-ray Absorption Fine Structure Spectra of Si-Containing Diamond-Like Carbon thin Films. *Jpn. J. Appl. Phys.* **2013**, *52*, 095504. [CrossRef]

39. Magnuson, M.; Lewin, E.; Hultman, L.; Jansson, U. Electronic structure and chemical bonding of nanocrystalline-TiC/amorphous-C nanocomposites. *Phys. Rev. B* **2009**, *80*, 235108. [CrossRef]

40. Kucheyev, S.O.; Buuren, T.V.; Baumann, T.F.; Satcher, J.H., Jr.; Willey, T.M.; Meulenberg, R.W.; Felter, T.E.; Poco, J.F.; Gammon, S.A.; Terminello, L.J. Electronic structure of titania aerogels: Soft X-ray absorption study. *Phys. Rev. B* **2004**, *69*, 245102. [CrossRef]

41. Tagawa, M.; Yokota, K.; Ochi, K.; Akiyama, M.; Matsumoto, K.; Suzuki, M. Comparison of Macro and Microtribological Property of Molybdenum Disulfide Film Exposed to LEO Space Environment. *Tribol. Lett.* **2012**, *45*, 349–356. [CrossRef]

metals

MDPI

Correction

Correction: Liss, K.-D., et al. Hydrostatic Compression Behavior and High-Pressure Stabilized β-Phase in γ-Based Titanium Aluminide Intermetallics. *Metals* 2016, 6, 165

Klaus-Dieter Liss [1,2,3,*] , **Ken-Ichi Funakoshi** [4], **Rian Johannes Dippenaar** [3], **Yuji Higo** [5], **Ayumi Shiro** [6,†], **Mark Reid** [2,3], **Hiroshi Suzuki** [1,‡], **Takahisa Shobu** [6,‡] and **Koichi Akita** [1,‡]

1   Quantum Beam Science Center, Japan Atomic Energy Agency, Tokai, Ibaraki 319-1195, Japan; suzuki.hiroshi07@jaea.go.jp (H.S.); akita.koichi@jaea.go.jp (K.A.)
2   Australian Nuclear Science and Technology Organisation, Lucas Heights 2234, Australia; mark.reid@ansto.gov.au
3   School of Mechanical, Materials & Mechatronic Engineering, Faculty of Engineering and Information Sciences, University of Wollongong, Northfields Avenue, Wollongong 2522, Australia; rian@uow.edu.au
4   Neutron Science and Technology Center, Comprehensive Research Organization for Science and Society (CROSS-Tokai), Tokai, Ibaraki 319-1106, Japan; k_funakoshi@cross.or.jp
5   SPring-8, Japan Synchrotron Radiation Research Institute, Kouto, Sayo, Hyogo 679-5198, Japan; higo@spring8.or.jp
6   Quantum Beam Science Center, Japan Atomic Energy Agency, Kouto, Sayo, Hyogo 679-5148, Japan; shiro.ayumi@qst.go.jp (A.S.); shobu@sp8sun.spring8.or.jp (T.S.)
*   Correspondence: kdl@ansto.gov.au or liss@kdliss.de; Tel.: +61-2-9717-9479
†   Present Address: Quantum Beam Science Research Directorate, National Institute for Quantum and Radiological Science and Technology, Kouto, Sayo, Hyogo 679-5148, Japan.
‡   Present Address: Materials Sciences Research Center, Japan Atomic Energy Agency, Tokai, Ibaraki 319-1195, Japan and Kouto, Sayo, Hyogo 679-5148, Japan.

Academic Editor: Hugo F. Lopez
Received: 15 August 2017; Accepted: 22 August 2017; Published: 7 September 2017

The authors would like to apologize for any inconvenience regarding misleading errors and inconsistencies in some of the units and one number, and wish to make the following corrections to this paper [1]:

Page 11: unit should be GPa in " ... up to 200 GPa [71] ... " (not MPa).

Page 12: unit should be GPa in " ... lying around 146 GPa ... " (not MPa).

Figure 9: units should be GPa for K_0 in all four occurrences (not MPa).

Table 1: add to caption: "a, c in (Å); V in (Å^3); K_0 in (GPa)."

Table 1: correct value for K_0 for α_2 to be "126" (not 116).

Page 16: unit should be GPa in "With values around 146 GPa ... " (not MPa).

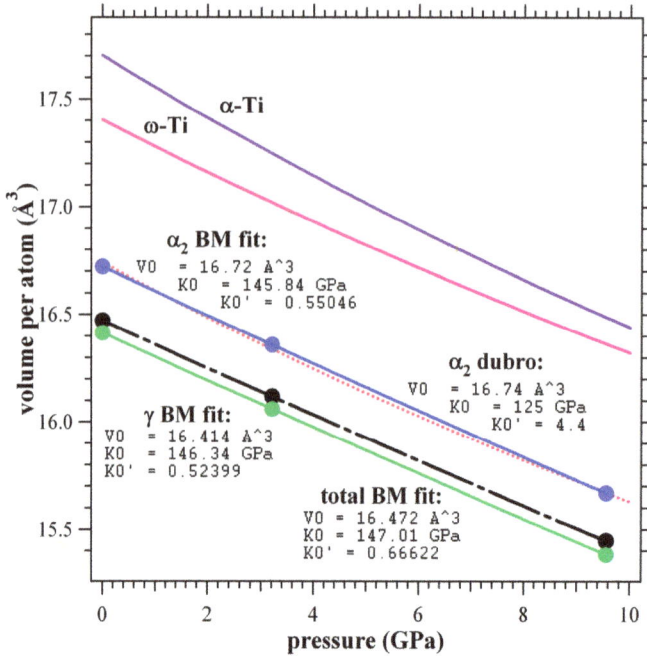

Figure 9. Atomic volumetric compression behavior of the investigated composition Ti-45Al-7.5Nb-0.25C with Birch–Murnaghan fits (experimental dots with continuous lines), as compared to α_2-single-phase compression, and α- and ω-titanium, reported by Dubrovinskaia [59] and Errandonea [48], respectively.

Table 1. Compilation of experimental lattice parameters a_0 and c_0 under ambient conditions, as well as the derived quantities; their axis ratios and volume per atom V_A, compression parameters K_0, K_0' (first three rows) and data from the literature. The α-phase lattice is given in α_2 cell notation, and therefore, $2c/a$ is noted. The first VA column is computed from a_0 and c_0, while the second results are from the fit of pressure data to Equation (6). The original data of Yeoh's publication [31] has been re-visited to extract the listed values at 300 K. Literature values are reported from their experimental findings, in addition to Ghosh's first-principles study [61]. Further listed references are Dubrovinskaia [59], Errandonea [48], Asta [60], Zhang [55], JCPDS [56], and Menon [78]. a, c in (Å); V in (Å3); K_0 in (GPa).

Phase	a_0	c_0	Axis Ratio	V_A	V_A	K_0	K_0'	Reference
γ	4.01867	4.06542	1.0116332	16.4138371	16.414	146.34	0.52399	this work
α/α_2	5.76803	4.64241	1.60970383	16.7201111	16.72	145.84	0.55046	this work
total				16.4720291	16.472	147.01	0.66622	this work
α_2-Ti-33.3Al	5.7763	4.6348	1.6047643	16.7406041	16.74	125	4.4	Dubrovinskaia
α_2-Ti-28.4Al	5.7829	4.6388	1.60431617	16.7933623	16.79375	131	3.6	Dubrovinskaia
α_2-Ti-24.0Al	5.8083	4.6563	1.60332627	17.0051191	17.005	133	2.6	Dubrovinskaia
α-Ti			1.583		17.7013462	117	3.9	Errandonea
ω-Ti			0.609		17.4024491	138	3.8	Errandonea
γ			1.012			128		Asta
α_2			1.698			126		Asta
γ	3.9814	4.0803	1.02484051	16.1697657	16.181	112.1	3.91	Ghosh
α_2	5.7372	4.6825	1.63232936	16.6847003	16.584	111.9	3.83	Ghosh
α-Ti			1.5868			114	4	Zhang
α-Ti	5.901	4.6826	1.58705304	17.651391				JCPDS
γ-Ti-50Al	3.9973	4.0809	1.02091412	16.3015706				Menon
γ-Ti-45Al-7.5Nb-0.5C	4.02421	4.07335	1.01221109	16.4912285				Yeoh
α_2-Ti-45Al-7.5Nb-0.5C	5.77568	4.65646	1.61243698	16.8152283				Yeoh

Reference

1. Liss, K.-D.; Funakoshi, K.-I.; Dippenaar, R.J.; Higo, Y.; Shiro, A.; Reid, M.; Suzuki, H.; Shobu, T.; Akita, K. Hydrostatic Compression Behavior and High-Pressure Stabilized β-Phase in γ-Based Titanium Aluminide Intermetallics. *Metals* **2016**, *6*, 165. [CrossRef]

MDPI AG

St. Alban-Anlage 66

4052 Basel, Switzerland

Tel. +41 61 683 77 34

Fax +41 61 302 89 18

http://www.mdpi.com

Metals Editorial Office

E-mail: metals@mdpi.com

http://www.mdpi.com/journal/metals

www.ingramcontent.com/pod-product-compliance
Lightning Source LLC
Chambersburg PA
CBHW051720210326
41597CB00032B/5545